THE IDEA OF THE BRAIN

A HISTORY

Matthew Cobb

D1644588

P

PROFILE BOOKS

This paperback edition published in 2021

First published in Great Britain in 2020 by
PROFILE BOOKS LTD
29 Cloth Fair
London EC1A 7JQ

www.profilebooks.com

1 3 5 7 9 10 8 6 4 2

Typeset in Palatino by MacGuru Ltd
Printed and bound in Great Britain by
CPI Group (UK) Ltd, Croydon CR0 4YY

A CIP catalogue record for this book is available from the British Library.

ISBN 978 1 78125 590 2
eISBN 978 1 78283 225 6

In memory of Kevin Connolly (1937–2015),
Professor of Psychology at the University of Sheffield,
who set me on the road to here.

The brain being indeed a machine, we must not hope to find its artifice through other ways than those which are used to find the artifice of the other machines. It thus remains to do what we would do for any other machine; I mean to dismantle it piece by piece and to consider what these can do separately and together.

Nicolaus Steno, *On the Brain*, 1669

CONTENTS

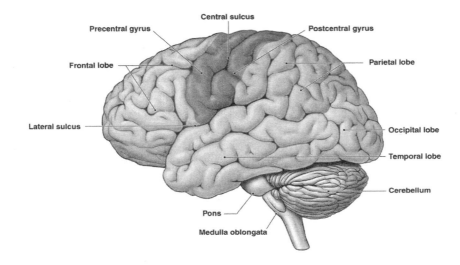

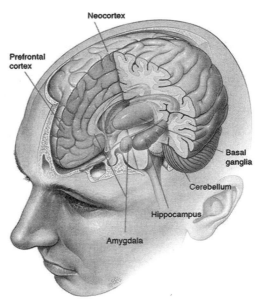

Key areas of the human brain.

INTRODUCTION

In 1665 the Danish anatomist Nicolaus Steno addressed a small group of thinkers gathered together at Issy, on the southern outskirts of Paris. This informal meeting was one of the origins of the French Académie des Sciences; it was also the moment that the modern approach to understanding the brain was set out. In his lecture, Steno boldly argued that if we want to understand what the brain does and how it does it, rather than simply describing its component parts, we should view it as a machine and take it apart to see how it works.

This was a revolutionary idea, and for over 350 years we have been following Steno's suggestion – peering inside dead brains, removing bits from living ones, recording the electrical activity of nerve cells (neurons) and, most recently, altering neuronal function with the most astonishing consequences. Although most neuroscientists have never heard of Steno, his vision has dominated centuries of brain science and lies at the root of our remarkable progress in understanding this most extraordinary organ.

We can now make a mouse think it is somewhere it is not, turn a bad mouse memory into a good one and even use a surge of electricity to change how people perceive faces. We are drawing up increasingly detailed and complex functional maps of the brain, human and otherwise. In some species we can change the brain's very structure at will, altering the animal's behaviour as a result. Some of the most profound consequences of our growing mastery can be seen in our ability to enable a paralysed person to control a robotic arm with the power of their mind.

We cannot do everything: at least for the moment, we cannot artificially create a precise sensory experience in a human brain (hallucinogenic drugs do this in an uncontrolled way), although it appears that we have the exquisite degree of control required to perform such an experiment in a mouse. Two groups of scientists recently trained mice to lick at a water bottle when the animals saw a set of stripes, while machines recorded how a small number of cells in the visual centres of the mice's brains responded to the image. The scientists then used complex optogenetic technology to artificially recreate that pattern of neuronal activity in the relevant brain cells. When this occurred, the animal responded as though it had seen the stripes, even though it was in complete darkness. One explanation is that, for the mouse, the pattern of neuronal activity was the same thing as seeing. More clever experimentation is needed to resolve this, but we stand on the brink of understanding how patterns of activity in networks of neurons create perception.

This book tells the story of centuries of discovery, showing how brilliant minds, some of them now forgotten, first identified that the brain is the organ that produces thought and then began to show what it might be doing. It describes the extraordinary discoveries that have been made as we have attempted to understand what the brain does, and delights in the ingenious experiments that have produced these insights.

But there is a significant flaw in this tale of astonishing progress, one that is rarely acknowledged in the many books that claim to explain how the brain works. Despite a solid bedrock of understanding, we have no clear comprehension about how billions, or millions, or thousands, or even tens of neurons work together to produce the brain's activity.

We know in general terms what is going on – brains interact with the world, and with the rest of our bodies, representing stimuli using both innate and acquired neural networks. Brains predict how those stimuli might change in order to be ready to respond, and as part of the body they organise its action. This is all achieved by neurons and their complex interconnections, including the many chemical signals in which they bathe. No matter how much it might go against your deepest feelings, there is no disembodied person floating in your head looking at this activity – it is all just neurons, their connectivity and the chemicals that swill about those networks.

However, when it comes to really understanding what happens in a brain at the level of neuronal networks and their component cells, or to being able to predict what will happen when the activity of a particular network is altered, we are still at the very beginning. We might be able to artificially induce visual perception in the brain of a mouse by copying a very precise pattern of neuronal activity, but we do not fully understand how and why visual perception produces that pattern of activity in the first place.

A key clue to explaining how we have made such amazing progress and yet have still barely scratched the surface of the astonishing organ in our heads is to be found in Steno's suggestion that we should treat the brain as a machine. 'Machine' has meant very different things over the centuries, and each of those meanings has had consequences for how we view the brain. In Steno's time the only kinds of machine that existed were based on either hydraulic power or clockwork. The insights these machines could provide about the structure and function of the brain soon proved limited, and no one now looks at the brain this way. With the discovery that nerves respond to electrical stimulation, in the nineteenth century the brain was seen first as some kind of telegraph network and then, following the identification of neurons and synapses, as a telephone exchange, allowing for flexible organisation and output (this metaphor is still occasionally used in research articles).

Since the 1950s our ideas have been dominated by concepts that surged into biology from computing – feedback loops, information, codes and computation. But although many of the functions we have identified in the brain generally involve some kind of computation, there are only a few fully understood examples, and some of the most brilliant and influential theoretical intuitions about how nervous systems might 'compute' have turned out to be wrong.

Above all, as the mid-twentieth-century scientists who first drew the parallel between brain and computer soon realised, the brain is not digital. Even the simplest animal brain is not a computer like anything we have built, nor one we can yet envisage. The brain is not a computer, but it is more like a computer than it is like a clock, and by thinking about the parallels between a computer and a brain we can gain insight into what is going on inside both our heads and those of animals.

Exploring these ideas about the brain – the kinds of machine we have imagined brains to be – makes it clear that, although we are still far from fully understanding the brain, the ways in which we think about it are much richer than in the past, not simply because of the amazing facts we have discovered, but above all because of how we interpret them.

These changes have an important implication. Over the centuries, each layer of technological metaphor has added something to our understanding, enabling us to carry out new experiments and reinterpret old findings. But by holding tightly to metaphors, we end up limiting what and how we can think. A number of scientists are now realising that, by viewing the brain as a computer that passively responds to inputs and processes data, we forget that it is an active organ, part of a body that is intervening in the world and which has an evolutionary past that has shaped its structure and function. We are missing out key parts of its activity. In other words, metaphors shape our ideas in ways that are not always helpful.

The tantalising implication of the link between technology and brain science is that tomorrow our ideas will be altered yet again by the appearance of new and as yet unforeseen technological developments. As that new insight emerges, we will reinterpret our current certainties, discard some mistaken assumptions and develop new theories and ways of understanding. When scientists realise that how they think – including the questions they can ask and the experiments they can imagine – is partly framed and limited by technological metaphors, they often get excited at the prospect of the future and want to know what the Next Big Thing will be and how they can apply it to their research. If I had the slightest idea, I would be very rich.

*

This book is not a history of neuroscience, nor a history of brain anatomy and physiology, nor a history of the study of consciousness, nor a history of psychology. It contains some of these things, but the history I tell is rather different, for two reasons. First, I want to explore the rich variety of ways in which we have thought about what brains do and how they do it, focusing on experimental evidence – this is rather different from telling the story of an academic discipline. It also means that the book does not deal solely with how we have thought about the human brain – other brains in other animals, not all of them mammals, have shed light on what is happening in our heads.

The history of how we have understood the brain contains recurring themes and arguments, some of which still provoke intense debate today. One example is the perpetual dispute over the extent to which functions are localised in specific areas of the brain. That idea goes back thousands of years, and there have been repeated claims up to today that bits of the brain appear to be responsible for a specific thing, such as the feeling in your hand, or your ability to understand syntax or to exert self-control. These kinds of claims have often soon been nuanced by the revelation that other parts of the brain may influence or supplement this activity, and that the brain region in question is also involved in other processes. Repeatedly, localisation has not exactly been overturned, but it has become far fuzzier than originally thought. The reason is simple. Brains, unlike any machine, have not been designed. They are organs that have evolved for over five hundred million years, so there is little or no reason to expect they truly function like the machines we create. This implies that although Steno's starting point – treating the brain as a machine – has been incredibly productive, it will never produce a satisfying and full description of how brains work.

The interaction between brain science and technology – the thread that runs through this book – highlights the fact that science is embedded in culture. So an element of this story reveals how these ideas have reverberated through the works of Shakespeare, Mary Shelley, Philip K. Dick and others. Intriguingly, cultural history shows that metaphors can flow both ways – in the nineteenth century, just as the brain and the nervous system were thought of as a telegraph network, so too the flow of Morse Code messages down the telegraph wires and the responses they evoked in their human readers were seen in terms of nervous activity. Similarly, at its birth the computer was seen as a brain – biological discoveries were used to justify John von Neumann's plans to build the first digital computer, rather than the other way around.

The second reason why this is not simply a history can be seen from the contents page – the book is divided into three parts: Past, Present and Future. The conclusion of the 'Present' section, which deals with how our understanding of the brain has developed over the last seventy years or so under the computational metaphor, is that some researchers sense we are approaching an impasse in how we understand the brain.

This might seem paradoxical – we are accumulating vast amounts of data about structure and function in a huge array of brains, from the tiniest to our own. Tens of thousands of researchers are devoting massive amounts of time and energy to thinking about what brains do, and astonishing new technology is enabling us to both describe and manipulate that activity. Every day we hear about new discoveries that shed light on how brains work, along with the promise – or threat – of new technology that will enable us to do such far-fetched things as read minds, or detect criminals, or even be uploaded into a computer.

In contrast to all this exuberance, there is a feeling among some neuroscientists, as shown by think-pieces in academic journals and books over the last decade or so, that our future path is not clear. It is hard to see where we should be going, apart from simply collecting more data or counting on the latest exciting experimental approach. That does not mean that everyone is pessimistic – some confidently claim that the application of new mathematical methods will enable us to understand the myriad interconnections in the human brain. Others favour studying animals at the other end of the scale, focusing our attention on the tiny brains of worms or maggots and employing the well-established approach of seeking to understand how a simple system works, and then applying those lessons to more complex cases. Many neuroscientists, if they think about the problem at all, simply consider that progress will inevitably be piecemeal and slow, because there is no Grand Unified Theory of the brain lurking around the corner.

The problem is twofold. Firstly, the brain is mind-bogglingly complicated. A brain – any brain, not just the human brain, which has been the focus of much of the intellectual endeavour described here – is the most complex object in the known universe. The astronomer Lord Rees has pointed out that an insect is more complex than a star, while for Darwin the brain of an ant, which is so tiny but which can produce such diverse behaviour, was 'one of the most marvellous atoms of matter in the world, perhaps more so than the brain of a man'. That is the scale of the challenge before us.

Which leads to the second aspect. Despite the tsunami of brain-related data being produced by laboratories around the world, we are in a crisis of ideas about what to do with all that data, about what it all means. I think that this reveals that the computer metaphor, which has served us so well for over half a century, may be reaching its limits, just as the idea of a brain as a telegraph system eventually exhausted its power in the nineteenth century. Some scientists are now explicitly challenging the usefulness of some of our most basic metaphors about the brain and nervous systems, such as the idea that neuronal networks represent the outside world, through a neuronal code. This suggests that scientific understanding may be chafing at the framework imposed by our most deeply held metaphors about how the brain works.

It may prove to be that even in the absence of new technology, developments in computing, in particular relating to artificial intelligence and neural networks – which are partly inspired by how brains do things – will feed back into our views of the brain, giving the computational metaphor a new lease of life. Perhaps. But, as you will see, leading researchers in deep learning – the most fashionable and astonishing part of modern computer science – cheerfully admit that they do not know how their programs do what they do. I am not sure that computing will provide enlightenment as to how the brain works.

One of the most tragic indicators of our underlying uncertainty about the brain is the very real crisis in our understanding of mental health. From the 1950s, science and medicine embraced chemical approaches to treating mental illness. Billions of dollars have been spent developing drugs, but it is still not clear how, nor even if, many of these widely prescribed treatments work. As to future pharmaceutical approaches to major mental health problems, there is nothing on the horizon – most of the large drug companies have abandoned the search for new drugs to treat conditions such as depression or anxiety, considering that both the costs and the risks are far too great. This situation is not surprising – if we do not yet properly understand the functioning of even the simplest animal brains, there does not seem much prospect of responding effectively when things apparently go awry in our own heads.

A great deal of energy and resources are being devoted to describing the myriad connections between neurons in brains, to create what are called connectomes, or more crudely and metaphorically, wiring diagrams. There is currently no prospect of creating a cell-level connectome of a mammalian brain – they are far too complex – but lower-definition maps are being established. Such efforts are essential – we need to understand how bits of the brain are connected – but on their own they will not produce a model of what the brain does. Nor should we underestimate how long this might take. Researchers are currently drawing up a functional connectome that includes all 10,000 cells in a maggot brain, but I would be amazed if, in fifty years' time, we fully understand what those cells and their interconnections are doing. From this point of view, properly understanding the human brain, with its tens of billions of cells and its incredible and eerie ability to produce the mind, may seem an unattainable dream. But science is the only method that can reach this goal, and it will reach it, eventually.

There have been many similar moments in the past, when brain researchers became uncertain about how to proceed. In the 1870s, with the waning of the telegraph metaphor, doubt rippled through brain science and many researchers concluded it might never be possible to explain the nature of consciousness. One hundred and fifty years later we still do not understand how consciousness emerges, but scientists are more confident that it will one day be possible to know, even if the challenges are enormous.

Understanding how past thinkers have struggled to understand brain function is part of framing what we need to be doing now, in order to reach that goal. Our current ignorance should not be viewed as a sign of defeat but as a challenge, a way of focusing attention and resources on what needs to be discovered and on how to develop a programme of research for finding the answers. That is the subject of the final, speculative part of this book, which deals with the future. Some readers will find this section provocative, but that is my intention – to provoke reflection about what the brain is, what it does and how it does it, and above all to encourage thinking about how we can take the next step, even in the absence of new technological metaphors. It is one of the reasons this book is more than a history, and it highlights why the four most important words in science are 'We do not know'.

Manchester, December 2019

PAST

THE HISTORY OF SCIENCE is rather different from other kinds of history, because science is generally progressive – each stage builds upon previous insights, integrating, rejecting or transforming them. This produces what appears to be an increasingly accurate understanding of the world, although that knowledge is never complete, and future discoveries can overthrow what was once seen as the truth. This underlying progressive aspect leads many scientists to portray the history of their subject as a procession of great men (and it generally has been men), each of whom is given approval if they are seen as having been right, or criticised – or ignored – if they were wrong. In reality, the history of science is not a progression of brilliant theories and discoveries: it is full of chance events, mistakes and confusion.

To properly understand the past, to provide a full background to today's theories and frameworks, and even to imagine what tomorrow may hold, we must remember that past ideas were not seen as steps on the road to our current understanding. They were fully fledged views in their own right, in all their complexity and lack of clarity. Every idea, no matter how outdated, was once modern, exciting and new. We can be amused at strange ideas from the past, but condescension is not allowed – what seems obvious to us is only that way because past errors, which were generally difficult to detect, were eventually overcome through a great deal of hard work and harder thinking.

Where people in the past accepted mistaken or what now appear to be unbelievable ideas, the challenge is to understand why. Often, what now might be taken as ambiguity or lack of clarity in an approach or set of ideas in fact explains why those ideas were accepted. Such imprecise theories may allow scientists with different views to accept a common framework, pending the arrival of decisive experimental evidence.

We should never dismiss past ideas – or people – as stupid. We will be the past one day, and our ideas will no doubt seem surprising and amusing to our descendants. We are simply doing the best we can, just as our forebears did. And, like previous generations, our scientific ideas are influenced not only by the internal world of scientific evidence, but also by the general social and technological context in which we develop those ideas. Where our theories and interpretations are wrong or inadequate, they will be proved so by future experimental evidence and we will all move on. That is the power of science.

ONE

HEART

The scientific consensus is that, in ways we do not understand, thought is produced by the activity of billions of cells in the most complex structure in the known universe – the human brain. Surprising as it may be, this focus on the brain seems to be a relatively recent development. Virtually all we know from prehistory and history suggests that for most of our past we have viewed the heart, not the brain, as the fundamental organ of thought and feeling. The power of these old, pre-scientific views can be seen in our everyday language – words and phrases like 'learn by heart', 'heartbroken', 'heartfelt', and so on (similar examples can be found in many other languages). These phrases still carry the emotional charge of the old world-view that we have supposedly discarded – try replacing the word 'heart' by 'brain' and see how it feels.

Our earliest written artefacts show the importance of this idea to past cultures. In the *Epic of Gilgamesh*, a 4,000-year-old story written in what is now Iraq, emotions and feelings were clearly based in the heart, while in the Indian Rigveda, a collection of Vedic Sanskrit hymns written around 3,200 years ago, the heart is the site of thought.[1] The Shabaka Stone, a shiny grey slab of basalt from ancient Egypt, now in the British Museum, is covered in hieroglyphs that describe a 3,000-year-old Egyptian myth focused on the importance of the heart in thinking.[2] The Old Testament reveals that at around the same time as the Shabaka Stone was carved, the Jews considered the heart to be the origin of thought in both humans and God.[3]

Heart-centred views also existed in the Americas, where the great empires of Central America – the Maya (250–900 CE) and the Aztecs (1400–1500 CE) – both focused on the heart as the source of emotions and thought. We also have some insight into the beliefs of those peoples from North and Central America who did not develop extensive urban cultures. In the early years of the twentieth century, US ethnographers worked with indigenous peoples, documenting their traditions and beliefs. Although we cannot be certain that the recorded views were typical of the cultures that existed before the arrival of Europeans, most of the peoples who contributed to these studies considered that something like a 'life-soul', or an emotional consciousness, was linked to the heart and to breath. This view was widespread, from Greenland to Nicaragua, and was held by peoples with ecologies as diverse as the Eskimo, the Coast Salish of the Pacific north-west, and the Hopi of Arizona.[4]

These views are remarkably congruent with the account of the Swiss psychoanalyst Carl Jung, who in the early decades of the twentieth century travelled to New Mexico. On the roof of one of the white adobe buildings built by the Pueblo people on the high Taos plateau, Jung talked with Ochwiay Biano of the Taos Pueblo. Biano told Jung that he did not understand white people, whom he considered cruel, uneasy and restless – 'We think that they are mad', he said. Intrigued, Jung asked Biano why he thought this:

'They say they think with their heads,' he replied.
'Why, of course, what do you think with?' I asked him in surprise.
'We think here,' he said, indicating his heart.[5]

Not all cultures have shared this widespread focus on the heart. On the other side of the planet, a key aspect of the outlook of the Aboriginal and Torres Strait Islander peoples in Australia was (and is) their link with the land, which extends to ideas about mind and spirit. Locating the seat of thought within the body appears not to have been part of their world view.[6] Similarly, traditional Chinese approaches to medicine and anatomy were primarily focused on the interactions of a series of forces, rather than localisation of function. However, when Chinese thinkers did seek to identify the roles of particular organs, the heart was the key.[7] The Guanzi, a document originally written by the Chinese philosopher Guan Zhong in the seventh century BCE, argued that the heart was fundamental for all functions of the body, including the senses.

Heart-centred views correspond to our everyday experience – the heart changes its rhythm at the same time as our feelings change, while powerful emotions such as anger, lust or fear seem to be focused on one or more of our internal organs, and to course through our bodies and change our way of thinking as though they are transported in, or simply are, our blood. This is why those old phrases about being 'down at heart' and so on have persisted – they correspond to the way we perceive an important part of our inner life. Just as with the appearance that the sun goes around the earth, everyday experience of being human provided a simple explanation of where we think – our hearts. People believed this idea because it made sense.

*

Even though the heart was widely seen as the centre of our inner life, certain cultures recognised that the brain had some kind of function, even if this could only be detected through injury. For example, in ancient Egypt a number of scribes created a medical document known as the Edwin Smith Papyrus.[8] The manuscript includes a brief description of the convolutions of the brain and the recognition that damage to one side of the head could be accompanied by paralysis on the opposite side of the body, but for these writers, as for all ancient Egyptians, the heart was nevertheless the seat of the soul and mental activity.

The first recorded challenge to our global heart-centred view occurred in ancient Greece. In the space of about three and a half centuries, between 600 and 250 BCE, Greek philosophers shaped the way that the modern world views so many things, including the brain. The early Greeks, like other peoples, considered that the heart was the origin of feelings and thought. This can be seen in the epic oral poems now attributed to Homer, which were created sometime between the twelfth and eighth centuries BCE; similarly, the ideas of the earliest recorded philosophers were focused on the heart.[9] In the fifth century BCE the philosopher Alcmaeon took issue with this view. Alcmaeon lived in Croton, a Greek town in the 'foot' of Italy, and is sometimes presented as a physician and as the father of neuroscience, although everything we know of him and his work is hearsay. None of his writings survive – all that remain are fragments quoted by later thinkers.

Alcmaeon was interested in the senses, and this naturally led him to focus on the head, where the key sense organs are grouped. According to subsequent writers, Alcmaeon showed that the eyes, and by extension the other sense organs, were connected to the brain by what he called narrow tubes. Aetius, living 300 years after Alcmaeon, is reported as having said that, for Alcmaeon, 'the governing facility of intelligence is the brain'. It is not clear how exactly Alcmaeon arrived at this conclusion – subsequent writers imply that he based his ideas not simply on introspection and philosophical musings, but also on direct investigation, although there is no evidence of this. He may have dissected an eyeball (not necessarily a human one) or he may have witnessed the culinary preparation of an animal's head, or he may simply have used his fingers to see how the eyes, tongue and nose were connected to the inner parts of an animal's skull.[10]

Despite these insights, the earliest unambiguous statements about the centrality of the brain were written several decades after Alcmaeon died; they came from the school of medicine on the island of Kos, whose most famous member was Hippocrates. Many of the works produced by the Kos school of medicine are attributed to Hippocrates, although the actual authors are unknown. One of the most significant of these documents was *On the Sacred Disease*, which was written around 400 BCE for a non-specialist audience and dealt with epilepsy (why epilepsy was considered a sacred or divine disease is unclear[11]). According to the author(s):

It ought to be generally known that the source of our pleasure merriment, laughter, and amusement, as of our grief, pain, anxiety, and tears, is none other than the brain. It is specially the organ which enables us to think, see, and hear, and to distinguish the ugly and the beautiful, the bad and the good, pleasant and unpleasant ... It is the brain too which is the seat of madness and delirium, of the fears and frights which assail us, often by night, but sometimes even by day, it is there where lies the cause of insomnia and sleep-walking, of thoughts that will not come, forgotten duties, and eccentricities.[12]

The argument in *On the Sacred Disease* was based partly on some pioneering but rudimentary anatomy ('the brain of man, as in all other animals, is double, and a thin membrane divides it through the middle', the author(s) stated), but it also revealed a great deal of confusion. For example, the document claimed that 'when a person draws in air by the mouth and nostrils, the breath goes first to the brain', arguing that the veins transport air around the body. Epilepsy was explained by the idea that a humour or fluid called phlegm entered the veins, preventing the air from getting to the brain and so causing the fit. Some people took the implications of localising epilepsy to the brain very seriously. Aretaeus the Cappadocian, a Greek physician who lived around 150 BCE, treated it by trepanation – drilling holes in the skull – a tradition that lived on in European medical manuals until the eighteenth century.[13] Aretaeus did not invent this operation – the earliest traces of any medical intervention are holes that were drilled or scraped into people's skulls and which can be found all over the planet, sometimes from over 10,000 years ago.[14] Although it is tempting to view prehistoric trepanation as an early form of psychosurgery (it is often suggested that trepanation was performed to let out 'evil spirits'), the global dominance of heart-centred ideas about the origins of thought suggests this is unlikely. There are more credible justifications for such a dangerous operation, including relief of painful subcranial bleeding or removal of bone fragments following a head injury.

Despite the arguments of Alcmaeon and of the Kos school, in the absence of any evidence to prove that the brain is the site of thought and emotion, there was no reason to prefer this claim to the obvious explanation that the heart plays this role. This led one of the most influential Greek philosophers, Aristotle, to dismiss the idea that the brain played any significant part in thinking or movement. As he wrote in *Parts of Animals*:

> And of course, the brain is not responsible for any of the sensations at all. The correct view [is] that the seat and source of sensation is the region of the heart ... the motions of pleasure and pain, and generally all sensation plainly have their source in the heart.

Aristotle's argument for the centrality of the heart was based on apparently self-evident principles, such as the link between movement, heat and thought. Aristotle noted that the heart clearly changed its activity at the same time as emotions were felt, whereas the brain apparently did nothing; he also affirmed that the heart was the source of blood, which is necessary for sensation, while the brain contained no blood of its own. Furthermore, all large animals have a heart, whereas – he claimed – only the higher animals have a brain. His final argument was that the heart is warm and shows movement, both of which were seen as essential features of life; in contrast, the brain is immobile and cold.[15] Given there was no actual proof of any link between thought and the brain, Aristotle's logical arguments were just as valid as those to be found in the writings of the Kos school. There was no way to choose between them. Elsewhere around the planet, things continued as before: for the vast majority of people, the heart was what counted.

*

After Aristotle's death, insight into the role of the brain emerged from Alexandria, at the western edge of the Nile Delta, in Greek-ruled Egypt. With a grid system of streets, underground plumbing and a multicultural population, Alexandria was one of the most significant centres of the Graeco-Roman world. Among those who benefited from this fertile intellectual atmosphere were the two leading Greek anatomists of the period, Herophilus of Chalcedon and Erasistratus of Ceos, both of whom worked in Alexandria.[16]

None of the writings of Herophilus and Erasistratus have survived, but subsequent writers claimed that they made important discoveries in the structure of the brain. The reason these breakthroughs came about in Alexandria was that, for a brief period, and apparently for the first time in history, the dissection of human bodies was permitted. It is even said that criminals who were condemned to death were vivisected under what must have been appalling circumstances. Exactly why dissection was allowed in Alexandria but not elsewhere is unclear, but whatever the case, physicians in the city made substantial anatomical advances relating to the liver, the eye and the circulatory system. They even described the heart as a pump.

The direct study of human anatomy enabled Herophilus and Erasistratus to make significant discoveries with regard to the brain and the nervous system. Herophilus supposedly described the anatomy of two key parts of the human brain – the cortex (the two large lobes of the brain), and the cerebellum, at the back of the brain, which he considered to be the seat of intelligence – as well as showing the origin of the spinal cord and how the nerves branch. He is said to have distinguished between nerves that were linked to the sense organs and the motor nerves that guide behaviour, developing a theory of sensation in which the optic nerve was hollow and some kind of air moved through this space.[17] Erasistratus apparently took a different approach, comparing the human brain with the brains of stags and hares, concluding that the greater complexity of the human brain, as shown by its convolutions, was responsible for our greater intelligence.

Despite the accuracy of their descriptions, the work of Herophilus and Erasistratus did not settle the issue of whether the heart or the brain is the site of thought and feeling. They merely showed that the brain was complicated. Aristotle's heart-centred view remained enormously influential, partly because of his immense prestige, but above all because it corresponded to everyday experience.

It was another 400 years before decisive evidence about the role of the brain was obtained, through the work of one of the most influential thinkers in the history of Western civilisation: Galen. A Roman citizen, Galen was born in 129 CE to a wealthy family in the city of Pergamon in what is now western Turkey.[18] Although today Galen is principally known as a writer on medical matters – his ideas shaped Western medicine and culture for 1,500 years – he was in fact one of the major thinkers of the late Roman world, producing millions of words of philosophy, poetry and prose.[19]

Galen travelled and studied throughout the eastern Mediterranean, including Alexandria, but the key years of his life were spent in Rome. He arrived there in 162 CE, aged thirty-two, following a four-year stint as physician to the gladiators in Pergamon, during which time he learned much about the human body by treating the fighters' wounds. He soon became a fashionable Roman physician, attending some of the leading figures in the city, including Emperor Marcus Aurelius, and gaining a reputation as a brilliant anatomist who had a taste for polemical argument. To demonstrate his discoveries Galen used 'lecture-commentaries' in which he simultaneously described his new knowledge and showed it in an animal. In these lectures the audience was invited to witness Galen's performance, and thereby to validate his claims – this was part of Galen's emphasis on the importance of experience in understanding. (The following explanation of how Galen came to some of his conclusions is rather grisly. If you are squeamish, you might prefer to skip the next three paragraphs.)

One of the key issues that interested Galen was the role of the brain and the location of thought and the soul – he was convinced that the brain was fundamental to behaviour and thought, and that he could prove it by experimentation on animals. All this at a time where there were no anaesthetics. Galen was not immune to the horror that he was inflicting – he counselled against using monkeys as their facial expressions during the experiment were too disturbing. Although Galen disagreed with those who argued that animals lacked part of the soul relating to anger and desire, he said nothing about pain – pain is not to be found in his descriptions of his work.[20]

One of Galen's most decisive experiments focused on the role of the nerves in the production of the voice; this was done on a pig because 'the animal that squeals the loudest is the most convenient for experiments in which the voice is harmed'.[21] With the poor pig strapped down on its back, its muzzle bound tightly shut, Galen cut into the flesh and revealed the recurrent laryngeal nerves that run either side of the carotid artery in the neck. If he tied a thread tightly round the nerves, the muffled squealing of the animal ceased; if he loosened the ligature, the voice would return. Although squealing was clearly produced by the larynx, something appeared to be moving down the nerves from the brain.

This insight was reinforced by one of Galen's most remarkable demonstrations, in which he proved the importance of the brain by directly confronting opponents with the implications of their heart-centred views. Having cut open a living animal, Galen obliged his contradictor to squeeze the beast's heart and prevent it from beating. Even with the heart stopped, the poor animal continued its muted whimpers, showing that the movement of the heart was not necessary for the animal to make sounds. But when Galen opened the skull and made his rival press on the brain, the animal immediately stopped making a noise and became unconscious. When the pressure was released, Galen reported, 'the animal returns to consciousness and can move again'. This must have been quite astonishing for the audience. As the historian Maud Gleason has put it, 'Galen's anatomical performances look less and less like an intellectual debate and more like a magic show.'[22]

On the basis of this evidence – supported by many other anatomical descriptions and surgical interventions, including on patients – Galen became certain that the brain was the centre of thought. He argued that the brain produced a special kind of air or *pneuma* that leaked out if the brain was injured, producing unconsciousness; when enough of this air accumulated, consciousness returned. Movement of the body was a consequence of the air produced by the brain moving down the apparently hollow nerves, Galen said. His anatomical work – most of it done on animals rather than on humans – showed that all nerves came from the brain, not the heart as Aristotle had claimed.

Despite the evidence that Galen presented, the authority of thinkers such as Aristotle and the power of everyday experience prevented brain-centred views from driving out the old ideas, even in Rome. Galen left an immense volume of work – around 400 treatises, of which over 170 survive, covering the whole range of medicine and natural science – but the decline and fall of the Roman Empire led to a collapse of the intellectual environment that could have permitted further discoveries. Simply thinking about where thought came from would never resolve the issue – as Galen's work indicated, it would require anatomical and experimental investigation, which in turn could occur only in a context of intellectual openness and knowledge of past successes and failures through the circulation of ideas. Those conditions would not be repeated for centuries.

*

Much of the cultural heritage of Rome and Greece was preserved in the libraries of the eastern Roman Empire, centred on Byzantium (modern-day Istanbul). From the seventh century onwards, the appearance of various caliphates associated with the rise of Islam led to a culture that spread to France in the west, to Bulgaria in the north and to Turkmenistan and Afghanistan in the east. This Islamic society placed a high value on knowledge and technical skills, and to meet the appetites of the new dominant classes and ruling groups, bridges and canals were constructed, horoscopes were cast, paper and glass were made. All this required rediscovering old wisdom or developing new understanding.[23]

First there was a wave of translations of the Greek and Roman texts that could be found in Persian or Byzantine libraries – this trend was centred on Baghdad and was sponsored by the caliphs and rich merchants. The ideas in these documents were soon extended as thinkers developed whole new areas of knowledge such as algebra, astronomy, optics and chemistry. But medicine and anatomy remained firmly anchored in Greek and Roman views, tied to the texts that were translated. In particular, the arguments about the roles of the heart and the brain that had existed since the time of Aristotle and Hippocrates were transmitted down the centuries more or less intact.

One of the leading physicians and philosophers of this period was Ibn-Sīnā, known in the West as Avicenna. Born in what is now Uzbekistan in 980, Avicenna lived in what is now Iran and wrote hundreds of books. His work combined Greek and Arabic thinking as well as treatments and diagnoses from as far afield as India; translated into Latin in the twelfth century, it exerted a profound influence on Western medicine for 500 years. Avicenna accepted Galen's claim that nerves arise from the brain or the spinal cord, but insisted, like Aristotle, that the primary source of all movement and sensation was nevertheless the heart.[24] This view also fitted with the Qur'an, which often refers to the heart as the source of understanding and, like the Bible, contains no mention of the brain at all.

Another route by which Galen's ideas were transmitted during this period was through the work of the tenth-century physician 'Alī ibn al-'Abbās Maǧūsī, known in the West as Haly Abbas – a historian has described him as 'a Persian who took an Arab name and wrote in the language of the Qu'ran, a Zoroastrian who was imbibed with Greek traditions, a thinker from the Islamic world who was adopted by the Western Latin community less than a century after his death'. To emphasise the cosmopolitan mix of this period, his work was subsequently translated into Latin in Italy by a Christian monk who had been a Muslim refugee from North Africa.[25]

Among the writings of Galen that Haly Abbas translated were those covering the structure and the role of the brain: 'The brain is the principal organ of the psychical members. For within the brain is seated memory, reason and intellect, and from the brain is distributed the power, sensation and voluntary motion.'[26]

Haly Abbas also put forward an idea that was not present in Galen – he claimed that the three cavities or ventricles in the brain were full of animal spirits* that were created in the heart and transported in the blood. Each of the ventricles, he said, had a different psychological function: 'Animal spirit in the anterior ventricles creates sensation and imagination, animal spirit in the middle ventricle becomes intellect or reason, and animal spirit transmitted to the posterior ventricle produces motion and memory.'

Despite the lack of evidence for this idea, it was widely held throughout Europe and the Middle East for over a millennium.[27] It had first appeared in the writings of the fourth-century Bishop

---

*The term 'animal' has the same root as in 'animated', and refers to *animus*, the Latin word for heart and mind.

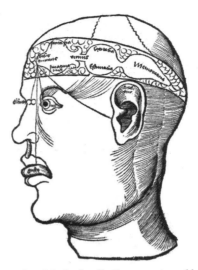

The theory of ventricular localisation as portrayed by Gregor
Reisch, from 1504. Perception and imagination are located in the
front, cognition in the centre and memory at the back.

Nemesius of Emesa in Syria, and a few decades later was briefly
mentioned by Saint Augustine, thereby acquiring a patina of religious approval that helped maintain its popularity.[28] For over 1,200
years, ventricular localisation was widely accepted as self-evident
– between the fourth and sixteenth centuries, at least twenty-four
different versions were put forward.[29] Among those who unques-
tioningly accepted this theory were some of the greatest thinkers
of Europe and the Arab world, including Leonardo da Vinci, Roger
Bacon, Thomas Aquinas, Averroes and Avicenna.

*

By the beginning of the thirteenth century, Latin translations of
Avicenna's writings, including its uneasy alliance of ventricular
localisation and a heart-centred origin for all thought and emotion,
became dominant in the new universities of Europe. Although Haly

Abbas's version of Galen's brain-centred views had been spread by the medical school of Salerno, south of Naples, Avicenna's ideas were eventually favoured because they were based on Aristotle's philosophy. Aristotle's ideas came to dominate thinking in Europe partly through the writings of the Dominican monk Thomas Aquinas, who towered over Western intellectual life for centuries. Aquinas sought to render Aristotle's ideas compatible with Christianity, fusing Christian dogma with the contradictory ideas of the ancient pagans. Areas of understanding that should have been the focus of empirical investigation, such as anatomy, became shrouded in a fog of religiosity, with theologians playing a decisive role in transmitting knowledge and in determining what was acceptable.

Readers of these newly available texts were well aware of the difference between the heart-centred view of Avicenna and Aristotle, the brain-oriented conceptions of the Salerno school and Galen, and the various attempts to reconcile them. In the thirteenth century, for example, Albertus Magnus squared the circle by simply arguing that Galen was wrong, and that all nerves did indeed have their origin in the heart, as Aristotle said.[30] The modern response to such contradictory claims would be to make direct observations; the solution in the Middle Ages was scholastic and theoretical – thinkers tried to resolve the contrasting views of their revered predecessors by close textual analysis, not by experimentation.

But at the beginning of the fourteenth century, medieval scholasticism's grip on anatomical knowledge was loosened slightly in the Bologna medical school, where Mondino de Luzzi was Professor of Medicine and Anatomy. Mondino wrote a manuscript entitled *Anatomia Mundini* (Mondino's Anatomy), which was based on his experience of dissecting the human body – the first such account since the time of Erasistratus and Herophilus in Alexandria over 1,500 years earlier.

The moral and sociological changes of the early 1300s that allowed Mondino to carry out dissections are not clear. The cadavers he dissected appear to be those of criminals (his instructions begin, matter-of-factly: 'The human corpse, killed through decapitation or hanging, is placed in the supine position.'[31]) There were some precedents – animal dissections were carried out in Salerno in the twelfth century, and post-mortem investigations had taken place in Bologna in the previous decades, apparently in order to determine the cause of death. Mondino's integration of dissection into the training of physicians may therefore have felt more like an obvious development, rather than a daring innovation.[32] It did not mark a break with religious teaching – dissection was not forbidden by either Christian or Islamic theology. Some Arabic texts from the ninth and twelfth centuries refer to dissection, but in general it seems that when scholars discovered and translated the writings of Galen and Aristotle, they were satisfied with the knowledge they contained, and did not seek to compare the views of the ancients with their own observations.[33] That now began to change. And unlike the brief period in Alexandria over 1,500 years earlier, this shift in attitude to dissection became permanent, in western Europe at least.

The decisive point was not that Mondino looked inside a dead body, but that in so doing he revealed the importance of personal investigation. The implication – that claims about the human body could be tested, and that knowledge could be independently acquired and not simply copied from the ancients – would eventually prove revolutionary. However, although Mondino's method was radical, his conclusions were not – he simply reiterated Galen's views with regards to anatomical structures and added an Aristotelian interpretation of their function, according to which the heart was the origin of movement, including the voice.[34]

Mondino's book showed that dissection was a potential tool for understanding, but his work did not have any major impact. In a world before print, ideas moved slowly. Textual evidence from the ancients was seen as decisive – starting with the Bible and including all the texts that Aquinas and other church leaders had integrated into their theology. Faith, not fact, was still the essence of knowledge and formed the framework of European intellectual life.

*

From the fifteenth century onwards, the pace of cultural and technological change in Europe suddenly accelerated, in successive periods that are traditionally called the Renaissance and the Scientific

Revolution; historians are still arguing about what might have caused these moments, or even whether they occurred at all. The European invention of printing (several hundred years after the Chinese produced movable type) transformed the distribution of knowledge, while translations of the Bible into the vernacular and the rise of Protestantism encouraged the idea that knowledge of the world could be acquired directly by individuals, rather than being necessarily mediated through authority. Revolutions in the Netherlands and England overthrew the old aristocratic powers, allowing political, social and economic space for new classes, with more radical views about the world. Meanwhile, the discovery of the Americas by Europeans, and the appearance of new diseases such as syphilis, undermined faith in ancient texts that were of little use in trying to understand these developments. Finally, the invention of the telescope and the microscope revealed hitherto unimagined worlds, while technological developments such as piston pumps and clockwork provided suggestive new metaphors that seemed to explain how everything, from the movement of the stars to the human body, might work.

In 1543 two books were published that, on very different scales, helped transform how we look at the universe and its inhabitants. The first, *De Revolutionibus Orbium Coelestium* (On the Revolutions of the Heavenly Spheres) by Nicolaus Copernicus, outlined a mathematical model in which the earth rotated around the sun, using theorems developed over two centuries earlier by Arabic astronomers. The second was *De Humani Corporis Fabrica* (On the Fabric of the Human Body) by Andreas Vesalius. Running to over 700 pages in seven books, the *Fabrica* combined knowledge and aesthetics, presenting its readers with the most accurate description of human anatomy ever assembled. Vesalius used the power of the new printing technology to the full, enriching his text with more than 200 striking woodcut illustrations, all based on his dissections of human bodies. Vesalius, who was Professor of Medicine at Padua, had produced a work that was truly revolutionary, not only because of the knowledge it contained, but also because of the way that knowledge had been obtained and how it was presented to the reader.

In previous decades other authors, such as Jacopo Berengario da Carpi, had published illustrated accounts of human anatomy based

on dissection, but these provided neither graphic impact nor anatomically accurate detail.[35] Even in terms of brain dissection, there
were precedents: in 1517, the German military surgeon Hans von
Gersdorff produced a sheet showing six small images of the cerebral
cortex in various stages of dissection, while in 1538 Johannes Dryander of Marburg published eleven woodcuts portraying the dissection
of the brain in relatively simple terms.[36] Vesalius's 1543 masterpiece
was in a completely different league. Nothing like it had been seen
before.

Each of the books in the *Fabrica* dealt with a different body
system (bones, muscles, internal organs and so on). The final book,
sixty pages long, was devoted to the brain and contained eleven
figures of the open skull, apparently taken from at least six individuals.[37] Although the brain engravings seem to be naturalistic, accurate
illustrations, like the rest of the work they are highly selective and
posed representations of what can be seen.[38] Nevertheless, the *Fabrica*
represented an immense step forward in anatomical knowledge. For
example, Vesalius reported that he was unable to observe the *rete
mirabile* – the network of blood vessels that Galen claimed enabled

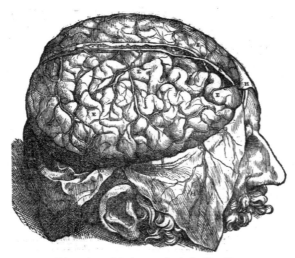

Dissection of the human brain by Vesalius.

the animal spirits to enter the brain. Vesalius audaciously – and accurately – concluded that Galen was wrong and that this structure did not exist in humans.[39] Students, he argued, should attend an autopsy, look closely, 'and in the future have less faith in anatomy books'.[40] Vesalius turned his objection to Galen's claim about this structure into a rallying-cry for a new way of studying the body.

Vesalius also grappled with the mystery of what it all might mean, how the body – and in particular the brain – might actually work. And at this point, his scalpel understandably failed him. Careful dissection of the human body could reveal structures, but apart from trivial cases (bones, sinews and nerves), it could not provide any real insight into function. These difficulties of interpretation were at their greatest when it came to understanding the origins of behaviour in humans and of our differences with animals. The problem, Vesalius explained, was that when he looked, he discovered that 'there is no difference at all in the structure of the brain in the parts I have dissected in the sheep, goat, cow, cat, monkey, dog, and birds when compared with the human brain'.[41]

Although the brain was proportionately much larger in humans than in other animals, Vesalius could find no qualitative difference between the structure of a human brain and that of other vertebrates. Whatever produced the evident behavioural and psychological differences between us and animals, Vesalius could not see it. Although his dissections were unable to provide an explanation of how the brain worked, they did suggest that the dominant ventricular theory of psychology might be wrong – the ventricles appeared to be 'nothing more than cavities or passages'. Without a better explanation of how the brain worked, Vesalius concluded by saying that 'nothing should be told about the locations in the brain of the faculties of the supreme spirit', lashing out at the theologians who dared to localise them, describing their ideas as 'lies and monstrous falsehoods'. Strong stuff.

The whole of Vesalius's study of the brain was based on the idea that it, rather than the heart, was the origin of thought and movement. The evidence for this assumption was in fact rather poor – the only experimental proof came from Galen over 1,200 years earlier. Three decades after the death of Vesalius, André du Laurens, Professor at the University of Montpellier and physician to Henri IV of France, could do no more than assert his belief in the role of the brain:

> I say then that the principall seate of the soule is in the braine, because the goodliest powers thereof doe lodge and lye there, and the most worthie actions of the same doe there most plainly appeare. All the instruments of motion, sence, imagination, discourse and memorie are found within the braine, or immediatly depending thereupon.[42]*

As to the role of the ventricles, du Laurens cautiously avoided the question, merely stating that it 'is not fully resolved upon'.

All these hesitant steps towards understanding the role of the brain in generating thought show that there was no single 'brain-centric moment' when thinkers realised that the brain, not the heart, was the key organ. The evident complexity of the brain compared with the heart strongly suggested where thought and emotion might be located, but the weight of tradition and the power of everyday experience meant that contradictory ideas were held by some of the greatest thinkers of the sixteenth and seventeenth centuries. The confusion many people felt was nicely summed up by Shakespeare, in one of the songs from Act 3 of *The Merchant of Venice*:

> Tell me where is fancy bred,
> Or in the heart or in the head?[43]

---

*Du Laurens described the widespread 'glass delusion' in which patients were convinced they were made of glass and feared they might shatter. Reporting of these symptoms declined in frequency from the early nineteenth century, showing that some mental health symptoms are tightly linked to social context. This is even true for post-traumatic stress disorder in soldiers – very different symptoms are reported now compared to those in the First World War. Speak, G. (1990), *History of Psychiatry* 1:191–206.

TWO

FORCES

17TH TO 18TH CENTURIES

During the seventeenth century, European thinkers became increasingly convinced that the answer to Shakespeare's question was most definitely 'the head' and, more precisely, the brain. The shift in attitude was slow and complex – there was no single experiment or dissection that resolved the question in favour of the brain. Instead there was a gradual accumulation of ideas and knowledge, all of which suggested a role for the brain, although old and new ideas coexisted. For example, in the 1620s William Harvey showed that 'the heart is simply a muscle', as the Danish anatomist Nicolaus Steno put it some decades later.[1] Although Harvey recognised the brain's complexity, calling it 'the organ of sensation' and 'the richest member of the body', he also felt that Aristotle was right, and that the blood carried some mysterious spirit that was generated by the heart. What now appears as a lack of clarity in Harvey's thinking reflected the absence of decisive evidence at the time.

One of the most influential figures to emphasise the significance of the brain was the French thinker René Descartes, who made dissections of the brain in the 1620s and 1630s. Although he decided not to publish his ideas following the condemnation of Galileo by the Catholic Church in 1633, they finally appeared posthumously in 1662.[2] Like many other thinkers, Descartes dismissed the suggestion that the heart was the seat of the passions as 'not worth serious consideration'.[3] His view of the brain was much more novel. For Descartes, animal bodies appeared to function as though they were machines – he even considered animals to be *bêtes machines* (animal machines or, more dramatically, beast machines) – with the brain playing a fundamental role.[4] Humans differed from other animals above all by their possession of a soul and their use of language, while the key anatomical difference between the brain of a human and an ape, say, was to do with the pineal gland, a pea-size structure at the base of the brain. Descartes claimed the pineal gland was unique to humans and that it generated the animal spirits from the blood that was supplied to it by the heart, thereby allowing for the interaction between the mind and the body. This was the place where, according to Descartes, the two fundamental parts of the universe – *res extensa* (the material thing) and *res cogitans* (the thinking thing, mind or spirit) – interacted.

This focus on the pineal gland was based on a mixture of assertion and doubtful anatomical evidence – Descartes claimed that nerves that projected upward into the cerebral cortex enabled the pineal to swing, and thereby to respond to the perception of various objects by moving 'in as many different ways as there are perceptible differences in the objects'.[5] No such nerves exist, and as soon as his claim became known in the 1660s, anatomists easily showed that this supposedly uniquely human structure is found in virtually all vertebrates.

One aspect of Descartes's ideas that had a lasting influence was his explanation of how the animal spirits moved in the nerves. Like many others, he thought that these spirits were fluid, and moved rapidly. But unlike previous thinkers, Descartes had an explanation for how this could produce behaviour: he had seen it in action in the shape of the hydraulic automata that were fashionable in French royal gardens at the time. These moving statues would eerily appear from the vegetation, play instruments or even speak as water and air were forced through their metal bodies. Descartes drew an explicit parallel between such automata and how humans and animals behaved:

Descartes's view of how movement occurs.

Indeed, one may compare the nerves of the machine I am
describing with the pipes in the works of these fountains, its
muscles and tendons with the various devices and springs
which serve to set them in motion, its animal spirits with the
water which drives them, the heart with the source of the
water, and the cavities of the brain with the storage tanks.[6]

Descartes used this model to describe the origin of simple
behaviours – what we would call reflexes.[7]* He presented a figure
showing what looks like a giant baby pulling its foot away from a
fire, because the spirits had moved from the foot, along the nerve, up
into the brain and then back down again to the muscles in the leg.
All this represented a decisive step forward from previous rather
vague explanations of behaviour and nerve function. For thousands
of years, thinkers had suggested that the spirits moved like a fluid
or the wind – the rapidity and intangibility of these forms of motion
made them attractive analogies. The organisation of hydraulic power
in automata was a far more compelling metaphor, but despite its

---

*Real and imagined automata were around in antiquity, but for reasons that
are unclear had not provoked the same kind of reflection among philosophers.

significance, there was still widespread disagreement as to what the spirits in nerves were made of, and Galen's confusing ideas about nervous air or *pneuma* were not much help.* Steno outlined the problem in 1665:

> Could they be a special substance separated from … glands? Might not serous substances be their source? There are some who compare them to spirit of wine, and it is suspected that they are in fact of a substance similar to light. In short, our standard dissections cannot clarify any of these difficulties concerning animal spirit.[8]

Steno's confident dismissal of all existing descriptions of nerve function was based partly on work using the latest technology – the microscope. His friend the Dutch microscopist Jan Swammerdam and the Italian anatomist Marcello Malpighi had both studied the contents of nerves, and agreed that there was neither fluid nor air within them – Swammerdam described such ideas as idle and absurd.[9] Experimental evidence against Descartes's hydraulic view of nerve function came when Swammerdam showed that if he stroked the outside of a dissected frog nerve with a pair of scissors, the attached muscle would contract, a result that he claimed 'applied to all the motions of the muscles in Men and Brutes'. Whatever was happening in nerves to produce behaviour, it was nothing like the moving water in Descartes's hydraulic automata – the same thing happened even if the end of the nerve was cut, thereby allowing any liquid or gaseous spirit to escape. As Swammerdam explained: 'a simple and natural motion or irritation of the nerve alone is necessary to produce muscular motion, whether it has its origin in the brain, or in the marrow, or elsewhere'.

Although Swammerdam was convinced that the true explanation

---

* According to the historian Erica Daigle, 'the nature of the *pneuma* has always perplexed readers of Galenic theory, whose mystified ranks include Vesalius, Thomas Willis, Descartes, and other interested parties of the seventeenth century'. She can add me to that list. Daigle, E. (2009), 'Reconciling Matter and Spirit: The Galenic Brain in Early Modern Literature', PhD thesis, University of Iowa, USA, p. 7, http://ir.uiowa.edu/etd/286.

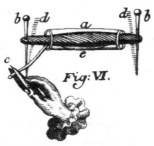

Swammerdam's experiment showing that touching a frog nerve (c) with metal scissors causes the muscle (a) to contract, pulling the pins (b) to position (d).

of nerve function lay 'buried in impenetrable darkness', he was pre-pared to speculate, using a new metaphor for how nerves work:

> it cannot be demonstrated by any experiments, that any matter of sensible or comprehensible bulk flows through the nerves into the muscles. Nor does any thing else pass through the nerves to the muscles: all is a very quick kind of motion, which is indeed so rapid, that it may be properly called instantaneous. Therefore the spirit, as it is called, or that subtle matter, which flies in an instant through the nerves into the muscles, may with the greatest propriety be compared to that most swift motion, which, when one extremity of a long beam or board is struck with the finger, runs with such velocity along the wood, that it is perceived almost at the same instant at the other end.

Swammerdam's experiments showed that neither the pneumatic nor the hydraulic models of nerve function were correct; instead, it seemed that some kind of intangible motion was involved – irrita-tion of the nerve produced an almost instantaneous response in the muscle, like a vibration. Swammerdam was groping for appropriate metaphors, but the key point was that he had shown that previous explanations were wrong, and that it was possible to produce move-ment artificially, by physically stimulating the nerve.

At the same time as the basis of nerve function was being explored, new studies of the brain were undertaken as anatomists responded to Descartes's ideas. Probably the most significant contribution came from Thomas Willis, a well-connected physician from Oxford. His gossipy contemporary John Aubrey sketched him in half a sentence – 'middle stature: darke brindle haire (like a red pig) stammered much'.[10] Influenced by Robert Boyle, the intellectual leader of the newly founded Royal Society of London, in the early 1660s Willis began to sketch out materialist explanations of mental health problems, which he saw as having their origins in the brain.[11]

In 1664 Willis published a book in Latin describing the anatomy of the brain, which was beautifully illustrated by his friend Christopher Wren. Over the following two decades the book went through eight editions and was published in Amsterdam and Geneva as well as London. The English translation, which appeared in 1684, is hard to understand, partly because of the archaic language but also because, according to the curmudgeonly historian of comparative anatomy F. J. Cole, Willis's Latin was 'elegant but involved'. For Cole, Willis lacked 'the gift of clear and intelligible expression' and was prone to 'the subtleties of speculative disputation'.[12] To put it bluntly, Willis was unclear about what exactly he thought.

Willis described the results of a massive programme of dissection that far surpassed that of Descartes – as well as human brains, his study involved 'hecatombs' of animals: horses, sheep, calves, goats, pigs, cats, foxes, hares, geese, turkeys, fish and a monkey.[13] Through his dissections, and the use of stains injected into blood vessels to reveal connections between regions of the brain, Willis came to the conclusion that it was the substance of the brain itself that enabled thought, not the ventricles, which were merely 'a vacuity resulting from the folding up of its exterior border'.[14] As Vesalius had suggested, they were nothing more than fluid-filled spaces.

For Willis, the structural complexity of the matter that made up the brain – he unhelpfully described it as having an 'anfractuous or broken crankling frame' – reflected its functional organisation. Memory was to be found in the convolutions of the cortex, he claimed, while the cerebellum was involved in involuntary acts such as heartbeat and was common to most vertebrates. Willis reached these conclusions mainly on the basis of extensive comparative anatomy and the connections he observed between these regions and various parts of the body. In humans the surface of the brain was highly complex, showing many convolutions, while it was simpler in cats and simpler still in fish and birds. Willis correlated these differences with different mental abilities – 'These folds or rollings about are far more and greater in a man that in any other living Creature, to wit, for the various and manifold actings of the superior Faculties.'

In the case of visual perception, Willis argued that a 'sensible impression'* created in the eye would be carried by an 'undulation or waving of waters' to the cerebrum where perception would 'arise', a memory of the image would be localised in the outer layers or the cortex while, for reasons that were unclear, imagination would be found in the corpus callosum (the structure that links the two hemispheres of the brain). The animal spirits, argued Willis, were created in the cortex of the brain, which transformed something in the blood into spirit. He assumed that blood and the heart were the sources of the essential features of life, just as thinkers had argued for thousands of years. As to how these spirits produced behaviour, he was vague: they 'enter into other manner of motions, and divers ways of emanations', Willis wrote, and would 'unfold', 'diffuse' and 'go forward', and eventually 'produce the acts of the Imagination, Memory, Appetite, and other superior Faculties of the Soul'.

Despite Willis's anatomical precision, his ideas about how the brain worked were pure speculation. A few months after Willis's book appeared, Steno visited Paris at the invitation of his patron, the rich and influential French bibliophile and ex-spy, Melchisedec Thévenot.[15] At the beginning of 1665, the brilliant but intense Dane, who was only twenty-seven, gave a lecture on the brain at Thévenot's country house in Issy, just south of Paris. He addressed Thévenot's

---

*The term 'impression', which began to be used as a synonym for 'sensation' in the seventeenth century, expressed the idea that perception involves a physical trace, altering the shape or function of the nerves, through some kind of pressure.

small circle of intellectual friends, which was one of the precursors of the French Académie des Sciences, and was forthright in describing contemporary ignorance about the brain: 'Instead of promising to satisfy your curiosity in what concerns the anatomy of the brain, I do confess here sincerely and publicly that I know nothing of this matter.'[16]

For Steno, as for Willis, the organisation of the brain was presumed to reflect its function, and yet, as Steno emphasised, that organisation was unfathomable. Steno not only dismissed ventricular localisation but was also scornful of Willis's gratuitous identification of various bits of the brain with different activities. With regards to the corpus callosum, which Willis claimed was the site of imagination, Steno pointed out that so little was known about this structure that anyone could 'say what he pleases about it'.[17] As Steno repeatedly underlined, most of what had been written about the brain was characterised by 'very obscure terms, metaphors and inappropriate comparisons'.

Unlike Descartes, Steno was not interested in the location of the soul, beyond accepting that the brain 'is certainly the main organ of our soul and the tool with which the soul carries out admirable tasks'. Steno was a profoundly religious man – he would soon convert to Catholicism, abandon science and become a bishop – but his investigations told him nothing about the location of the soul within the brain, so he did not speculate.

Steno argued that thinkers had first to accurately describe the components of the brain – this should include precise drawings, and comparative studies of animals, including at different developmental stages. Then, in a dramatic turn of phrase, Steno made a bold suggestion not only as to how we should think about the brain, but also how we should go about investigating it:

> The brain being indeed a machine, we must not hope to find its artifice through other ways than those which are used to find the artifice of the other machines. It thus remains to do what we would do for any other machine; I mean to dismantle it piece by piece and to consider what these can do separately and together.[18]

Steno himself did not carry out this research programme – soon afterwards he left for Tuscany, where, in a brief period of time, he founded geology, declared that women have eggs and found out how muscles work, before becoming a priest in 1675. Nevertheless, Steno's insight into how we should investigate brain function was profound – this approach of taking it apart and attempting to identify the functions of those parts is more or less what we have been doing ever since.

*

Steno's view that the brain was not simply like a machine, but actually was some kind of device, was part of a shift in view that took place in seventeenth-century Europe. Philosophers and physicians became comfortable about using mechanical metaphors when thinking about the body (the same view was also extended to the whole universe, with the regularity of celestial mechanics being seen in terms of some cosmic clockwork).[19] For example, in 1641 the philosopher Thomas Hobbes asked rhetorically: 'For what is the Heart, but a Spring; and the Nerves, but so many Strings; and the Joynts, but so many Wheeles, giving motion to the whole.'[20]

The analogies Hobbes drew between technology and anatomy corresponded reasonably well to the physical function of many body parts – the heart is indeed a pump (or a spring), and so on. The brain was rather different, both in its apparent lack of internal organisation that could be understood in terms of physical components, and in the absence of any machine, beyond a clock, that could provide an appropriate metaphor. Because of the lack of any decisive experimental evidence about brain function, the seventeenth- and eighteenth-century debates over the link between brain and mind focused on metaphysical aspects of how such a link might exist, rather than using contemporary machines as illuminating metaphors or providing anything concrete in the way of evidence. These philosophical debates laid the foundation for most subsequent views of the link between brain and mind.

A strictly materialist approach was represented by Hobbes, who dismissed Descartes's contradictory ideas about the soul being an 'immaterial substance' and instead simply argued that the thing that is thinking must be composed of matter. Thinking matter. Hobbes's approach was shared by the extraordinary Margaret Cavendish, Duchess of Newcastle.[21] In 1664 Cavendish argued that 'sensitive and rational matter … makes not onely the Brain, but all Thoughts, Conceptions, Imaginations, Fancy, Understanding, Memory, Remembrance, and whatsoever motions are in the Head, or Brain'. She went on to challenge those who believed in a non-material mind:

> I would ask those, that say the Brain has neither sense, reason, nor self-motion, and therefore no Perception; but that all proceeds from an Immaterial Principle, and an Incorporeal Spirit, distinct from the body, which moveth and actuates corporeal matter; I would fain ask them, I say, where their Immaterial Ideas reside, in what part or place of the Body?[22]

Princess Elizabeth of Bohemia had similarly expressed her incomprehension at Descartes's views in a private letter to the French philosopher, written in 1643: 'I have to say that I would find it easier to concede matter and extension to the soul than to concede that an immaterial thing could move and be moved by a body.'[23]

For the Princess, it was more straightforward to imagine the existence of thinking matter than to accept Descartes's suggestion that an immaterial substance – whatever that was – somehow interacted with the physical world.

A few decades later the radical Dutch philosopher Baruch Spinoza was convinced that 'mind and body are one and the same thing', but he also accepted that, given the knowledge of the time, this identity could not be proved:

> Again, no one knows how or by what means the mind moves the body, nor how many various degrees of motion it can impart to the body, nor how quickly it can move it. Thus, when men say that this or that physical action has its origin in the mind, which latter has dominion over the body, they are using words without meaning, or are confessing in specious phraseology that they are ignorant of the cause of the said action, and do not wonder at it.[24]

Many philosophical heavyweights were opposed to materialist explanations of mind. In one of his last writings, in 1712, Gottfried Leibniz expressed the commonly held view that there was no such thing as thinking matter, because it was impossible to imagine how it might work:

> If we pretend that there is a machine whose structure enables it to think, feel and have perception, one could think of it as enlarged yet preserving its same proportions, so that one could enter it as one does a mill. If we did this, we should find nothing within but parts which push upon each other; we should never see anything which would explain a perception.[25]

This argument became known as Leibniz's Mill, and suitably technologically updated versions of it have been used down the centuries, including in present-day arguments over how the brain works.

*

Philosophical debates over the possible existence of thinking matter grew in intensity following the appearance of John Locke's 1689 *Essay Concerning Human Understanding*.[26] Now remembered as a philosopher, Locke trained as a physician and was a close friend of Richard Lower, who had helped Willis carry out his dissections at the beginning of the 1660s; he was also a member of the Royal Society (Boyle was his patron). Although the *Essay* was initially well received – it was soon taught at Oxford – by the end of the century it was subject to a growing number of attacks, because of the way Locke dealt with the question of thinking matter. Locke's views, or what people took to be his views – philosophers are still arguing about exactly what he meant – shaped much of eighteenth-century Western thinking about mind, soul and self.

Surprisingly, given this long-term influence, Locke's direct contribution to the thinking matter debate was fairly minimal. In the third part of his *Essay*, Locke briefly put forward two possible explanations for the origin of thought, which he considered to be equally likely. Either God could have created matter such that it was able to think, or he could have fixed onto inert matter some immaterial substance that was thought. As Locke explained in his typically tortuous prose:

> We have the ideas of *matter* and *thinking*, but possibly shall never be able to know whether any mere material being thinks or no: it being impossible for us, by the contemplation of our own *ideas*, without revelation, to discover where Omnipotency has not given to some system of matter, fitly disposed, a power to perceive and think, or else joined and fixed to matter, so disposed, a thinking immaterial substance: it being, in respect of our notions, not much more remote from our comprehension to conceive that God can, if he pleases, superadd to matter a faculty of thinking, than that he should superadd to it another substance with a faculty of thinking.[27]*

Despite being far less assertive than the arguments of Hobbes or Cavendish, Locke's mild suggestion that thinking matter might be possible outraged many conservative thinkers, who perceived in it the following blasphemous chain of argument: if matter could think, that implied the soul must be material, in which case logic suggested it could not be immortal. One Irish theologian accused Locke's work of being 'in all probability the last great Effort of the Devil against Christianity'.[28]

Another thread of opposition to thinking matter flowed from the growing conviction that the universe is composed of particles. The argument went like this: given that all matter is made of atoms, then the atoms involved in thinking matter must have some special quality; but all atoms must be fundamentally identical, so the stuff that makes up the brain cannot be special in any way. This paradox was seen by many as a killer argument against thinking matter – either all matter could think, or none of it could. According to Richard Bentley, in a 1692 lecture to the Royal Society entitled

---

*My colleague Professor Helen Beebee reassures me that I am not alone in finding Locke's writing hard to understand: 'Your view of Locke accords with philosophy undergrads everywhere.'

Matter and Motion Cannot Think, belief in thinking matter led to 'monstrous absurdities': 'Every Stock and Stone would be a percipient and rational Creature ... every single Atom of our Bodies would be a distinct Animal, endued with self-consciousness and personal sensation of its own.'[29]

Some thinkers embraced this possibility – the English physician Francis Glisson argued that a fundamental feature of all matter was irritability (responsiveness might be a modern synonym), which was also the basis of perception and implied that the whole universe was in some way sentient. This view is known as panpsychism and continues to reverberate in some modern neuroscientific debates about the nature and origin of consciousness.[30]

For Bentley, thinking matter of any kind was simply impossible. He was even prepared to deny the omnipotence of the Creator by rejecting Locke's tentative suggestion that God could have created thinking matter: 'Omnipotence it self cannot create cogitative Body, And 'tis not any imperfection in the Power of God, but an incapacity in the Subject: the Idea's of Matter and Thought are absolutely incompatible'.[31] One of Locke's defenders, a theologian called Matthew Smith, rightly said of Bentley's argument: 'the substance of all his reasons amounts to no more than, this, we cannot conceive how meer matter and motion should produce sensation'.[32]

After Locke's death, the debate over thinking matter was summarised in the publication of a series of letters written in 1706–8 by Anthony Collins, a wealthy English freethinker who was a friend of Locke, and the philosopher Samuel Clarke, who was deeply opposed to Locke's ideas. For Clarke, as for Richard Bentley a decade earlier, if one part of the human body is conscious, then every particle of it must be, because every quality of any material system must reside in all its component parts.[33] Collins replied with an attempt to explain how the organisation of particles in the brain might give rise to consciousness, through what we would call an emergent property:

> it may be conceived that there is a power in all those particles that compose the brain to contribute to the act of thinking before they are united under that form, though while they are disunited they have no more of consciousness than any being

which produces sweetness in us ... has a power to produce sweetness in us when its parts are disunited and separated.[34]

Ultimately, the argument revolved around the nature of matter itself, and in particular the possibility, as explained by Collins, that the whole is characterised by something that is not possessed by each of its parts.[35] These were big questions that could not be resolved merely by arguing about them.

One of the suggestions about thinking matter that particularly irked many thinkers was that it implied that there was no fundamental difference between humans and machines. Parallels between humans and machines were generally considered to be highly immoral, because they were seen as questioning free will. If human choices somehow flowed from an underlying material process, rather than the spirit, then morality would collapse, went the argument. Many critics suspected that materialists would use the machine analogy to inveigle naive youngsters into rampant sexual behaviour. According to one John Witty, the cunning scheme of the materialists would be '*first* to Argue themselves into mere Machines; and afterwards in Letters to the Ladys; to persuade 'em, for what ends 'tis not difficult to determine, out of their Immaterial and Immortal Souls'.[36]

This convoluted conviction was widely held – materialism was considered a real threat to sexual morality. For example, the English mathematician Humphry Ditton clearly felt that the world was going to hell in a handcart and that this was all down to belief in thinking matter, which, he claimed, had the aim of undermining 'the very Foundations of Christianity' and was behind 'the Whole System of Modern Infidelity'.[37] Ditton's dramatic description of the consequences of accepting the existence of thinking matter showed how keenly he felt about the question: 'They have divested us of all our Intellectual Powers, and made up our very Souls of Wheels and Springs, so that we are only a Set of Moving pratling Machines.'[38]

In France, as might be expected, there was less concern about the implications of thinking matter with regards to morality, sexual and otherwise, and Locke's tentative ideas were generally better received on that side of the Channel. For example, in the early decades of the eighteenth century an anonymous manuscript entitled *L'Ame matériel*

(The Material Soul) circulated in French intellectual circles. This confused collection of texts included the claim that 'it is the matter of which the brain is composed that thinks, that reasons, that desires, that feels, and so on'.[39] The fact that the manuscript was never published reveals the official disapproval of such ideas, but the appetite among intellectuals to discuss them was real.

*

While philosophers worried away at the metaphysics of the mind, physicians and other investigators addressed the apparently simpler question of how perception and movement occurred.[40] Even Isaac Newton got in on the act: at the end of Book III of the second edition of his *Principia* (1713), Newton suggested that 'a certain most subtle Spirit' could be found in 'all gross bodies'. Bodily movement occurred through 'the vibrations of this medium, excited in the brain by the power of the will, and propagated from thence through the solid, pellucid and uniform capillaments of the nerves into the muscles for contracting and dilating them'.[41] Newton's views were not based on any specific physiological knowledge, but rather on his assumptions about how the universe worked. In the absence of any experimental data they were nothing more than hand-waving.

Many of the most influential eighteenth-century ideas about the link between the brain and bodily movement were spread through the teaching of Herman Boerhaave, Professor of Medicine at the University of Leiden. Boerhaave was probably the most significant physician of his time – between 1715 and 1776 nearly a hundred editions of his writings or commentaries upon them were published in Britain alone, while his students became some of the leading anatomists and physiologists of the age. Although Boerhaave knew that Swammerdam and Glisson's work showed there was no nervous fluid – in his final years he assembled Swammerdam's masterpiece, *The Book of Nature*, for publication – he continued to state that nerves contained 'Juice', 'the swiftest and easiest of any'.[42] This 'subtile Fluid' was formed from the blood, claimed Boerhaave, as people had argued since Galen. Boerhaave dismissed Swammerdam's experiments on frogs as being of little consequence to the understanding of humans:

these are in Reality no Objections to the Existence of a nervous Fluid, for the first two Experiments make nothing against us, and the rest only shew that the Fabric of the Nerves in cold amphibious Animals is different from that of the Nerves in Quadrupeds and hot Animals; so that no argument of Force can be thence drawn to make any Conclusions with regard to the human Body.

Boerhaave's conception of movement and nervous function was a souped-up version of Descartes's hydraulic view and may have been reinforced by the work of Giorgio Baglivi, who in 1702 claimed that the pulsations of the brain produced the circulation of a nervous fluid (in fact these brain movements are the consequence of arterial activity).[43]

In 1752 one of Boerhaave's students, the austere Swiss Calvinist Albrecht von Haller, outlined a new way of looking at the functioning of the nerves and the brain. Haller described two fundamental properties of living tissues – irritability and sensibility. He argued that movement is produced by irritability (he took the term from Glisson), which can be observed when muscles contract, and was carried by what he called the *vis insita* (contractile force), which continued to exist after death, as seen in Swammerdam's experiments on frogs' legs. Nerves, on the other hand, showed sensibility, which was carried by the *vis nervosa* (nervous force). This force ceased with death, said Haller, and his experiments indicated that it could be suppressed by tying a nerve, injuring the brain or treating a patient with opium. Extensive experimentation suggested that these two fundamental forces were completely separate: 'the most irritable parts are not at all sensible, and vice versa, the most sensible are not irritable', wrote Haller.[44]

Haller later argued instead that nerves must contain some kind of fluid, produced in the cortex of the brain, which moved down the 'small tubes of the nerves'. This 'nervous liquor', he argued, 'which is the instrument of sense and motion, must be exceedingly movable, so as to carry the impressions of sense, or commands of the will, to the places of their destination, without any remarkable delay'.[45] Despite Haller's claim that knowledge should be based on experimentation rather than analogy, in the end his understanding of nervous function was not dissimilar to that which had dominated thinking for centuries – Haller's fluid was no different from Galen's animal spirits.

Other thinkers were bolder. In 1749 the Yorkshire physician David Hartley published a book in which he proposed that vibrations run along nerves 'as sound runs along the surface of rivers'. This view was opposed by Alexander Monro, Professor at the University of Edinburgh and another of Boerhaave's students, who was convinced that there were fluids in nerves and that 'the nerves are unfit for vibrations because their extremities ... are quite soft and pappy'.[46] Hartley retorted that he did not believe that 'the nerves themselves should vibrate like musical strings' for the obvious reason that nerves are not taut, but nevertheless he could not explain how a vibration moved down a soft and pappy nerve.[47]

Despite these problems, Hartley extended his vibrational view to encompass the whole brain. Perception, he argued, somehow induced vibrations in the brain that would be essentially identical in different individuals. Furthermore, the location of such vibrations could explain learning:

> That when two or more Objects present themselves at the same time, the Impressions on the Sensory caused by them lying so near each other that in turning to that Part of the Sensory, the Mind cannot view one without the other, and so the Ideas answering to those Objects ever after keep in Company together.[48]

Hartley's idea, later called associationism, implied that sensations that were physically linked in the brain could form a memory.[49] Hartley also distinguished between 'automatic motions' – such as those shown by the heart and the bowels – and voluntary movements.[50]

Both Haller and Hartley's views were opposed by Robert Whytt, a Scottish physician from Edinburgh and yet another of Boerhaave's pupils. Whytt argued that there was an immaterial 'sentient principle' that operated via the nerves and the brain, which enabled bodies to move. In 1751 Whytt attacked Haller's suggestion that there was a force that led to muscular contraction as 'no more than a refuge of ignorance', and argued instead that irritability was simply a power of the soul.[51] Appropriately irritated, Haller responded by pointing out that Whytt could not explain how a muscle removed from the body still contracted if stimulated, unless he imagined that the soul was somehow present in every part of the body.[52] The two men carried on arguing in print until Whytt's death in 1766, and even afterwards – Haller continued to pursue his defunct opponent for over a decade.[53]

Whytt was so hostile to the possibility of a material basis to behaviour that he would not use the term 'automatic' to describe involuntary motions. As he warned, it could suggest that the body was 'a mere inanimate machine, producing such motions purely by virtue of its mechanical construction'.[54] However, he perceptively noted that some involuntary motions can in fact be influenced by the mind, showing these were not truly mechanical responses: 'Thus the sight, or even the recalled *idea* of a grateful food, causes an uncommon flow of spittle into the mouth of a hungry person.'

Whytt's work built upon the ideas of Jean Astruc, Professor of Medicine at Montpellier and Paris. Astruc was an extraordinary scholar – as well as writing the first book on venereology, he was also one of the pioneers in applying textual analysis to the Bible, suggesting that the Book of Genesis was written by more than one author. Astruc claimed that involuntary behaviours such as blinking, ejaculation and breathing were produced by animal spirits that flowed down the nerves, arrived at the brain and were then, following Descartes's idea, were 'reflected' back to produce the appropriate movement in the relevant organ. This is the origin of Astruc's coinage 'reflex'.[55] Nearly a century after it had first been described by Descartes, the reflex had a name.

Whytt's key contribution was to explore the physical basis of reflex movements. He showed that a spinal cord was required to make them happen, and that specific reflexes were associated with different parts of the cord – the movement of the lower limbs was produced by the lower part of the spinal cord, and so on.[56] Like Astruc, Whytt interpreted these facts in terms of a link between nerves that were stimulated and those involved in movement, which seemed to be located at the point where they met in the spinal cord or in the brain.[57] Although Whytt's view of the basis of thought was resolutely anti-materialist, his work suggested that certain behaviours could be explained by some kind of nervous connection between different parts of the body.

*

The most notorious contribution to the eighteenth-century debate over thinking matter came from another of Boerhaave's students, the Frenchman Julien Offray de La Mettrie. In 1747 La Mettrie published *L'Homme machine* (Machine Man), a manifesto for a new way of looking at the human mind and body, in which all the workings of the body and the mind could be explained by matter.[58] As La Mettrie wrote: 'all the soul's faculties depend so much on the specific organisation of the brain and of the whole body that they are clearly nothing but that very organisation'.[59] There was such a thing as thinking matter, claimed La Mettrie, and it was the brain.

According to his patron, Frederick the Great of Prussia, La Mettrie's realisation that 'the ability to think is merely the consequence of the organisation of the machine' came during a fever in 1744. In 1746 La Mettrie tentatively outlined this idea in print; his work was immediately condemned by the French authorities and he astutely fled to the Netherlands.

Undaunted, La Mettrie thought even harder about the material basis of mind; the outcome was *L'Homme machine* which was written and published – anonymously – in Leiden in 1747. This book had everything needed to make it a best-seller: it dealt with a daring idea but was written in an easy, conversational style, it made jokes and poked fun at the powerful, and it contained mild sexual references. The book was immediately banned in France, thereby inevitably encouraging the clandestine circulation of printed and manuscript versions. Even in supposedly tolerant Amsterdam the book was proscribed and publicly burned by the hangman. Despite – or, more likely, because of – its sulphurous reputation, La Mettrie's enterprising Leiden publisher quickly produced two further editions.[60]

Many of La Mettrie's ideas sound very modern – he suggested that we might be able to teach great apes to use sign language, because from 'animals to man there is no abrupt transition ... What was man before he invented words and learnt languages? An animal of a particular species.'[61] Over a century before Darwin, he also claimed that 'the form and composition of the quadruped's brain is more or less the same as man's ... man is exactly like animals both in his origin and in all the points of comparison'.

Interestingly, La Mettrie's starting point in *L'Homme machine* was mental health, and how it is affected by the state of the body. Some of the symptoms identified by La Mettrie look odd today – 'those who imagine they have been changed into werewolves, cocks, or vampires' – but he described, with evident sympathy, various forms of emotional turmoil, the terrible effects of insomnia or the tragedy of phantom limb syndrome in amputees. This was typical of a slow shift in attitudes towards mental illness that took place in Europe during the second half of the eighteenth century, reinforced by the madness of George III of Great Britain, which was diagnosed in 1788. But although some physicians expressed a more caring attitude to patients with mental health problems, they had little idea how to treat physical illnesses and their grasp of how to understand and treat mental health was even flimsier.[62] La Mettrie, despite his compassion and his modern-sounding ideas, was no different.

Behind La Mettrie's apparent modernity there were some rather older ideas. His explanation of how the brain might work focused on involuntary movements, which he called the 'springs of the human machine', but which he could only describe vaguely with analogies about clocks.[63] Unable to explain how matter could think, La Mettrie fell back on the assumption that it was a consequence of some unknown force that is specific to life: 'organised matter is endowed with a motive principle, which alone distinguishes it from unorganised matter ... that is enough to solve the riddle of substances and of man'. This produced a remarkable image of the human brain and body which La Mettrie saw as 'a machine that winds its own springs – the living image of perpetual motion ... man is an assemblage of springs that are activated reciprocally by one another'.[64] As modern commentators have recognised, these vitalist views suggest that despite the dramatic title of his book La Mettrie did not fully embrace a materialist approach.

In February 1748, when it was clear that *L'Homme machine* would land La Mettrie in a great deal of legal hot water in the Netherlands, he fled to Berlin, having accepted an invitation from Frederick the Great of Prussia. He became physician to the king, and joined Voltaire and other radical thinkers in the court. Frederick, who was extremely liberal when it came to philosophical questions, shared La Mettrie's views about thinking matter ('thought and movement ... are attributes of the animated machine, formed and organised as man', the king wrote to Voltaire[65]).

La Mettrie was a cheerful, vivacious man – his portrait makes him look like the kind of bloke you would like to chat with in the pub – and he became notorious for his relaxed attitude to court conventions. He would throw himself down on the palace couches and fall asleep; when it was hot, he would drop his wig on the floor, remove his collar and unbutton his jacket.[66] His contemporaries were not impressed – the conservative Haller disavowed him, while the French philosopher Denis Diderot described him as 'lunatic', 'dissolute, impudent, a fool, a flatterer'.[67] In November 1751 La Mettrie died suddenly and mysteriously, aged only forty-two. According to Voltaire, the cause was a meal of 'pâté of eagle disguised as pheasant ... well mixed with bad lard, chopped pork and ginger'.[68]

For the first half of the nineteenth century La Mettrie was forgotten, and the recent revival of interest in his work is primarily due to the parallels with modern conceptions of brain and behaviour, rather than any influence on subsequent thinking.[69] However, in broader terms, La Mettrie's work was significant – his suggestion that humans are machines soon penetrated into popular culture, in exactly the place that some of Locke's critics had predicted – pornography.

One of the most scandalous books ever published in English is *Memoirs of a Woman of Pleasure*, popularly known as *Fanny Hill* after its main character. The book was published a year after *L'Homme machine*, and within twelve months its author, John Cleland, was accused of corrupting the king's subjects and the book was banned.

So explicit was its content that it was only in 1970 that an unexpurgated version could finally be sold in the UK. In the book, the young Fanny repeatedly uses the term 'machine' to describe the various penises she encounters (there are many), while erection is often said to be due to 'irritation'. Several characters are described as 'a machine' or 'the man-machine' when engaging in piston-like sexual acts, while a central theme of the book is the link between the body and mind, as seen through the omnipresent prism of sexual desire.[70] Cleland may have read and been impressed by *L'Homme machine*, or he may have cynically added a zest of forbidden philosophy to spice up his piece of one-handed reading. Whatever the case, the cultural impact of the new view of humans as machines was real.

No matter how vague its exposition, the machine analogy at the heart of La Mettrie's work chimed with a growing interest in intricate machines and automata. Technological development, in particular miniaturisation, meant that Descartes's hydraulic statues had long been surpassed by clockwork mechanisms that were eerily lifelike. In 1738 the French inventor Jacques Vaucanson amazed Parisians with his mechanical flute player, followed a year later by a piper that accompanied itself on a drum, and a device known as the *Canard digérateur* ('Digesting Duck') that could move, eat and defecate.[71] In London, the clock-maker James Cox had an entire gallery devoted to his automata, including his beautiful mechanical silver swan, which can be seen at the Bowes Museum in County Durham. Perhaps the height of this period of ingenuity was The Writer, an automaton composed of nearly 6,000 parts that was created in the 1770s by the Swiss watchmaker Pierre Jaquet-Droz. This extraordinary device – now on display in Neuchâtel – could write letters with a quill pen, the glass eyes flicking back and forth, following the movement of the automaton's hand as though it was concentrating.

Nobody suggested these automata were alive, or that they were thinking, but their uncanny ability to reproduce aspects of behaviour suggested that their ticking innards might somehow shed light on how squishy brains and bodies might work.

*

Throughout the eighteenth century, the fundamental role of the brain
became increasingly rooted in the academic and popular imaginations. In 1734 the English writer Samuel Colliber proclaimed: 'That
the Brain is the Seat of Sensation (which we have observ'd to be one
sort of Thinking) is at present universally agreed.'[72] This was a slight
exaggeration, but that was clearly the way things were going. Nearly
half a century later Joseph Priestley, the great British chemist and
dissenting clergyman, who was heavily influenced by David Hartley,
proclaimed that thought 'is a property of the *nervous system*, or rather
of the *brain*'.[73] As he put it with typical Yorkshire bluntness: 'In my
opinion there is just the same reason to conclude that the brain *thinks*,
as that it is *white* and *soft*.'[74] Priestley even gave some good evidence
to back up his conviction:

> as far as we can judge, the faculty of thinking, and a certain
> state of the brain, always accompany and correspond to one
> another; which is the very reason why we believe that any
> property is inherent in any substance whatever. There is no
> instance of any man retaining the faculty of thinking, when his
> brain was destroyed; and whenever that faculty is impeded,
> or injured, there is sufficient reason to believe that the brain is
> disordered in proportion; and therefore we are necessarily led
> to consider the latter as the seat of the former.[75]

However, over the course of the eighteenth century there was a
slow shift in scientific thinking, from a universe ruled by mechanical explanations to one in which forces and sensitivities seemed to
dominate. The vitalism that had apparently been driven out by the
mathematisation of the universe in the seventeenth century was
making a comeback. Mechanistic views that had proved so successful in the hands of Newton and others had also revealed their limits
– Newton's theory of gravity had immense predictive power, but no
one was sure how gravity worked.* Gravity was real, but it could
only be observed, not captured or broken down into its component
parts. In the field of physiology, attempts to explain heat in the body

---

* We still do not know.

by mechanical models failed the test of experimentation, and by the mid-1700s, more vitalist interpretations were put forward, suggesting that there was something special about the processes taking place inside a living body, as La Mettrie suggested.[76] Similarly, ideas about nerve function and the nature of the mind had been dominated by mechanical analogies, but these seemed inadequate when faced with the newly identified forces of irritability and sensibility.

Furthermore, the expression of these forces in the nerves was not like some kind of hydraulic force that emerged as a consequence of pressure. Instead, it was conditional, and could be observed only under certain circumstances. In 1784 the Austrian physiologist Georg Prochaska claimed: 'As the spark is latent in the steel or flint, and is not elicited, unless there be friction between the flint and steel, so the *vis nervosa* [nervous force] is latent, nor excites action of the nervous system until excited by an applied stimulus.'[77]

This conditional, non-mechanical, view raised the question of how any known force could fulfil such a role. Neither water nor air nor vibration seemed to fit the bill. But there were exciting hints about what this latent force might be, hints that came from a new phenomenon with dramatic, terrifying effects on bodies, and which seemed to be linked with life itself – electricity.

THREE

ELECTRICITY

18TH TO 19TH CENTURY

At the beginning of April 1815 the Tambora volcano in Indonesia erupted with astonishing violence. A hundred cubic kilometres of rock were pulverised and thrown high into the sky, thick gases and microscopic debris circulated in the atmosphere for months on end and the whole planet's climate was severely affected. In Europe, the following year became known as 'the year without a summer' – crops failed, diseases spread, and in Switzerland four British tourists on the banks of Lake Geneva found themselves housebound by the 'wet, ungenial summer, and incessant rain'.[1] To pass the time, they decided that each of them would write a ghost story. One of those travellers was eighteen-year-old Mary Shelley, and the story she wrote was called *Frankenstein*. As Shelley later explained, her idea of Doctor Frankenstein assembling body parts and bringing them to life had its origin in experiments that had been conducted a few years earlier, as the bodies of newly executed criminals were stimulated with electricity, making their muscles twitch in a travesty of life.[2]

Interest in electricity had grown throughout the eighteenth century, and by the 1750s public displays of electrical phenomena had become commonplace in Europe.[3] These demonstrations were performed by 'electricians' who generated static electricity by rubbing a piece of glass or amber with some wool, or, better still, by using a custom-made machine in which a hand-cranked flywheel spun a glass object against some felt or wool, generating an electrical charge. The results were sometimes uncanny – St Elmo's Fire could be conjured up in a glass sphere, while in a trick known as 'the Hanging Boy', an unfortunate youth was suspended from the ceiling and electrified with a static charge by being rubbed with a glass tube; light objects such as feathers and metal flakes would then magically fly through the air and stick to him.

A key moment came in 1746, when Pieter van Musschenbroek of Leiden University invented a way of capturing and storing electricity.[4] When a silk thread was passed between a generator and a glass jar, which in its earliest form was filled with water, the jar accumulated an electric charge (it was soon discovered that metal foil linings on the inside and the outside of an empty jar would do the trick even better). If wires connected to the inside and the outside of the jar were touched at the same time there would be a massive shock as the jar instantaneously discharged (they can store over 30,000 volts). If someone else held on to the brave soul who was connecting the two parts of the device, they too would be shocked. This could be taken to surreal lengths: the French philosopher Jean Nollet persuaded 200 hapless monks to hold hands in a chain over 400 metres long and made them all involuntarily leap into the air as a charge passed through them, much to the delight of onlookers.[5]

Electricity was also used to treat various kinds of paralysis. Among the itinerant practitioners who toured Britain with their generating devices and what were called Leyden jars, offering to cure the sick, were the founder of Methodism, John Wesley, and a future leader of the French revolution, Jean-Paul Marat. So apparently successful was this therapy that from the 1780s onwards a number of European hospitals installed generators and Leyden jars.[6]

It was soon realised that electricity could have an effect on the body of any animal, even if it was dead. In 1753 Professor Giambatista Beccaria of Turin showed that violent contractions could be induced in 'the muscles of the thigh of a strong cock' if stimulated with electric sparks.[7] In Bologna, Marc Caldani removed the rear legs from a frog, and then brought an electrified rod close to the limbs: 'we always saw the muscles of the lower extremities make a movement.

This occurred solely through the power of electricity.'[8] Joseph Priestley also studied the effects of electricity on frogs and showed that a
shock from a Leyden jar could make a dead animal's lungs inflate.
Priestley's explanation of why he did not do more experiments to
investigate the matter shows that these researchers were not all hard-
hearted brutes: 'I would have given the shock to toads, serpents,
fishes etc., and various other exanguious animals, but I had not the
opportunity. Besides, it is paying dear for philosophical discoveries,
to purchase them at the expense of humanity.'[9]

In 1749 David Hartley linked the growing fascination with electricity to Newton's sketchy ideas about nerve function and suggested
that 'Electricity is also connected in various Ways with the Doctrine
of Vibrations'.[10] Six years later, the Swiss thinker Charles Bonnet took
a further step, wondering if 'the animal spirits are of an analogous
nature to that of light or electric matter'. Bonnet was perhaps the first
to use a word that is now fundamental to our understanding of how
nerves function – 'transmission': 'Are nerves simply threads devoted
to the transmission of this matter that is so marvellously quick?' he
asked.[11] In 1760 Bonnet again reflected on the link between electricity
and nerve function, suggesting that the nerves contained 'a fluid that
approached light in its subtility and mobility'. Bonnet was careful to
make clear that he had no evidence for this suggestion:

> We do not know the nature of the Animal Spirits: they are
> even more out of the range of our senses and of our instru-
> ments than the Vessels that filter or prepare them. It is only by
> way of reason that we are led to accept their existence, and to
> suspect some kind of analogy between these Spirits and the
> Electric Fluid. This analogy is principally based on some sin-
> gular properties of this Fluid; in particular the rapidity and the
> freedom with which it moves, along one or more threads, or
> through a body of water, even if it is moving.[12]

From the late 1750s, Bologna – location of the world's oldest
university – became the stage for an intense series of debates over the
role of electricity in nerve function. Marc Caldani and Felice Fontana
supported Haller's view that irritability was the explanation, while

others defended the traditional notion of animal spirits, but followed Bonnet in imagining that they were a form of electricity.[13] The fact that the dispute could not be resolved showed that knowledge had reached an impasse: new evidence of a different kind was needed to make progress.

*

For millennia, it was known that some fish produce a weird shock. In Europe, people knew of the benumbing effects of a small ray called the crampfish or torpedo ('torpedo' comes from the Latin *torpere*, meaning to be stiffened or paralysed), the ancient Egyptians drew pictures of the Nile catfish, which has similar powers, while the peoples of the Amazon Basin were well aware of the ability of the electric eel to paralyse animals.[14] However, the exact nature of the dramatic shock produced by these beasts was unclear – those who had investigated the effect, such as Francesco Redi in the seventeenth century or Réaumur in the eighteenth, considered that it was produced by a rapid movement of the fish.

In 1757 the French explorer Michel Adanson concluded that the effects of a shock from a freshwater catfish from Senegal were the same as those produced by a Leyden jar.[15] A decade later the naturalist Edward Bancroft showed that the shock produced by the torporific eel of Guyana (it is not, in fact, an eel) could be transmitted through a fishing line and thence to a chain of a dozen people, 'exactly similar to that of an electric machine'. Further investigations of the torpedo, inspired by the naturalist John Walsh and involving the physicist Henry Cavendish and the anatomist John Hunter, suggested that the organs responsible for the creation of the shock – large structures on either side of the upper surface of the fish – might function like a series of Leyden jars.

In 1775 Walsh was eventually able to obtain a spark from the charge produced by the fish, showing that it could generate electricity. Suggestions that the animal spirits might be electrical had encountered a major problem: the spirits were clearly limited to the nerves, but electricity flowed easily throughout the whole body. The torpedo showed that electricity could be contained within a particular organ, suggesting that nerves might be able to do something similar.[16] The French physicist Pierre Bertholon concluded that all animals had 'their own electricity', produced by the friction created by movements such as breathing, blood circulation and so on, acting like artificial electrical generators.[17] That electricity, he claimed, was the basis for all movement, stimulating the muscles via the nerves.

A few years later Luigi Galvani, a physician from Bologna, began to explore how animals respond to electricity from Leyden jars, mainly by studying the movement of isolated frogs' legs, following the work of Priestley and others three decades earlier. In 1791 Galvani revealed the acute sensitivity of nerves to electricity when he accidentally discovered that even the charged atmosphere that occurs on thundery days could induce muscular contractions.[18] Galvani's most challenging discovery was his observation of contractions in the absence of any external source of electric charge. Over a century earlier, Swammerdam had shown that if he stroked a frog nerve with a scalpel, the attached muscle would contract, an effect he imputed to irritation. Galvani found a similar effect, but noticed that a frog muscle contracted if it was placed on an iron plate and the associated nerves were then touched with another metal such as silver. Galvani concluded that there was some kind of innate electricity in the nerve, which passed via the metals into the muscle.[19] This effect was not limited to frogs. In May 1792 Galvani attended a double amputation performed by Professor Gaspar Gentili in St Ursula's hospital in Bologna; immediately after the operation, Galvani took the poor patient's amputated arm and leg and 'in the presence of the aforesaid Professor and other physicians and men of learning' made the fingers of the hand move and the muscles of the leg contract, simply by touching a nerve with a piece of tinfoil, the muscles with a piece of silver, and then allowing the two metals to come into contact.[20]

Galvani claimed that these experiments revealed the existence of what he called animal electricity, which was 'contained in most parts of animals; but manifests itself most conspicuously in muscles and nerves' and was essentially the same stuff as could be observed in the torpedo and other similar fish.[21] Animal electricity, claimed Galvani, was generated by the cortex and was then extracted from the blood and entered the nerves. In a way, nothing much had changed – this

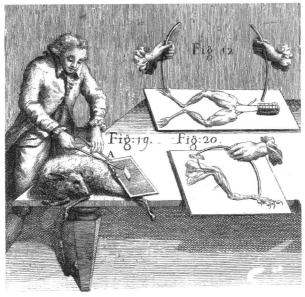

Galvani's experiment on frogs' legs. The man on the left is generating a static electric charge by rubbing the wool of the sheep.

was similar to the ideas about the generation of animal spirits from centuries earlier.

As to how electricity in the nerve caused the muscle to contract, Galvani could not say, although he wondered whether it might be due to a vapour, or to irritation. Despite these problems of understanding how the simplest of movements might occur, Galvani was prepared to speculate about the most complex question of all – the link between mind and movement:

> perhaps the mind, with its marvellous power, might make some impetus either into the cerebrum, as is very easy to believe, or outside the same, into whatever nerve it pleases, wherefrom it will result that neuro-electric fluid will quickly flow from the corresponding muscle to that part of the nerve to which it was recalled by the impetus.[22]

In 1793 the Turin physician Eusebio Valli enthusiastically supported and supplemented Galvani's claims, arguing that the old animal spirits had been replaced by the new idea of animal electricity.[23] Valli realised that if nerves functioned on the basis of electricity, then like the torpedo's electric organ, they must have some special structure, completely unlike that found in other tissues: 'the brain, spinal marrow, and nerves have a specific constitution, and that it is upon this that the mode of electricity in them depends'.

A few months later, the Edinburgh physician Richard Fowler put his finger on a problem: Galvani's animal electricity effect seemed to occur only if the tissue was touched by two different metals.[24] This criticism was also at the heart of the work of Alessandro Volta of the University of Pavia, who showed that the mere contact of two different metals generated a weak electrical current, which in turn induced contraction in the frog muscle. He summarily dismissed Galvani's claim to have discovered an innate electricity in animals – muscular contraction was simply a response to the electrical stimulation produced by the contact of two different metals.[25]

Stung by Volta's criticisms, Galvani, along with his nephew Giovanni Aldini, carried out experiments showing that muscle contraction could be obtained simply by allowing a nerve to touch a bared muscle, with no metal involved. This result was confirmed two years later by Alexander Humboldt.[26] Volta was not impressed, arguing that even in these cases, some external component, such as liquids on the outside of the tissues, was involved in inducing contraction.[27] The body, Volta claimed, was entirely passive, responding to an external electrical stimulus that was generated in some unknown way by the interaction of what he called heterogeneous substances.

This was not so far off the mark – we now know that the results of Galvani's initial bimetallic experiments were caused by a differential affinity for electrons shown by the two kinds of metal, thereby leading to current flow, while Galvani and Humboldt's metal-free experiments produced what is called an injury current, whereby injured tissue has a negative charge compared with the rest of the body.[28] But Galvani was fundamentally correct in claiming that there is a kind of electricity present in animals, and that what he called disturbed equilibria lie at the root of the flow of electrical current. The deeper explanation, which would not become completely clear for nearly 150 years, was that in the body electrical charges have a chemical basis – nerves transmit their signals electrochemically.

Not everyone was convinced that these experiments revealed anything about how movement occurs. In 1801 the English physician Erasmus Darwin (the grandfather of Charles) wrote: 'I do not think the experiments conclusive, which were lately published by Galvani, Volta and others, to shew a similitude between the spirit of animation, which contracts the muscular fibres, and the electrical fluid.'[29] Darwin soon found himself in a minority as new experiments seemed to settle the matter.

These new insights were based on a truly revolutionary discovery by Volta, made when he decided to focus on one of the strongest arguments in favour of animals having some kind of inherent electricity – the torpedo's electric shock. In the autumn of 1799, following the ideas of the English chemist and inventor William Nicholson, Volta began to investigate whether the repetitive structure of the torpedo's electric organ was at the root of its ability to generate electricity.[30] To test his hypothesis, Volta created what he described as an artificial electric organ based on the torpedo's anatomy and composed of alternating discs of zinc and copper, interspersed with pieces of cardboard soaked in diluted acid. This was called a pile, after the pile of discs that it was made of – the term persists in French, but in English we now say 'battery'.*

Amazingly, this device created a continuous current of electricity by the interaction of its components. Volta's argument with Galvani had led to a new source of energy. This momentous finding was announced to the world in a letter to the Royal Society, written in March 1800 and published in June of the same year.[31] The age of chemical electricity was born, and soon physicists and chemists all over Europe were using batteries in their research, spellbinding the public with demonstrations of the new form of power, such as at Humphry Davy's famous lectures in London. In 1812 a teenage girl

---

*One exception to this is 'atomic pile', which is still used for a nuclear reactor. 'Atomic battery' would sound weird.

is thought to have attended one of Davy's dramatic demonstrations of electricity; her name was Mary Godwin, but she would be better known by her married name, Shelley.[32]

In his letter to the Royal Society, Volta described the power of external electricity to stimulate nerves in the absence of any supposed charge from muscles, explaining how connecting the artificial electric organ to various parts of his head induced taste in his tongue, lights in his eyes, and sounds in his ears. The only sense that he could not artificially stimulate was his sense of smell – sending an electric current through the inside of his nose produced only a tingling sensation.* Remarkably, Volta did not address how nerves actually functioned. He argued that responses to electricity were always due to an external stimulus, but he did not explain how nerves worked in the absence of such stimulation. Galvani had argued that in its normal state, the brain somehow released electrical charges through the nerves; Volta had nothing to say.

*

Although he lived until 1827, Volta made no further contribution to the investigation of animal electricity. However, in one of history's ironies it was his invention of the battery that led to the popularisation of Galvani's ideas about the importance of electricity in the body, through the work of Galvani's nephew and collaborator, Aldini. In the early years of the nineteenth century Aldini carried out a series of gruesome experiments in various European cities, in which he used Volta's batteries to show the power of electricity to produce movement in animal bodies, and, most dramatically, in dead humans.[33] The most widely known of these events took place in London, in January 1803, when Aldini experimented on the body of George Forster, who had been hanged an hour earlier for drowning his wife and child in a canal.[34] In front of a small audience of physicians at the Royal College of Surgeons, Aldini placed electrodes on

---

* Volta got no effect from his nose because he did not actually stimulate the olfactory neurons, which lie high up in the nasal cavity, at about eye level, dangling down through the base of the skull. Do not try this at home.

the corpse's head, causing Forster's left eye to open and his face to grimace.[35] According to a brief account that appeared in *The Times*: 'In the subsequent part of the process, the right hand was raised and clenched, and the legs and thighs were set in motion. It appeared to the uninformed part of the bystanders as if the wretched man was on the eve of being restored to life.'[36]

Aldini's experiments, which he performed throughout Europe, astonished hardened medical observers, because continuous current from a battery was able to produce eerie life-like coordinated behaviour – very different from the brief spasms induced by the single shocks from Leyden jars. This suggested that electricity was more than a mere irritant and was in fact the nervous source of complex behaviour.[37] Aldini's own accounts of his experiments are often grotesque and unpleasant. Here are two of the milder examples. His studies of animals showed that passing a current through a dead ox's head produced 'a commotion so violent in all the extremities of the animal, that several of the spectators were much alarmed, and thought it prudent to retire to some distance'; sending a current through a decapitated cow's body led to violent contractions of the diaphragm and the expulsion of faeces.[38] Following an experiment in France, the Parisian Institut National reported:

> The head of a dog being cut off, Aldini subjected it to the action of a strong pile, by which means the most frightful convulsions were produced. The mouth opened, the teeth gnashed, the eyes rolled in their orbits; and, if the imagination had not been restrained by reason and reflection, one might have almost believed the animal was restored to life, and in a state of agony.

Not all of Aldini's experimentation was sordid or heartless. He used electrical stimulation to make a cicada sing and a glow-worm glow, and with an insight that would not bear fruit for nearly 200 years, he even wondered whether it would be possible to use this technique 'to gain a more precise knowledge of the organisation of insects'. Aldini also used batteries to carry out some pioneering therapy. He described the case of Louis Lanzarini, a twenty-seven-year-old farmer who suffered from 'a deep melancholy'. Following

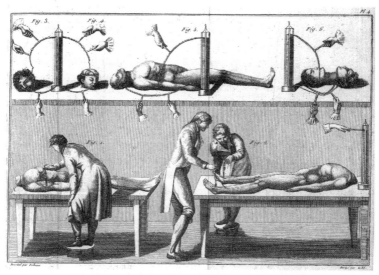

Aldini's experiments using electricity on human bodies.

a series of electric shocks, first to the face, then to the skull, Lanza-rini's symptoms eventually diminished. Aldini followed his patient for some months and reported that he was able to 'enjoy good health, and to exercise his usual employment'.[39]

Although Aldini was far from being Frankenstein, some other people came quite close. The German physician Karl August Wein-hold produced a series of extravagant observations that could have come from the pages of Mary Shelley's masterpiece, including the claim that bimetallic electricity could actually restore life.[40] The faint-hearted reader should skip the next paragraph.

In an 1817 book chillingly entitled *Versuche über das Leben und seine Grundkräfte* (Experiments on Life and its Basic Forces), Wein-hold said that dissimilar metals could act as an artificial brain. He claimed to have removed the brain from a living kitten, and then inserted a mixture of zinc and silver into the empty skull; the animal began to move and for twenty minutes 'it raised its head, opened its eyes, looked straight ahead with a glazed expression, tried to creep, collapsed several times, got up again, with obvious effort, hobbled about, and then fell down exhausted'.[41] In Frankenstein style, Weinhold concluded that he had shown he could 'create a complete physical life'.[42] All this should be taken with more than a pinch of salt: a few decades later the young German physician Max Neuburger described Weinhold's work as 'most bizarre' and suggested that his experiments 'illustrate the fantasy of his thinking and observation'.[43] The reason for Neuburger's scorn was simple: Weinhold's claims are impossible.

Despite – and, frankly, because of – such dramatic but doubtful evidence, the idea that electricity lay at the root of how the brain worked became commonplace. Many thinkers agreed with the German chemist and physicist Johann Ritter, who in 1805 concluded that animal spirits and the animal electricity observed by Galvani were functionally identical.[44] The general public soon caught up with these ideas, as practical demonstrations of what were known as galvanism became a form of entertainment. On 28 September 1804, *The Times* announced a lecture by one Mr Hardie, to be given in the Lyceum theatre in London, which he promised would involve 'exciting the actions of crawling, kicking, leaping, &c. in the detached limbs of dissected animals. The functions of smelling, biting, chewing, swallowing, drinking, and other voluntary motions, will be induced in the head of a sheep, ox, or other large animal, long after separation from the body.'

The significance of these discoveries for how nerve and brains function was explained to the general reader in the 1827 edition of the *Encyclopaedia Britannica*. The physician Peter Mark Roget – later the author of the eponymous *Thesaurus* – explained that nerve function bore 'a greater resemblance to the transmission of the electric agency along conducting wires, than to any other fact we are acquainted with in nature'.[45] Similar ideas were spread by the growing self-improvement movement that flourished in the United Kingdom in the nineteenth century. In 1832 a young woman called Eliza Sharples, who had scandalously entered a 'moral marriage' with the radical pamphleteer Robert Carlile*, gave a series of lectures in London in

---

*In 1819 Carlile played a key role in publicising the Peterloo massacre in

Carlile's Blackfriars Rotunda theatre in which she dressed as various figures from antiquity and mythology.[46] In 'The Seventh Discourse of the Lady at the Rotunda', given in March 1832, Sharples explained to her audience that the brain was merely 'an electrical pile giving pulsation to the heart and accounting for all the phenomena of the body'.[47]

Probably the strongest sign that the ordinary person knew there was a link between brain, mind and electricity was its appearance in *Vestiges of the Natural History of Creation*, the widely read mid-nineteenth century work of popular science.[48] This book, which was published anonymously in 1844, was written by the Scottish writer and geologist Robert Chambers, and became an international best-seller. The section dealing with the brain boldly underlines 'the absolute identity of the brain with a galvanic battery', although the evidence cited in support of this claim was Weinhold's fraudulent description of his experiments on kittens.[49] If the brain is a battery, then thought could be merely electricity, the author of *Vestiges* suggested, and 'if mental action is electric', then its speed could be measured. The most recent calculations of the speed of light suggested that it zipped along at 192,000 miles per second, and it could be assumed that electricity, and therefore mental action, moved at a similar speed.[50]

*

Although there was a growing consensus about the link between nerve action and electricity, the experimental evidence for this view was remarkably weak. Despite nearly half a century of investigations into the role of electricity in nerve action and muscular contraction, no one had shown that the only thing that travelled down a nerve was an electric current, nor had anyone been able to explain how that current might be conducted. As the French physician François-Achille Longet put it in 1842: 'there is no direct proof in support of the hypothesis that there are electric currents in the nerves'.[51]

Manchester; his writings were a source for Shelley's famous poem *The Masque of Anarchy*.

The difficulties involved in coming to a decisive conclusion were shown in the work of Italian physiologist Carlo Matteucci, whose experimental results led him to repeatedly change his mind about whether or not there was a link between electricity and nervous action. In 1838 Matteucci studied muscle contraction using a galvanometer, which measures the strength and direction of an electrical current, and he found that muscular contraction was always linked with a flow of electrical current.[52] Within four years, faced with some complicated results, Matteucci reversed his opinion, arguing that electricity was not the cause of contraction, which was produced by something called nervous force.[53] By the end of the decade, new experimental evidence led Matteucci to alter his views yet again: he now claimed that 'the cause of these contractions is evidently an electrical phenomenon'.[54] This kind of flip-flopping from one of the leading researchers in the field did little to encourage confidence in any explanation.

The breakthrough came as a result of work inspired by one of the nineteenth century's greatest scientists, Johannes Müller of the University of Berlin.[55] Müller was particularly interested in the nature of nervous activity and its links with mind and perception – in his mid-twenties he noticed that if you stimulated a particular kind of nerve (say, the nerves in your retina by pressing your eyeball), then the stimulus was perceived not in terms of its physical nature (in this case, pressure), but in terms of the sense that the nerve normally communicated (vision). Müller called this effect 'the law of specific nerve energies': he imagined that each peripheral nerve carried a particular kind of energy, depending on the sensory organ it was connected to.

One of the reasons Müller adopted this position was that he did not accept that nerves transmitted electricity. Instead, he considered that organisms contained a vital principle that kept them alive and which was involved in the working of the mind and the production of behaviour. This vitalist view was typical of the Romantic movement in early nineteenth-century Europe and was one of the threads that contributed to Mary Shelley's *Frankenstein*. For Müller, all talk of electricity in organisms was mere metaphor:

To speak, therefore, of an electric current in the nerves, is to use
quite as symbolical an expression as if we compared the action
of the nervous principle with light or magnetism. Of the nature
of the nervous principle we are as ignorant as of the nature of
light and electricity; but with its properties we are nearly as
well acquainted as with those of light and other imponderable
agents.[56]

Müller was not only uncertain about the nature of nervous activity, he also thought that its speed precluded it from ever being fully
understood: 'We shall probably never attain the power of measuring
the velocity of nervous action; for we have no opportunity of comparing its propagation through immense space, as we have in the
case of light.'

Müller had a relatively brief academic career – he died in 1858,
apparently having committed suicide at the age of fifty-seven – but
he attracted a remarkable array of brilliant students and researchers,
including some of the greatest figures in nineteenth-century science.
These included Hermann von Helmholtz and Ernst Haeckel, as well
as less well-known but equally significant individuals such as Rudolf
Virchow and Emil du Bois-Reymond.[57] These young men, imbued by
Müller with a taste for applying the methods and outlook of physics
to the study of physiology, formed part of the long academic tradition of students trying to prove their teacher wrong. In this case,
they rejected Müller's vitalism in favour of a consistently materialist
approach. As du Bois-Reymond and Ernst Brücke put it in a manifesto they wrote in 1842: 'no forces operate in the organism other
than those common to physics and chemistry'.[58]

In 1841 Müller prompted du Bois-Reymond to investigate Matteucci's contradictory findings on the role of electricity in nerves and,
if possible, get to the bottom of the nature of nervous action. By the
end of the 1840s, du Bois-Reymond had demonstrated that there was
nothing mysterious about the way nerves functioned – it was indeed
on the basis of electricity. He showed that what he called an action
current of electricity flowed down the nerve, and that tissues were
polarised – they contained both negative and positive particles, in
different proportions. The fundamental feature of the action current,

he argued, was what he called negative variation, whereby changes in polarity led to the flow of current. Although du Bois-Reymond would turn out to be mistaken in many details, in 1848 he claimed, in terms that echoed *Frankenstein*: 'I have succeeded in restoring to life in full reality that hundred-year-old dream of the physicist and physiologist, the identity of the nerve substance with electricity.'[59]

Not everyone agreed with him. Nearly forty years later, the argument was still raging in some quarters – in 1886 the Dean of Harvard Medical School, Henry Bowditch, wrote an article in *Science* in which he rejected du Bois-Reymond's claim. One of Bowditch's pieces of evidence was the well-known but misunderstood fact that a tied nerve could not stimulate a muscle but was nevertheless able to conduct electricity.* He also argued that the production of an electric charge in the nerve was expected to create heat, but precise experimental measures revealed no such effect. Bowditch was sure that electricity was not involved and instead harked back to old ideas, suggesting that 'the nerve-force is transmitted from molecule to molecule by some sort of vibratory action, as sound is transmitted through a stretched wire'.[60]

Another of Müller's students, Hermann von Helmholtz, investigated the speed of the nervous impulse, something that Müller thought was impossible.[61] In 1849 Helmholtz devised an apparatus consisting of a frog's leg with a circuit-breaker attached to one end; when the muscle contracted, the circuit was broken, and the change in the readings on a galvanometer revealed the time that had passed between the beginning of stimulation and the breaking of the circuit. A simple calculation based on the length of the nerve made it possible to calculate the speed of transmission. The answer was surprisingly slow – slower even than the speed of sound, and nothing like the light-speed that had been imagined by Müller or the author of *Vestiges of the Natural History of Creation*. Whatever kind of electricity was in nerves, it seemed to behave in a different way from that found in wires. To confirm this surprising finding, Helmholtz asked human subjects to indicate when they perceived a mild electric

---

* This can be explained by the presence of conducting fluids on the outside of the nerve.

shock. By calculating the distance from the shock point to the brain, he worked out the speed of sensory nerve action to be about thirty metres/second. He eventually showed that human motor nerves reacted with a similar speed. Helmholtz also invented a new term to describe what passes down the nerve – action potential – that is still in use today.

These surprisingly slow speeds posed two problems. First, as Helmholtz realised, it had consequences for perception, as it would mean the brain could only respond to events in the past. Helmholtz dismissed this as a cause of any major problems in the real world: 'Happily, the distances our sense-perceptions have to traverse before they reach the brain are short, otherwise our consciousness would always lag far behind the present.'[62] Despite Helmholtz's breezy confidence, the implication is indeed that we live – ever so slightly – in the past; we never perceive the world instantaneously.

The second problem was more fundamental: an explanation was needed as to why the speed of electrical activity in nerves was far slower than that in wires. Although du Bois-Reymond and Helmholtz had shown that nervous systems functioned according to physical principles, they were not able to show how nervous electrical activity was propagated. For Helmholtz, as for many other nineteenth-century thinkers, the obvious technological metaphor for the nervous system was the telegraph network, which was spreading throughout Europe.* Indeed, the links between the two were not simply metaphorical – early neurophysiologists, including Helmholtz, made use of telegraphic devices in their experiments on nervous activity.[63] In 1863 Helmholtz drew a parallel and pointed out that nerves, like telegraph wires, could lead to all sorts of functions:

> Nerves have often and not unsuitably been compared to telegraph wires ... according to the different kinds of apparatus with which we provide its terminations, we can send

---

*The metaphor went both ways – not only was the nervous system described as being like a telegraph, the telegraph system was described as the nervous system of the country. In the language of the time, both telegraph and nerves communicated intelligence nearly instantaneously, and they both enabled action.

telegraphic dispatches, ring bells, explode mines, decompose water, move magnets, magnetise iron, develop light, and so on. So with the nerves.[64]

What telegraphs could not do, but nerves could, was produce sensation and perception. How that happened was not clear.

*

The most far-reaching but now forgotten mid-century attempt to explore the link between brain, thought and electricity was made by Alfred Smee, a brilliant polymath and inventor. At the age of twenty-two, Smee was given a near-sinecure as Surgeon to the Bank of England (the post was created especially for him), and the following year he was elected a Fellow of the Royal Society. He had a wide range of interests, from aphid-transmitted diseases in potatoes (this got him a name-check in a Drury Lane pantomime) to inventing a new kind of battery, and in the middle of the nineteenth century he used electricity to explain everything about brain function, from the senses to memory.[65] In his 1849 book *Elements of Electro-Biology*, Smee claimed the brain was made up of hundreds of thousands of tiny batteries, each of which was connected to a part of the body. Desire, he argued, was merely an expression of electrical charge in the brain; once a desire was gratified and the charge released, the battery required a certain amount of time to recharge itself, and for desire to be sensed once again.[66] Smee even applied his theory to the nature of mind, suggesting that ideas and consciousness were the product of combinations of batteries in the brain.[67]

Smee was a prolific writer, and a year later he produced a popular version of his theory in a book called *Instinct and Reason*. Some of his ideas appear remarkably prescient. Taking as his starting point the assumption that 'light falling upon the nerve determines a voltaic current which passes through the nerves to the brain', Smee suggested it would be possible to make an artificial eye, by aggregating 'a number of tubes communicating with photo-voltaic circuits'. All you had to do was to repeat these structures over and over and then 'there is no reason why a view of St Paul's in London should not

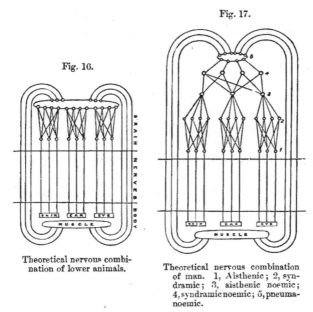

Fig. 17.

Fig. 16.

Theoretical nervous combination of lower animals.

Theoretical nervous combination
of man. 1, Aisthenic; 2, syndramic; 3, Aisthenic noemic;
4, syndramic noemic; 5, pneuma-
noemic.

Smee's schema of animal and human brains.

be carried to Edinburgh through tubes like the nerves which carry
the impression to the brain'.[68] Similar approaches were possible with
the other senses, he argued. If sensation was electric, then it should
be possible to build devices that could imitate it.

Smee produced a complex diagram showing how nerves from
the muscles and nerves in the sense organs each converged on a
central battery in the brain, with the sensory nerves showing intricate interactions. This kind of structure, he claimed, could explain
how 'the idea of a nest may be implanted in the bird, of a comb in
the wasp or bee, or a web in the spider, and upon this supposition
we have a complete explanation of instinctive operations'. These
rigid, inborn connections were the way instinctive behaviour was
represented in the animal brain. Humans, being more complicated,
required an extra two layers of nervous combination, which would
enable what Smee called a general law to appear by combinations of

simple expressions – 'man is made up of a great number of voltaic elements, so arranged as to form one whole'. All brains and bodies worked along the same general principles, which were analogous to the most sophisticated technology of the time:

> In animal bodies we really have electro-telegraphic communication in the nervous system. That which is seen, or felt, or heard is telegraphed to the brain … and, from the whole of our previous ideas being included in the circuit, the act determined takes place momentarily.

Despite sounding modern, Smee's attempt to reveal the principles of the human mind using his battery theory contained no great insights beyond those that had been discovered by philosophers and investigators in previous centuries. Probably for that reason, Smee's views were described as 'fatuous' by one reviewer, while another dismissed them as 'crude, unphilosophical, and unsupported by facts'.[69] These criticisms stung – Smee later complained that his books had received 'unmeasured abuse' from certain quarters.[70]

In 1851 Smee outlined a device that was based on the fact that 'every idea, or action on the brain, is ultimately resolvable into an action on a certain combination of nervous fibres'. He was trying to make a machine that could think. Having initially described a primitive cypher, whereby a concept or word might be represented, Smee claimed it was all relatively simple:

> By taking advantage of a knowledge of these principles it occurred to me that mechanical contrivances might be formed which should obey similar laws, and give those results which some may have considered only obtainable by the operation of the mind itself.

Although Smee produced a diagram of part of such a machine, composed of a series of metal plates linked by hinges, and even claimed to have made some prototypes ('I have before me, whilst I write, seven or eight varieties of these contrivances'), it is not clear how they were supposed to work. All he would say was that it required 'movements to move upon other movements' and involved some new principle that could not be seen in any other machine in London. As to representing the full workings of the human brain in his apparatus, Smee thought that it would be impossible:

> When the vast extent of a machine sufficiently large to include all words and sequences is considered, we at once observe the absolute impossibility of forming one for practical purposes, inasmuch as it would cover an area exceeding probably all London, and the very attempt to move its respective parts upon each other, would inevitably cause its own destruction.

He therefore focused on two components of his hypothetical device, the first of which he called a relational machine, which would produce a predetermined response to a given stimulus and could therefore be used for mathematical calculations. Smee claimed: 'this mechanism gives an analogous representation of the natural process of thought, as perfectly as a human contrivance can well be expected to afford'. An illustration of the relational machine was published in 1875; it shows a complex fan-like hierarchical structure, but there is no indication of how it might work. The second component, which remains even more mysterious, was called a differential machine. It would 'exemplify the laws of judgement' through a system of pins of different sizes, which would enable the apparatus to give one of four responses (yes/probably/possibly/no) to the potential existence of a relation between what Smee called a set of facts or principles.

Smee closed his discussion with some confident words, which seem to imply he was a precursor of modern attempts to get machines to mimic thought:

> By using the relational and differential machines together, we are enabled to obtain the bearing of any facts, or to arrive at any conclusion to which the mind by itself is competent. From any definite number of premises the correct answer may be obtained, by a process imitating, as far as possible, the natural process of thought.

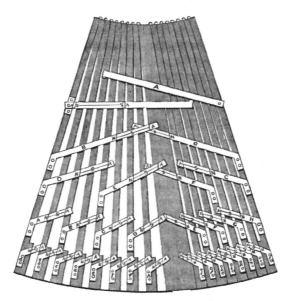

Smee's mysterious relational machine.

Remarkably, Smee's writings contain no reference, not even a hint, of Charles Babbage's earlier attempts to build mechanical calculators – first the Difference Engine (1820s) and then the Analytical Engine (1830s onward). These devices had far more limited ambitions, but the affinity with Smee's later work seems obvious. Babbage was still very active at the time that Smee was thinking about his machines, and both men were members of the Royal Society, but there is no evidence they ever met.

The significant limits of Smee's ideas are shown by the fact that although he claimed his initial concept of brain function was entirely based on electricity, it could equally have functioned on the basis of hydraulic power, like Descartes's statues. Even though he used the analogy of the telegraph and photovoltaic cells to describe the workings of the nervous system, this had no consequences for his model, nor did it give any insight into brain function. And when it came to actually building a machine to represent his vision, the language of batteries and electricity was replaced by that of hinges and metal. Smee's contrivances, which he claimed were representations of brain and thought, were purely mechanical.

Although Smee is now forgotten by all but a handful of historians, and he had no influence on our understanding of brain function or even on the history of computing, his bold ambition of representing thought in the activity of a machine was remarkable.[71] He took the suggestion that brain, mind and electrical activity were intimately linked and, brimming with self-confidence, suggested that if the brain was thinking matter, then a machine could think, or at least function along the same lines as a brain. Smee's approach was fatally flawed, not simply because the technology of the time was woefully inadequate to the task, but because he did not seem to suspect that some of the hundreds of thousands of batteries that he argued made up the brain might have specific functions and concomitant specific structures. In Smee's conception, there was no localisation of functions within the brain. And yet by the middle of the nineteenth century the assumption that brain structure was related to function – and indeed to human personality – had become deeply rooted in the popular imagination.

FOUR

FUNCTION

19TH CENTURY

In summer 1850 the Communist Labourers' Educational Club of London held a picnic, either at Hampton Court or Kew Gardens – memories differ. One of the guests was twenty-four-year-old Wilhelm Liebknecht, a German socialist who had recently been expelled from Switzerland for his revolutionary activity. The leading figure in the communist movement, Karl Marx, was also in attendance; this was the first time that the two had met. Liebknecht recalled that the thirty-two-year-old Marx 'examined me with questions, but also with his fingers, making them dance over my skull in a connoisseur's style'.[1] Like many other nineteenth-century Europeans and Americans, Marx was convinced that it was possible to determine an individual's personality by feeling the bumps on their head. At the other end of the political scale, Queen Victoria was also a believer in this guff, twice inviting a leading practitioner to read the skulls of her children.[2]

Known as phrenology ('study of the mind') this idea featured in novels such as Charlotte Brontë's *Jane Eyre* and Balzac's *Père Goriot* as well as in Arthur Conan Doyle's Sherlock Holmes stories – Moriarty makes a disdainful phrenological remark about Holmes when the pair first meet.[3] Virtually every cultural figure in the nineteenth-century English-speaking world, from Mark Twain to George Eliot, embraced phrenology at one point or another.* In continental Europe, leading thinkers such as the French founder of sociology, Auguste Comte, adopted the new way of looking at brain and behaviour.[4] Popular books on phrenology sold hundreds of thousands of copies in the UK alone. All this, despite being complete nonsense.

Initially known as cranioscopy, phrenology was the brainchild of Franz Gall, a Viennese physician.[5] During the 1790s Gall came up with the idea that human behaviour and personality could be divided up into a number of mental faculties, each of which were produced by a particular organ in the brain, and that it was possible to detect the relative size of these organs by feeling the shape of the skull. In 1800 Gall met Johann Spurzheim, a physician eighteen years his junior, who embraced his ideas. Over the next decade or so, Gall recalled, the pair toured Europe, presenting their ideas to 'sovereigns, ministers, intellectuals, administrators and all kinds of artists'.[6] Conservative forces viewed their ideas with scepticism, and sometimes downright opposition: the Holy Roman Emperor and the Catholic Church both condemned the theory. In 1807 Gall settled in Paris, where his presence was grudgingly accepted by Napoleon. Although Gall soon acquired a following among upper layers of society, he was never accepted by the French academic establishment – his attempts to join the Académie des sciences were repeatedly rebuffed and he never gained the intellectual recognition he craved.[7]

Despite being completely spurious, Gall's theory was significant because it was based on three insights that still form the basis of our understanding of the link between brain, mind and behaviour. Firstly, phrenology was focused on the brain – 'the brain is the organ of all the sensations and of all voluntary movements', said Gall.[8] Second, Gall assumed that there was a localisation of function, in that different parts of the brain were responsible for different aspects of thought and behaviour. Finally, Gall explained how humans shared most of their psychological faculties, and the underlying organs,

---

*Dickens was not entirely impressed. In *Great Expectations* the criminal Magwitch explains that he was subject to a phrenological examination, before providing a simpler explanation of his behaviour: 'they measured my head … they had better a measured my stomach'.

with animals. Only eight of the twenty-seven faculties were unique to humans – wisdom, poetry and suchlike. Gall claimed that this comparative approach enabled him to discover 'the laws of the organism', even though the link between behaviours in animals and in humans was sometimes tenuous – for example, the faculty of pride was taken to be identical to the propensity of mountain goats, birds and so on to live in high places.[9] Searching for biological principles by comparing different species has turned out to be a very powerful method in science, but Gall's phrenology was not based on an evolutionary view – he merely assumed that similar structures had similar functions.

Gall's idea was not entirely original – many of the psychological faculties he identified could be traced to the work of the eighteenth-century Scottish thinkers Thomas Reid and Dugald Stewart, while in the 1770s the old belief that facial features could be used to determine personality had been codified as 'physiognomy' by the Swiss pastor Johann Lavater.[10] Gall brought all these ideas together, using his own anatomical knowledge based on measurements of over 300 human skulls that he had collected.

In 1815 Spurzheim and Gall fell out and Spurzheim published his own version of the theory. At one level, the differences seemed trivial – Spurzheim described eight extra organs and their associated faculties and also introduced a different set of psychological terms.[11] But the dispute between the two men ran much deeper: in his version of phrenology, Spurzheim focused solely on humans, dramatically altering the social implications of the theory. Gall had argued that the faculties were innate and fixed, and that, if expressed in excess, many of them could give rise to less desirable behaviours such as lustfulness, fighting, or deceit, thereby showing the need for religion and punishment as ways of keeping humans in check.[12] For Spurzheim, however, 'all the faculties are themselves good and are given for a salutary aim'. Immoral or criminal behaviours were simply the consequence of experience; education could alter the size of organs, and therefore change behaviour (how such changes could be felt through the bony skull was never explained).[13] Spurzheim's more positive, even therapeutic, view of the link between brain, mind and behaviour was the version that began to capture the popular imagination in Europe and the USA. As Gall became less involved (he died in 1828, aged seventy), Spurzheim's version of phrenology gained the upper hand.

In Great Britain, Spurzheim's success was due partly to the ceaseless activity of George Combe, a Scottish lawyer. Combe not only helped create the first British phrenological society (in Edinburgh), he also wrote a large number of best-selling books, articles and pamphlets outlining his version of phrenology, which was focused on self-improvement.[14] From the 1820s onwards, phrenological societies popped up all over the UK. At first these groups were primarily composed of professional men and intellectuals, but they gradually interacted with the Mechanics Institutes and the Literature and Philosophical Societies that were a feature of the growing industrial cities and were aimed at the self-improvement of the working class. Combe and the phrenologists were no revolutionaries, but for those in power their twin tenets of materialism and self-improvement had disturbingly radical implications. This did not deter some religious leaders from embracing phrenology: in the 1830s, Richard Whateley, the Anglican Lord Archbishop of Dublin claimed that he was 'as certain that Phrenology is true as that the sun is now in the sky'.[15]

A similar process took place on the other side of the Channel. Napoleon I had eventually banned Gall's writings, but after the creation of a more liberal monarchy in 1830 some key physicians espoused phrenology, and King Louis Philippe showed an interest in the topic.[16] As in Britain, there was a real popular appetite for versions of phrenology that emphasised self-improvement, but despite this widespread interest, intellectuals and physicians were never completely at ease with the discipline. An early critic was the German philosopher Georg Hegel, who dismissed the new fad in 1807 – the lumps and bumps on a murderer's skull could not possibly reveal a murderous nature, said Hegel, not only because skulls have so many different lumps on them, but also because human behaviours, including murder, are complex phenomena. One murderer is not identical to another in their motivations and behaviour.[17] Napoleon was another doubter:

> Look at Gall's stupidity! He attributes to some lumps, tenden-
> cies and crimes that do not exist in nature, that in fact come
> from society and from human conventions. What would be the
> significance of the bump for theft if there were no property?
> Or of the bump for drunkenness if there were no fermented
> liquors, or of that for ambition if society did not exist?[18]

A more substantive and largely scientific criticism came from
Roget, who in the 1820s wrote a series of articles on phrenology for
the *Encyclopaedia Britannica*, some of which were later published as
pamphlets. Roget scoffed at what he called the 'metaphysical laby-
rinth of *the thirty-three special faculties* into which they have analyzed
the human soul' and dismissed the phrenologists' suggestion that
damage to the brain led to alterations to mental faculties – 'innumer-
able cases might be quoted in direct contradiction to this principle',
he claimed.[19] Although Roget accepted that the brain was 'the organ
of the mind', he insisted that 'nothing like direct proof has been given
that the presence of any particular part of the brain is essentially
necessary to the carrying on of the operations of the mind'. This was
particularly true in the case of those suffering from mental illness:
'the most accurate dissections have not taught us any thing with
regard to the seat of mental alienation'. Roget also rubbished Gall
and Spurzheim's fundamental claim that the skull provided access
to the shape of the brain, pointing out the fairly obvious problem
that the bony skull is thicker in some areas than others, and that
it is covered in muscles and skin that make it difficult to measure
its shape precisely. Roget's views were widely shared by that new
breed of intellectual – scientists.* In private, they could be even more
forthright: in 1845 the Cambridge professor of geology, the Reverend Adam Sedgwick, wrote a letter to his colleague Charles Lyell,
describing phrenology as 'that sinkhole of human folly and prating
coxcombry'.[20]

---

* William Whewell first used the word 'scientist' in print in 1834, in a review of
Mary Somerville's *On the Connexion of the Physical Sciences*. He had coined it a
few years earlier to describe the 'gentlemen' of the British Association for the
Advancement of Science. Whewell, W. (1834), *The Quarterly Review* 51:54–68,
p. 59.

From the late 1840s onwards, phrenology began to wane as a social force. The London Phrenological Society fell apart in 1846, and even George Combe eventually stopped writing about phrenological matters.[21] In France, the timid individually focused changes advocated by many phrenologists seemed completely inadequate as the wave of revolutions that swept through the continent in 1848 crashed over the country. The spectre that Marx and Engels claimed was stalking Europe – communism – offered more radical solutions to individual and social woes than self-improvement coupled with feeling bumps on the skull.

*

For the academic community, phrenology failed due to that ultimate arbiter of scientific thought: evidence. No matter how beautiful or logical or seductive or fashionable a theory may be, it will eventually be abandoned if it is not supported by experimental findings. In the case of phrenology, powerful experimental counter-evidence came in the shape of a series of studies by the French physician Marie-Jean-Pierre Flourens. Born in 1794, Flourens rose rapidly through the ranks of French academia; the protégé of the great naturalist Georges Cuvier, he became not only a member of the Académie des sciences, but also of the prestigious literary body the Académie française. The contrast with Gall's failure to be recognised by his adopted country could not be starker.

Flourens carried out a series of studies on various animals, surgically removing different parts of the brain and observing the animal's subsequent behaviour. Gall had criticised this approach, claiming that it was not possible to remove a particular part of the brain with sufficient precision to be confident that only that part was affected. Flourens recognised this danger, but as the historian Robert Young put it, his method was basically 'I removed this part, and the animal ceased to do that, so this must be the seat of the faculty of that.'[22]

For two decades, Flourens pursued this work on a depressingly wide range of birds, reptiles and amphibians, together with a few mammals. One of his clearest experimental findings involved the medulla oblongata, a structure common to all vertebrates that is found at the very top of the spinal cord, just beneath the brain. Flourens found that damage to this structure affected breathing and heartbeat – it seemed to represent a fundamental centre for the essential physiology of life. When Flourens damaged the next structure up – the cerebellum, at the base of the rear part of the brain – he found that animals showed poorly coordinated behaviour; he reported that a pigeon that had been subjected to cerebellar lesions behaved like a drunken man.[23]

The outermost layers of the brain, the cerebral lobes, produced a very different kind of result. If the cerebral lobes were removed, animals became completely unable to respond to stimuli – a frog could live for up to four months in such a condition, but 'in a state of complete stupidity ... it neither heard nor saw nor any longer gave any sign of will or intelligence'. Flourens concluded that the operation had led to 'the loss of all perception and all general intelligence, as well as all particular forms of intelligence that determine the unique behaviours of various species', including humans.

Flourens rejected the idea that there were large numbers of psychological faculties – he grouped all aspects of intelligence and will into one – and refused to recognise any anatomical subdivisions within the cortex. He argued that 'one of the most important results' he had discovered was that the cortex – the 'seat of intelligence' – was a unitary structure:

> not only do all perceptions, all will, all the intellectual faculties reside exclusively in this organ, but all of these faculties occupy the same space. As soon as one of these faculties disappears as a result of a lesion of a given part of the actual brain, they all disappear; as soon as one of them returns by the healing of that part, they all return.

In 1842 Flourens wrote a book-length counterblast against phrenology, and particularly against Gall, who had been dead for fourteen years. Flourens justified his place at the heart of the French academic establishment by attacking Gall not only with experimental findings, but also by invoking the quintessential French philosopher of mind,

Descartes, to whom the book was dedicated. Flourens's experimental studies suggested that the mind was, as Descartes had argued from philosophical premises, a unitary whole. Flourens showed that many of the higher behavioural functions associated with mind and perception, in animals and in humans, appear not to be highly localised, as Gall had argued, but instead to be broadly distributed in the cortex. Localisation applied only either to basic physiological functions, or to those related to motor coordination. Evidence from strokes, for example, showed that if the right side of the brain was damaged, then the patient suffered from paralysis of some or all of the left side of the body. But this kind of contralateral motor impairment was relatively trivial compared with the deep mysteries of human mental existence. The mind, it appeared, was distributed across the cortex.

Gall had claimed that each part of the brain produced its own kind of mental action – in French this was called *action propre*.[24] For Flourens, most of what the brain does seemed to be coordinated brain-wide action (*action commune*), which included an influence 'from each on all, and all on each'. The brain as a whole, even if it contained areas with specific physiological functions, formed 'nothing less than a single system', Flourens stated.[25]

Flourens had unwittingly opened a profound and long-running debate between those who argued that the brain functions as a whole, and those who claimed it contained localised areas that produce particular mental activities. For Flourens, only the simplest, most physiological or motor behaviours showed localisation. All that is specific to higher mental activity formed a unitary whole, and was somehow expressed throughout the brain.

*

The initial blow to this view came from studies of language. Gall's phrenology flowed from his youthful conviction that children with bulging eyes were best at rote learning; accordingly, he placed the faculty of language, along with other faculties associated with memory, at the front of the brain, just behind the eyes. In 1825 a young French physician called Jean-Baptiste Bouillaud presented a paper at the Académie royale de médecine in Paris, in which he attacked Flourens's claim that there was no cerebral localisation. Bouillaud argued that a number of pathological cases revealed that the brain contained an organ responsible for speech that was distinct from both verbal understanding and verbal memory. Bouillaud, who was an open advocate of phrenology, said that the cases of dozens of patients showed that Gall was correct, and that the organ of speech lay at the very front of the brain. Post-mortem studies of patients who were unable to speak, but could understand and remember words, always revealed damage to the frontal lobes of the brain, claimed Bouillaud.[26]

The link between Bouillaud's views and Gall's increasingly discredited phrenology, and the power of Flourens's experimental evidence, meant that Bouillaud's claim initially found few supporters. Furthermore, there were a number of clear counter-examples: in 1840, Gabriel Andral described fourteen cases of patients who had lost the power of speech but in whom post-mortem examination revealed no lesions to their frontal lobes, while many patients with damaged frontal lobes could speak quite normally. Andral concluded that it was 'at least premature' to suggest that specific parts of the brain were involved in speech.[27] These findings had little effect on Bouillaud, who was so confident that he offered 500 francs to anyone who could find a single patient with frontal lobe damage but no language impairment (he eventually paid up, in 1865).[28]

In February 1861 the Paris Société d'anthropologie hosted a series of debates on brain size and mental ability. The French surgeon Paul Broca argued that there was a clear link between brain size and intelligence, pointing to supposed differences between men and women and between races.[29] Broca's view was a development of the ideas put forward in 1839 by the American physician Samuel Morton, who used skull measurements from various ethnic groups to determine cranial capacity, which he then correlated with alleged differences in intellectual ability. Morton unsurprisingly found that what he termed Caucasians were intellectually superior to other 'races' and that this was reflected in the size of their skulls.*

---

*In 1981, Stephen Jay Gould argued that Morton's measurements were subtly

Opposing Broca in these debates was the French zoologist Louis-Pierre Gratiolet, who had used comparative anatomy to divide the brain into four lobes, corresponding to the bones that covered them – frontal, parietal, temporal and occipital, a nomenclature that is still in use today – and had also shown that the convolutions of the brain were consistent within a species.[30] Gratiolet argued that the mind and brain were both indivisible, with no localisation of function, and no simple link between cranial capacity and intelligence.

Responding to Gratiolet, Bouillaud's son-in-law, the physician Ernest Auburtin, provided some striking evidence for localisation in the shape of a Parisian patient who had attempted suicide with a pistol. The frontal lobes of the poor man's brain were laid bare by the shot, and, during treatment (which eventually failed), Auburtin carried out a gruesome experiment that recalled Galen's pig study of 1,700 years earlier:

> While the patient was speaking, the flat end of a large spatula was placed on the anterior lobes, which were gently pressed; *speech immediately ceased*, the word that had begun to be pronounced was cut off. Speech returned as soon as compression ceased. In this patient, compression, undertaken with great care, had no effect on the general function of the brain; *the only faculty that was abolished*, limited to the anterior lobes, was that of language.[31]

This strongly suggested that Bouillaud had been correct, and that the front part of the brain was indeed essential for speech.

Within two months a chance event offered Broca the opportunity of testing this idea. At a meeting of the Société d'anthropologie that

distorted, producing a result that suited his racist expectations. Thirty years later, this claim was criticised by a group of physical anthropologists; in 2014, that conclusion was in turn contested and Gould's position was apparently vindicated. Whatever the case, variation in skull size and cranial capacity reveals no more about the intellectual abilities of different groups than do the bumps on their heads. Gould, S. (1981), *The Mismeasure of Man* (New York: Norton); Lewis, J., et al. (2011), *PLoS Biology* 9:e1001071; Weisberg, M. (2014), *Evolution & Development* 16:166–78.

took place in April 1861, Broca showed his colleagues the brain of a recently deceased fifty-one-year-old man who for twenty-one years had been unable to speak – the only sound he could make, repeatedly, was 'tan, tan'. As a result, at the hospital where he had been an in-patient for over two decades, he was known as 'Tan'. Tan, whose real name was Louis Leborgne, had suffered from epilepsy throughout his life but was able to work normally as a cobbler's last-maker until he suddenly lost the power of speech, aged thirty.[32] Although when admitted to hospital he was classed as healthy and intelligent, he gradually became paralysed on his right side and his vision faded. On 12 April 1861, Leborgne was admitted to Broca's surgical ward with a severe case of gangrene – this was the first time that Broca had met him. Although he could not speak or write, Leborgne could tell the time, and indicate numbers by snapping his fingers; Broca's impression was that he was more intelligent than indicated by his inability to respond verbally. Five days later, poor Leborgne died. An autopsy revealed a series of lesions in his brain, which were concentrated on the left side of the frontal lobe. Broca concluded: 'Everything leads me to believe that, in this case, the lesion of the frontal lobe was the cause of the loss of speech production.'[33]

Broca soon published a more detailed description of his views, in which he linked Leborgne's case and the ideas of Bouillaud regarding the localisation of speech in the frontal lobes.[34] He also presented a detailed anatomical account of Leborgne's brain, drawing parallels between the gradual loss of function and the spread of lesions, strengthening the implication that some faculties were localised in the brain. Although Broca insisted that he was not trying to make Gall's phrenology scientifically respectable, the terminology he used was entirely phrenological – he claimed to be exploring the faculty of speech and its underlying organ in the brain.

A few months later, a second patient was referred to Broca, this time with a broken hip. This patient, Monsieur Lelong, had lost his power of speech five months earlier, and could pronounce only a handful of words, including 'Lelo' – his attempt at his surname. Nearly two weeks after going into hospital, Lelong died; an autopsy revealed a lesion in the left frontal lobe, in exactly the same region as had been observed in Leborgne.[35] Broca reported he felt

'astonishment bordering on stupefaction' when he made this discovery; nevertheless, he preferred to place his work in the context
of Bouillaud's theory, which suggested that the whole of the frontal
lobe was involved in speech. The fact that both patients showed the
same, highly localised lesion on the left side of the brain was pure
coincidence, Broca claimed.[36]

Broca's circumspection began to fade as he assembled eight
cases of speech loss or aphasia, all of which were linked to damage
to the same very specific area of the left frontal lobe. In April 1863,
Broca presented a paper describing these results, but caution still
prevailed, indicating the dominance of Flourens's anti-locationist
ideas at the time:

> Here we have eight cases where the lesion is situated in the
> posterior third of the third frontal convolution. This figure
> seems large enough to give rise to some strong presumptions.
> And, quite remarkably, in all of these patients, the lesion was
> on the left side. I do not dare to draw any conclusion from this,
> and I await more findings.[37]

No sooner had Broca published these results than he found
himself embroiled in a bitter priority dispute with one Gustave Dax,
a physician from Montpellier, who claimed that earlier in the century
his father, Marc Dax, had observed over forty cases in which loss
of language was linked with lesions to the left frontal lobe. Dax the
younger stated that his father had presented these results to a medical
conference in Montpellier in 1836, but no trace has ever been found.
In March 1863 Gustave Dax submitted two papers to the Académie
des sciences; one was his father's from 1836, the other was based on
his own observations, showing that lesions to the left frontal lobe
were linked with language disorders.[38]

It seems clear that Broca was unaware of the work of Dax *père
et fils*, and that he would have come to the inevitable conclusion that
speech is produced by a region of the left frontal lobe – now known
as Broca's area – even if he had never seen Dax's work. When Dax's
article did eventually appear, in April 1865, it provided powerful
support for Broca's claims. Although Dax the younger's anatomical

evidence was less precise than that of Broca, he had assembled a
massive data set in support of his hypothesis of functional localisation of speech. Reanalysing a series of case studies, including those
used by Bouillaud, Dax described 140 examples, eighty-seven of
which showed left frontal lobe lesions and loss of speech, while fifty-
three showed right frontal lobe lesions and no loss of speech. 'The
cerebral organ of speech is found,' concluded Dax.[39]

Rattled by Dax's claim to priority, Broca published a long paper
in which he showed that the Daxes' work had no influence on his
own ideas and provided new evidence to strengthen his case – precisely the kind of new facts he had argued were needed two years
earlier. He described how patients showing paralysis on the right
side of their body – which indicated a lesion in the left side of the
brain – tended to have speech difficulties. This was not due to motor
impairment in the mouth or throat, nor to an inability to understand
language, but was clearly linked to the ability to speak.

The implication – that there was a localised area of the brain,
on only one side, that was responsible for a very particular faculty –
posed Broca with a major problem. From an anatomical perspective,
the two hemispheres of the brain appeared perfectly identical, and
it was well known that where organs were paired or symmetrical,
both halves performed identical functions. Although there were no
discernible anatomical differences in the two sides of the brain, Broca
pointed out that in developmental terms the two hemispheres are not
strictly identical – in the embryo, the left side develops earlier than
the right, perhaps indicating a difference in function. Furthermore,
most people are right-handed, perhaps owing to the early development of the left side of the brain. Broca considered that other areas of
the brain might be involved in speech and that, in principle, it might
be possible to restore some function by training. Cautious as ever,
Broca concluded by insisting that 'this does not imply the existence
of a functional disparity between the two halves of the brain'.[40]

The complexity of what Broca had discovered was soon revealed
by a young German physician, Carl Wernicke. In 1874 Wernicke
described a woman patient who could speak – albeit in a confused
way – but was unable to understand language. 'The patient could
comprehend absolutely nothing that was said to her,' he reported.[41]

Wernicke concluded that it was 'highly improbable' that the whole faculty of language was located in the area of the temporal lobe identified by Broca; he argued that while speech production was located in Broca's area, other regions, including what is now known as Wernicke's area, to the back of the brain, were involved in comprehension. Wernicke was not simply locating a different component of speech in a different area; he was arguing that language comprehension as a whole was highly distributed.[42]

Part of the problem in finding any clearer evidence for the localisation of speech had been identified the previous year by a surgeon from Brest, Professor Ange Duval. After listing a series of cases that supported the localisation of speech to the left hemisphere, Duval underlined the methodological problem that everyone was facing:

> These facts are sufficiently numerous to form an indirect demonstration, but in physiology we rightly prefer direct demonstrations, provided by experiments on animals. But animals cannot be used in the study of a function that they do not possess. We therefore have to wait for chance accidents to produce in man lesions that are analogous to those that we might otherwise seek to produce by vivisection.[43]

Duval's moral scruples do him credit, but before the decade was out scientists became convinced of localisation of function in the brain, not because of damage caused by accident or illness, but partly through a gruesome and scandalous experiment on a human being.

*

On 26 January 1874, thirty-year-old Mary Rafferty was admitted to the Good Samaritan Hospital in Cincinnati. Mary was a slight woman who worked as a domestic servant. For some time she had suffered from a horrific ulcer on her scalp that had slowly eaten away parts of her skull, baring the brain beneath. Despite her condition, Mary was cheery and good-humoured, but infection had set in and in a world before antibiotics the prognosis was not good. Mary was seen by Professor Roberts Bartholow, a forty-two-year-old surgeon, who explained there was a procedure he would like to perform. After apparently obtaining Mary's consent, Bartholow introduced two slender electrodes just beneath the exposed surface of the left side of her brain, and then turned on a generator that produced a very slight electric current.

The result was dramatic: Mary's right arm and leg contracted and moved forward, while her fingers stretched out. When Bartholow inserted the electrodes into the rear part of Mary's brain, her eyes twitched and her pupils dilated and she described 'a very strong and unpleasant feeling of tingling' in her right leg and arm. Bartholow's report continued: 'Notwithstanding the very evident pain from which she suffered, she smiled as if much amused.' Undaunted, Bartholow pursued his experiment and put the electrodes into the other side of Mary's brain. What happened next was alarming – 'her countenance exhibited great distress, and she began to cry' – but Bartholow continued until the poor woman had a fit and became unconscious. After twenty minutes she came round, complaining of weakness and dizziness. Bartholow pressed on, repeating the electrical stimulation. This produced similar distressing effects, but no fit.

Two days later, Mary was again brought to Bartholow's consulting room where he again introduced electrodes into her brain, this time using direct current produced by sixty batteries housed in glass jars contained in an imposing wooden cabinet. The experiment was soon discontinued as Mary became pale and her lips turned blue; she complained of dizziness, and the right side of her body went into spasm. Apparently concerned for his patient, Bartholow asked his assistant to give her chloroform to relieve the pain. The following day, Mary was unable to get out of bed; in the evening she had a fit that led to total paralysis of her right side. Sometime afterwards, she died; Bartholow did not report exactly when.

A few weeks later, in April 1874, Bartholow published a brief account of his experiment. It immediately created a storm, not only because of its dramatic content, but also because of the apparently unethical way it had been performed. The American Medical Association criticised Bartholow's work because it involved experimentation on a human, while the *British Medical Journal* published a half-apology from Bartholow in which the physician regretted that his results 'were obtained at the expense of some injury to the patient'. He even acknowledged that to repeat such experiments 'would be in the highest degree criminal'.[44] The British brain physiologist David Ferrier warned that whatever the interest of Bartholow's findings, because 'the procedure is fraught with danger to life, it is not to be commended or likely to be repeated'.[45]

Leaving aside the grave ethical issues, scientists were fascinated by Bartholow's study because it provided dramatic support for some recent, highly controversial findings about brain function. Although at the beginning of the century Aldini had demonstrated that electrical stimulation of the outside of the head could induce movement, there was a general consensus that the cerebral hemispheres were completely unresponsive. Unlike the lower parts of the brain, no physical, chemical or electrical stimulation of these areas could induce a response.

But in 1870 two young German physicians, Gustav Fritsch and Eduard Hitzig, had shown that electrical stimulation of the outside of a dog's cortex could produce highly specific movements.[46] In his medical practice Hitzig used fashionable external electrotherapy to treat mild neuromuscular symptoms such as cramps and minor paralysis. In 1869 he gave a patient a mild electric shock by simultaneously placing electrodes on the ear and at the back of the skull; he was surprised to notice that the muscles around the eye contracted. Had the electrodes been on either side of the eye, Hitzig would have dismissed the effect as a classic example of an electric current stimulating a muscle. But instead he suspected that the electricity had gone into the brain and he had somehow stimulated a 'centralised feature' responsible for movement.*

A skilled electrophysiologist, Hitzig teamed up with Fritsch to see if it was possible to stimulate the exposed cortex of a dog and obtain a specific response. The experiment, which was done on a dressing table belonging to Frau Hitzig, used very delicate electrodes

---

* As Hitzig noted over 30 years later, he had not in fact stimulated a 'deep brain centre'. Instead, he inadvertently excited the vestibular nerve, which lies near the surface. As he ruefully pointed out: 'This is not the first or last time that a false premise will lead to the uncovering of the correct facts.' Hagner, M. (2012), *Journal of the History of the Neurosciences* 21:237–49, p. 243, note 1.

developed by du Bois-Reymond. It was part of a wave of invasive physiological studies that had become possible with the widespread adoption of anaesthetics from 1846 and Joseph Lister's 1867 discovery that simple antiseptic procedures could reduce the risk of post-surgical infection. Fritsch and Hitzig used very weak currents that were 'scarcely perceptible when applied to the tip of the tongue' to stimulate the thin outer layer of the front part of the anaesthetised animal's cortex.[47] When they did this, they found that various muscles on the opposite side of the dog would twitch. This effect was highly localised – stimulating one part of the brain moved the forelegs, another made the face twitch, and yet another moved the leg muscles.[48]

This discovery went against over a century of scientific certainty and suggested that functional localisation of the production of behaviour in the brain went far beyond Broca's discovery of a specific area for speech production in humans. Stereotypical reflexes, such as the knee-jerk, were known to take place without the involvement of the brain. But the movements induced by direct electrical stimulation of the brain were more like normal behaviours than the minor, repetitive movements of a reflex. Given that the cortex was widely considered to be the site of thought and the will, the implication of Fritsch and Hitzig's discovery was that they had localised the site of voluntary movement, although they were careful not to make this claim explicitly.

The two young men were aware of the significance of their breakthrough, and quickly published a paper describing their results, a large part of which was devoted to contrasting their findings with previous failures to produce a response from the cortex, going back to Haller in the eighteenth century. They also provided precise experimental details – in the end, they stated, it was their technique that explained why their results were radically different from those that had previously been obtained. Fritsch and Hitzig were confident that they had shown that 'individual mental functions' occurred 'in circumscribed centres of the cerebral cortex'.[49]

Fritsch and Hitzig's bombshell discovery immediately prompted David Ferrier, aged only twenty-seven, to carry out his own experiments.[50] Ferrier, like most scientists, accepted that the cerebrum was 'the seat of memory and perception', but it was not clear whether these mysterious capacities were localised to a specific area, or were instead to be found spread all over the surface.[51] As he put it, it was as yet unclear:

> whether the cerebrum, as a whole and in each and every part, contains within itself, in some mysterious manner inexplicable by experimental research, the possibilities of every variety of mental activity, or whether certain parts of the brain have determinate functions.

Ferrier set out to investigate this enigma through a series of rather sad experiments, in which he removed the cerebral hemispheres from frogs, fishes, birds and rabbits, much as Flourens had done four decades earlier. In each species, the same thing happened – if the animal was lucky enough to survive the surgery it would sit, immobile, responding only to stimuli such as pinching: 'If the animal be left to itself, undisturbed by any form of external stimulus, it remains fixed and immovable on the same spot, and unless artificially fed, dies of starvation.'

Ferrier concluded that 'ablation of the hemispheres abolishes certain fundamental powers of mind', which he suggested included volition, or the will to move.

More precise lesion studies revealed an intriguing inconsistency. In mammals where motor behaviour apparently involved an important learned component, destruction of the 'cortical motor centres' was more likely to cause paralysis than in the case of animals that relied on instinctive behaviours. This suggested that volition was more important in higher mammals and reinforced the suspicion that precise areas of cortex were involved in the voluntary movement of specific parts of the body.

The most striking set of findings were obtained by Ferrier's adaptation of Fritsch and Hitzig's technique, in which he stimulated the cortex of a monkey with slight electrical currents. Using batteries invented by Smee, Ferrier found that he was able to induce responses in some areas that his German competitors had claimed could not be excited. To summarise these findings, Ferrier drew a

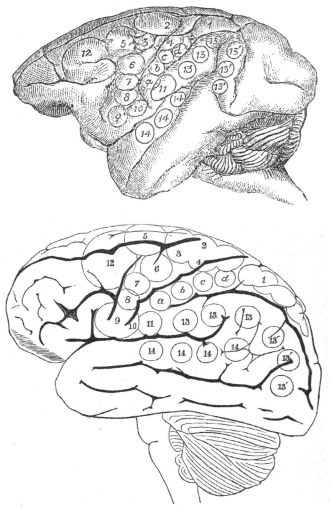

Figures from Ferrier showing areas in the monkey brain (top) and the human brain (bottom). The numbers show equivalent areas in the two species.

diagram showing the specific localisation of the production of particular movements in various parts of the monkey brain. For example, stimulation of area 3 (top centre) resulted in movements of the tail, whereas stimulation of the neighbouring area 5 produced extension of the opposite arm, and movement of the fingers and wrist, while stimulation of areas 9–14 produced precise and reproducible movements of the face and eyes.

Ferrier also experimented on dogs, cats, jackals (obtained from London Zoo), rabbits, guinea pigs, rats, pigeons, frogs and fishes. Each time he found that stimulation of specific areas of the cerebral hemispheres induced specific movements. The only exception was the frog, where the size of the brain made it difficult to obtain clear data. Ferrier was even able to apparently induce the illusion of hearing in his monkeys. When he stimulated area 14 he observed 'pricking of the opposite ear, head and eyes turn to the opposite side, pupils dilate widely', as though the animal had heard something.

Drawing on comparative anatomy and various reports of cerebral lesions in patients, as well the awful experiment on Mary Rafferty, Ferrier drew up a diagram of localised cerebral motor function in humans. But there was one part of the brain that did not seem to respond to the gently probing of his electrodes. In *The Functions of the Brain*, his 1876 book describing his discoveries, Ferrier reported that he failed to observe any responses from 'electrical irritation' of the frontal regions of the brain in the monkey, cat or dog. This fitted with his initial understanding of the case of Phineas Gage, an American railway worker who in 1848 suffered a terrible accident in which a metre-long iron tamping rod was explosively propelled through the front of his skull.[52] Gage miraculously survived, living another twelve years and even spending some years as a stagecoach driver in Chile. Gage's body was eventually exhumed and his severely damaged skull, along with the tamping iron, were displayed in the Harvard Medical Museum, where they can still be seen. During his lifetime and immediately afterwards, Gage was of interest to scientists because he survived, apparently unscathed.

Ferrier was an observant experimenter, and although he stated that removal of the front areas of the brain in one particular monkey 'caused no symptoms indicative of affection or impairment of the special sensory of motor faculties', he also noted that he perceived 'a very decided alteration in the animal's character and behaviour ... a considerable psychological alteration' characterised by a lack of interest and curiosity – as he put it, the poor beast had lost 'the faculty of attentive and intelligent observation'.

Intrigued, Ferrier re-examined the Gage case.[53] He was struck by a small detail in an 1868 report by John Harlow, the physician who attended Gage twenty years earlier, involving a brief description of Gage's behaviour before and after the accident. From having been a 'most efficient and capable foreman', Gage allegedly became 'fitful, irreverent, indulging at times in the grossest profanity', while friends said he was 'no longer Gage'.[54] These descriptions are now routinely reproduced in accounts of Gage's case, but until noticed by Ferrier they had gone unremarked. It should be noted that their provenance and their veracity are unknown. This vague and anecdotal account, published years after the event, is the only source that suggests there were any changes in Gage's personality or behaviour, but it was enough to convince Ferrier, who now highlighted the claim that after the accident 'Gage was no longer Gage' and that he purportedly became withdrawn and more impulsive.

In his 1878 book *The Localisation of Cerebral Function*, Ferrier boldly drew parallels between Gage's supposedly altered personality and his own observations of changes to the behaviour of monkeys with damaged frontal lobes. In many respects, the modern interpretation of Gage's injuries and their significance can be traced back to Ferrier's fusion of experimental, psychological and physiological insights – textbooks now frequently mention Gage, although they rarely get the complex story right.[55]

All this evidence suggested that various aspects of mental life linked to attention and behaviour were somehow localised in the frontal parts of the brain. Surprisingly, Ferrier even presented supporting evidence from phrenology, showing that these ideas persisted among some scientists late into the nineteenth century:

> The phrenologists have, I think, good grounds for localising the reflective faculties in the frontal regions of the brain, and there is nothing inherently improbable in the view that frontal

development in special regions may be indicative of the power of concentration of thought and intellectual capacity in special directions.[56]

Ferrier had evidence that movement and perhaps some higher psychological functions such as attention were situated in particular parts of the brain, but when it came to the most complex and immaterial aspect of brain function – thought – the evidence argued against localisation. Damage to one side of the human cortex led to loss of function in both sensation and motion on the opposite side of the body, but the ability to think was apparently unaffected, because the mind is present throughout the brain. As Ferrier explained:

> The brain as an organ of motion and sensation, or presentative consciousness, is a single organ composed of two halves; the brain as an organ of ideation, or representative consciousness, is a dual organ, each hemisphere complete in itself. When one hemisphere is removed or destroyed by disease, motion and sensation are abolished unilaterally, but mental operations are still capable of being carried on in their completeness through the agency of the one hemisphere. The individual who is para-lysed as to sensation and motion by disease of the opposite side of the brain (say the right), is not paralysed mentally, for he can still feel and will and think, and intelligently compre-hend with the one hemisphere.

Despite his discoveries, Ferrier had no model of how the brain worked. Indeed, he doubted it would ever be possible to know, arguing that even if it were possible to determine 'the exact nature of the molecular changes which occur in the brain cells when a sen-sation is experienced', 'this will not bring us one whit nearer the explanation of the ultimate nature of that which constitutes the sensation'. Ferrier, like many other scientists, was still unnerved by the problem identified by Leibniz in 1712 with his analogy of the brain working like a mill. Even if you could see inside the brain, and understand everything that was happening, that would not neces-sarily mean you would understand the nature of consciousness or thought. Ferrier was not the only thinker of the time to be uncertain about the power of science to understand the brain. Although great discoveries had been made, scientists were beginning to doubt.

FIVE

EVOLUTION

19TH CENTURY

In February 1838 the twenty-nine-year-old Charles Darwin sat down in his London lodgings, just opposite the top of Carnaby Street, and opened his copy of *Inquiries Concerning the Intellectual Powers and the Investigation of Truth*, the latest edition of a best-seller by a Scottish physician and philosopher called John Abercrombie. In the opening pages, Abercrombie bluntly proclaimed his complete ignorance of the precise relation between thought and the brain:

> The truth is, we understand nothing. Matter and mind are known to us by certain properties:— these properties are quite distinct from each other; but in regard to both, it is entirely out of the reach of our faculties to advance a single step beyond the facts which are before us. Whether in their substratum or ultimate essence, they are the same, or whether they are different, we know not, and never can know in our present state of being.[1]

Abercrombie was pointing to a fundamental problem, but this was of little concern to Darwin, who put two wavering pencil lines down the left side of the text and wrote at the bottom of the page: 'It is sufficient to point out close relation of kind of thought & structure of brain.'[2]

Darwin assumed that brain and mind were closely linked – as he put it pithily in one of his notebooks, 'brain makes thought' – but he was interested primarily in the implications of such a link rather than its precise nature.[3] Since returning from his long voyage on the survey ship HMS *Beagle* eighteen months earlier, Darwin had become preoccupied with understanding the origin of species and why they are so varied. Over time, Darwin became increasingly convinced of the role of natural selection in shaping organisms and explored the implications of this for the link between brain and thought.

In 1840 Darwin wrote in his copy of Müller's *Elements of Physiology*: 'The inherited structure of brain must cause instincts: this structure might as well be bred as any other adapted structure.'[4] Darwin realised that if brain makes thought, there must be a link between the structure of the brain and the kind of thought it produced; this in turn meant that natural selection could alter mind and behaviour by altering brain structure. This could explain the origin not only of instinctive behaviour, but also, in principle, of the human mind. From this point of view, brain and the behaviour it produced were no different from any other organ. Indeed, Darwin wrote in one of his notebooks that thought was 'a secretion of brain', 'as much function of organ, as bile of liver'.[5]*

For over twenty years, Darwin worked on what he called his 'big species book', collecting examples of natural and artificial selection and writing draft chapters, but without any particular sense of urgency or sign of finishing. Then, in June 1858, he received a shocking letter from Alfred Russel Wallace, a young explorer. Wallace sent Darwin an essay he had written in which he outlined the same mechanism of natural selection that Darwin had privately been exploring for the previous two decades. Aghast at the possibility of being

---

*This phrase was taken from a lecture given by the French physician Pierre-Jean-Georges Cabanis in the 1790s. It became notorious in the latter part of the nineteenth century, when the German zoologist Carl Vogt proclaimed that 'mental activities are merely functions of the brain, or, to put it crudely, thoughts are to the brain as bile is to the liver or urine to the kidneys'. Vogt, C. (1855), *Köhlerglaube und Wissenschaft: eine Streitschrift gegen Hofrath Rudolph Wagner in Göttingen* (Giessen: Rider), p. 32; Cabanis, J. (1815), *Rapports du physique et du moral de l'homme*, vol. 1 (Paris: Caille et Ravier), pp. 127–8.

scooped, Darwin forwarded the letter to his friends Joseph Hooker and Charles Lyell, who hastily found a solution that preserved Darwin's priority while recognising Wallace's insight. In a single session, the Linnean Society heard Wallace's letter and a note from Darwin, as well as an extract from an essay written by Darwin in 1844 that summarised his ideas. Wallace's letter finally pushed Darwin into action, and in November 1859 he published *On the Origin of Species by Means of Natural Selection*.[6]

Strikingly, Darwin's earth-shattering book avoided the thorny issues of human evolution and the link between behaviour, mind and brain. The word 'brain' appears only once in the first edition, which contains just one allusive comment about human evolution. Darwin later explained this choice:

> During many years I collected notes on the origin or descent of man, without any intention of publishing on the subject, but rather with the determination not to publish, as I thought that I should thus only add to the prejudices against my views. It seemed to me sufficient to indicate, in the first edition of my *Origin of Species*, that by this work 'light would be thrown on the origin of man and his history'; and this implies that man must be included with other organic beings in any general conclusion respecting his manner of appearance on this earth.[7]

The publication of *On the Origin of Species*, which was so destabilising for many readers, contributed to a period of uncertainty in Western intellectual life – according to the historian Owen Chadwick, in the 1860s 'Britain and France and Germany entered the age of Doubt, in the singular and with a capital D.'[8] One of the key questions that was both the focus of doubt and a contribution to it was the very issue that Darwin had so deftly avoided in his work – how (or if) consciousness emerged from the activity of the brain. In 1861 the Irish physicist and science educator John Tyndall – a supporter of Darwin's – explored this question in the pages of the weekly London newspaper *The Saturday Review*. Tyndall began with what might appear to be a straightforward materialist description of the question:

We believe that every thought and every feeling has its definite mechanical correlative in the nervous system – that it is
accompanied by a certain separation and remarshalling of the
atoms of the brain.

But as he showed, things soon got complicated when you explored
what 'correlative' and 'accompanied' actually meant:

[W]hen we endeavour to pass ... from the phenomena of
physics to those of thought, we meet a problem which tran-
scends any conceivable expansion of the powers we now
possess. We may think over the subject again and again – it
eludes all the intellectual presentation – we stand, at length,
face to face with the Incomprehensible.[9]

Nobody could even begin to describe how consciousness might
emerge from the activity of the brain.

That did not stop some scientists from speculating. In 1860 the
German physiologist Gustav Fechner made one of the boldest and
most remarkable predictions in the history of brain science. Fechner
argued that the apparent unity of mind flowed from the structural
integrity of the brain; the implication was that if you could divide
the brain hemispheres by cutting the structure that joins them, the
corpus callosum, then you would get two minds instead of one. Initially those minds would be identical, said Fechner, but they would
each gradually change with experience.[10] It would be over a century
before this dramatic hypothesis could be tested, following the use of
psychosurgery in the USA.

Some years later, Tyndall explained his position further, in two
influential lectures to the British Association, in 1868 and 1874, using
a modern version of Leibniz's Mill:

Were our minds and senses so expanded, strengthened, and
illuminated as to enable us to see and feel the very molecules
of the brain; were we capable of following all their motions,
all their groupings, all their electric discharges, if such there
be, and were we intimately acquainted with the corresponding

states of thought and feeling, we should be as far as ever from the solution of the problem, 'How are these physical processes connected with the facts of consciousness?'[11]

For Tyndall, as for Leibniz, it was impossible to explain thought on the basis of physical processes because the two classes of phenomena were qualitatively different. The terminology used by the two men was different, but their conclusions were the same.

There were two main ways of viewing this problem – either admitting that the answer was not yet known but could one day be understood, or insisting, like Tyndall, that it was inherently unknowable. Emil du Bois-Reymond agreed with Tyndall and in 1872 he explained forcefully that materialism could never provide any insight into the nature of thought: 'no imaginable movement of material particles could ever transport us into the realm of consciousness', he stated.[12] Mental processes, he said, 'lie beyond the law of causation, and are therefore unintelligible'.[13] Du Bois-Reymond closed his argument with a Latin phrase that for decades became famous in debates over the limits of scientific knowledge: '*Ignoramus et ignorabimus*' – 'We do not know and we will not know.'

*

This wave of scepticism about our ability to understand the brain infected the world of evolutionary biology, as some of Darwin's closest followers began to nuance their support for natural selection when it came to humans. In 1866, to Darwin's shock, Alfred Wallace argued that human evolution, and in particular the appearance of the human mind, could not be explained by natural selection – some supernatural force must have been involved. The immediate reason for this abrupt change was that Wallace had become obsessed with spiritualism after attending a seance in the winter of that year. During the performance, the medium, Miss Nichol, suddenly appeared on top of a table, apparently having floated there, and produced summer flowers still damp with dew.[14] Wallace was bewitched.

High on ectoplasm and the evidence of his own eyes, Wallace applied his newfound beliefs about the spirit world to the evolution of human beings, using Tyndall's doubts as support for his new view that humans were exempt from Darwin's theory:

> Neither natural selection or the more general theory of evolution can give any account whatever of the origin of sensational or conscious life … the moral and higher intellectual nature of man is as unique a phenomenon as was conscious life on its first appearance in the world, and the one is almost as difficult to conceive as originating by any law of evolution as the other.[15]

Humans were subject to very different rules from the rest of the natural world, argued Wallace, and the apparent impossibility of explaining the physical origin of thought was one element of this.[16]

Wallace was not the only one of Darwin's supporters to suggest that human evolution required some supernatural explanation. In his 1863 book *Geological Evidences of the Antiquity of Man*, the eminent geologist Charles Lyell brought together a range of evidence from palaeontology, geology and anthropology to argue that humans evolved from an ancestor that was shared with other primates.[17] This cheered Darwin, but he was less impressed by the closing section, in which Lyell argued that only divine intervention could explain the appearance of language in humans.[18]

With allies like Wallace and Lyell claiming that natural selection could not explain all aspects of human evolution, Darwin felt compelled to make his own views clear. In February 1871 he published *The Descent of Man*, which used the same approach that had been so effective in *On the Origin of Species*. Darwin presented examples of homology in anatomy and behaviour to reveal common ancestry and explained the origin of adaptations in terms of some original character that had a different function, rather than through their current apparent purpose. Darwin's conclusion was that there is no insuperable barrier between humans and other primates, including in brain, behaviour and morality. 'My object,' he wrote, 'is to show that there is no fundamental difference between man and the higher animals in their mental faculties'.[19] Darwin did not mean that apes were identical to humans, but that just as in their other physical characteristics, there was a continuum in the brain structure of different species, and therefore in their mental life.

Soon after the publication of *The Descent of Man*, Darwin realised that he needed more detailed evidence of changes in brain structure over evolutionary time, so he asked his friend and supporter Thomas Henry Huxley ('Darwin's bulldog') to provide an appendix to a second of edition of the book, dealing with comparative brain anatomy in humans and apes. Huxley obliged and concluded:

> Every principal gyrus and sulcus of a chimpanzee's brain is clearly represented in that of a man, so that the terminology which applies to the one answers for the other. On this point there is no difference of opinion ... There remains, then, no dispute as to the resemblance in fundamental characters, between the ape's brain and man's; nor any as to the wonderfully close similarity between the chimpanzee, orang and man.

The brains of humans and apes were not identical, but as Darwin was fond of saying, the difference was one of degree, not of kind. Summarising the evidence, Huxley wrote: 'this is exactly what we should expect to be the case, if man has resulted from the gradual modification of the same form as that from which other Primates have sprung'.* As Darwin explained, the behavioural and intellectual

---

*Huxley, like many of his contemporaries, including Darwin, combined opposition to slavery with views on the brain that veered between the naive and the racist. In 1865 Huxley claimed that white people had bigger brains than black people, and that this produced differences in mental ability. The consequence, he explained, was that 'no rational man, cognisant of the facts, believes that the average negro is the equal' of the average white man. He concluded that whatever might be done to remove barriers to advancement, 'The highest places in the hierarchy of civilisation will assuredly not be within the reach of our dusky cousins.' Huxley said that in women, too, differences in brain sizes supposedly led to differences in ability. He was in favour of removing all systematic discrimination, but that did not mean he thought that men and women, or white people and black people, had equal capabilities – he merely wanted to ensure 'that injustice is not added to inequality'. Huxley, T. H. (1898) *Collected Essays*, vol 3: *Science and Education* (London: Macmillan), pp. 66–75.

differences that existed between primate species must in some way reside in slight anatomical differences, not in some large structure present in one group and completely absent in another.

Darwin did not allow himself to be distracted by the doubts about the exact connection between brain and thought that concerned so many scientists and philosophers. He was confident there was such a link – in order for natural selection to operate on brain structure and thereby alter behaviour there had to be a causal connection between the two, not a mere correlation – but he was not preoccupied by its exact nature. As he had written over thirty years earlier, for his purposes he merely required that the link exist. So, with a pirouette on the pages of *The Descent of Man*, Darwin escaped from the epistemological quagmire that had trapped so many of his contemporaries: 'In what manner the mental powers were first developed is as hopeless an enquiry as how life itself first originated. These are problems for the distant future if ever they are to be solved by man.'[20]

*

While the precise link between brain and mind might be unknown, and perhaps unknowable, the more general connection between brain and behaviour in all animals was a question that absorbed Darwin. This issue was at the heart of *The Descent of Man*, which detailed the intricate behavioural adaptations or instincts shown by non-human animals and explained them in terms of natural selection. Darwin was particularly impressed by social insects, such as ants – their complex communication systems and ability to recognise nest-mates suggested the existence of memory. The explanation of this rich behavioural repertoire lay in the relatively large brains possessed by these insects, brains that Darwin said had 'extraordinary dimensions', given the size of the ants' bodies. When it came to explaining how an ant brain could pack all that behaviour into such a tiny space, Darwin's mind reeled:

> It is certain that there may be extraordinary mental activity with an extremely small absolute mass of nervous matter:

thus the wonderfully diversified instincts, mental powers, and affections of ants are notorious, yet their cerebral ganglia are not so large as the quarter of a small pin's head. Under this point of view, the brain of an ant is one of the most marvellous atoms of matter in the world, perhaps more so than the brain of a man.[21]

In the case of ants, there was apparently little need for metaphysical niggling about the link between brain and mind – no one disputed that their astonishing behaviour was simply produced by their brains. But this merely highlighted the fact that the link between brain and behaviour in tiny animals was just as marvellous and mysterious as that between brain and mind in humans.

In thinking about this question, Darwin drew an important conclusion about the significance of instinctive behaviour in most animals, and its apparently weaker role in humans. He suggested that if instinct played a relatively smaller role in the behaviour of a particular group of animals, then they should have a more complex brain structure:

Little is known about the functions of the brain, but we can perceive that as the intellectual powers become highly developed, the various parts of the brain must be connected by very intricate channels of the freest intercommunication; and as a consequence each separate part would perhaps tend to be less well fitted to answer to particular sensations or associations in a definite and inherited – that is instinctive – manner.

Darwin was suggesting that in more developed brains, there would be less localisation of function, as the 'freest intercommunication' would account in some unknown way for more complex intellectual powers.

Darwin had revealed a principle that could explain why the brains of different animals are different shapes – because they have evolved to produce different behaviours. Evolution by natural selection, together with patterns of common descent, could explain the origin of complex structures, in principle right up to human consciousness, whatever the metaphysical mysteries of how it worked. Darwin was convinced that consciousness of some kind extended deep into the animal lineage, and that the difference between humans and other animals was one of extent – we are merely more conscious than our ape relatives, rather than possessing a completely novel characteristic that requires some special explanation.

*

These debates about the origin of consciousness and its link to brain function were not restricted to academic circles. There was a huge appetite among the middle classes and educated workers to follow these arguments, in particular after a highly influential speech that Huxley gave to the 1874 meeting of the British Association in Belfast. Huxley's lecture was reprinted in *Nature*, and, in a modified form, in the widely read *Fortnightly Review*. There were hundreds of responses to it over the following years, including in the pages of *Nature*, in newspapers, and in magazines and novels.[22]

What made Huxley's lecture so provocative was his claim that animals – and humans – were 'conscious machines' or 'conscious automata':

> We are bound by everything we know of the operation of the nervous system to believe that when a certain molecular change is brought about in the central part of the nervous system, that change, in some way utterly unknown to us, causes that state of consciousness that we term a sensation.[23]

To explain the term 'conscious automata', Huxley presented a detailed re-examination of the views of Descartes, comparing them with the most recent scientific evidence, which showed that quite complex behaviours in animals, including swimming and jumping in frogs, were reflexes that could be produced without the involvement of the brain. While Descartes supposedly argued that animals were simply unfeeling machines, Huxley considered this 'a most surprising hypothesis' and instead followed Darwin's approach to show that there was no strict distinction between animals and humans:

the lower animals possess, though less developed, that part of the brain which we have every reason to believe to be the organ of consciousness in man; and as, in other cases, function and organ are proportional, so we have a right to conclude it is with the brain; and that the brutes, though they may not possess our intensity of consciousness, and though, from the absence of language, they can have no trains of thoughts, but only trains of feelings, yet have a consciousness which, more or less distinctly, foreshadows our own.[24]

Animal consciousness, Huxley argued, was a 'collateral product' of the working of the animal body, with no possibility of influencing behaviour, just as 'the steam-whistle which accompanies the work of a locomotive engine is without influence upon its machinery'. Consciousness in animals simply emerged out of nervous activity but was unable to affect behaviour, which was governed by inbuilt rules as though it were a machine.

When it came to applying this insight to humans, Huxley insisted that 'our mental conditions are simply the symbols in consciousness of the changes which take place automatically in the organism; and that, to take an extreme illustration, the feeling we call volition is not the cause of a voluntary act, but the symbol of that state of the brain which is the immediate cause of that act'. This was a major challenge to most readers' everyday experience, then and now. We all have the impression that we have volition; the implication of Huxley's argument was that free will, in terms of humans being able to think about alternatives and choose between them, is an illusion.

Huxley had form in this respect. In 1870 he had given a lecture on Descartes in which he explored the possibility that machines might be conscious:

I hold, with the Materialist, that the human body, like all living bodies, is a machine, all the operations of which will, sooner or later, be explained on physical principles. I believe that we shall, sooner or later, arrive at a mechanical equivalent of consciousness, just as we have arrived at a mechanical equivalent of heat.

This vision, which went even further than that of Smee two decades earlier, implied that whatever the apparently impenetrable nature of thought and its relation to the brain, ultimately thought would be explained by the creation of an appropriate machine. In Huxley's view, at least at this point in his intellectual development, matter could think.

After Darwin's death in 1882, evolutionary biologists appeared to lose confidence in the material link between brain and mind. The scientist who was widely seen as Darwin's successor, George Romanes (now forgotten except to historians), soon developed a view that was not far from panpsychism – the idea that all matter is somehow conscious – and abandoned natural selection as the motor force of biological adaptation. Not only did Romanes think that 'the association between mind and matter is one which is beyond the reach of human faculties to explain', he even questioned whether natural selection could explain complex instincts. He was particularly struck by the *Sphex* wasp, which digs a hole and then buries a paralysed caterpillar alongside her eggs. This caused Romanes to doubt that nature 'could ever have developed such an instinct out of merely fortuitous variations'.[25]

In contrast, the pioneer British psychologist Conwy Lloyd Morgan, writing in the 1890s and the early years of the twentieth century, was confident that such behaviours could emerge through natural selection. He showed that a chick will peck at grain without having learned to do so and he explained this behaviour by arguing that the young bird's nervous system 'is so organised that this stimulus produces that result through an organic coordination that is independent of conscious knowledge or experience'.[26] Lloyd Morgan's views of the nature of consciousness changed over time, but in 1901 he outlined what he called a 'double-aspect' theory of consciousness:

> the safest assumption is that what from a physical and physiological point of view is a complex molecular disturbance is at the same time from a psychological point of view a state of consciousness. The two are different aspects of one natural occurrence. Why such an occurrence should have two so different aspects we have not the faintest idea.

Some French philosophers were not convinced – are they ever? – and argued that whatever the brain might do, it is not responsible for the production of thought. Following Descartes, they argued that thought was an immaterial substance. In 1883 Henri Bergson claimed that 'if thought were in the head, it would occupy a place there, which by dissecting one could end up finding on the end of a scalpel ... But thought does not reside in the brain.'[27]

In 1872 the early psychiatrist Henry Maudsley looked at the crisis of confidence that was spreading among some scientists and tried to steady the ship:

> To say that it is inconceivable that matter, in however complex a state or organisation, should generate consciousness, should feel and think, is simply an appeal to the self-sufficiency of human intellect at the present day, and a sort of argument which, if logically carried through, would bar any new conception of what, from ignorance, is yet inconceivable to us.[28]

In other words, even though we may not currently understand a particular phenomenon, that does not mean we will never be able to understand it. To argue that there are things we can never understand is to undermine the whole point of science, which is to explain what is currently unexplainable.

Within a decade, Maudsley's confidence had evaporated and even he was caught up in the general mood of the time, speculating about the existence of 'an all-pervading mentiferous ether' that was beyond matter, but which could somehow interact with it. Perception occurred when the ether pervaded both the perceived object and the brain, and undulations produced by the object rippled through the ether to the brain, where they would produce consciousness. In 1883, Maudsley argued that mind was nothing more than 'the mentiferous undulations as conditioned by the convoluted and the exceedingly complicated and delicate structure of the brain'.[29] Maudsley modestly suggested that if his theory was properly elaborated it would 'without doubt, explain everything' about the universe. In the meantime, he could not actually use it to explain consciousness, beyond invoking what he called 'inconceivably rapid atom-quiverings'.

In this he might have been right, but there was no need to invoke a hypothetical 'mentiferous ether' that linked all matter. It added nothing, made no predictions that could be tested, and took Maudsley's views in a speculative, non-materialist direction. Around the same time, and in the same spirit, the neurologist John Hughlings Jackson argued 'we do not say that psychical states are functions of the brain, but simply that they occur during the functioning of the brain'.[30] Nothing seemed certain any more.

A decade earlier, Darwin had found no need for the kind of speculation that preoccupied Maudsley, nor was he seduced by the doubts that led Hughlings Jackson to distance himself from the identity of brain and mind. Instead, Darwin concentrated on showing that natural selection had operated on the brain and thereby on behaviour and psychological activity. Whatever link there was between brain structure and mental function, it was the focus of the action of natural selection, which by directly shaping organic forms could produce psychological and behavioural consequences. Furthermore, however the brain worked, there was a continuity between the mysterious phenomena of the human mind and the inner worlds of our animal relatives, both close and distant.

As the wave of doubt swept through Europe, these key lessons were forgotten and after Darwin's death the significance of his great insight faded. This was unfortunate, because a solid grounding in a Darwinian understanding would have reinforced the significance of a number of breakthroughs in our understanding of what the brain does, which had occurred in the 1860s. Each of these advances raised major problems with all aspects of existing explanations of brain function, from the vague mechanical metaphors, through the older hydraulic concepts, to the realisation that brain activity is based on electricity. Faced with these new ideas and discoveries, scientists were obliged to reconsider their ideas about what the brain does – the words they used, the metaphors they conjured up and the ways they represented those ideas.

SIX

INHIBITION

19TH CENTURY

Since the 1670s it had been known that artificial stimulation of a nerve could lead to muscle contraction. Nerves seemed to make things happen. But in the middle of the nineteenth century it became apparent that an equally fundamental property of some nerves is that they can stop things happening.[1] In 1845 Ernst and Eduard Weber, two brothers working in Leipzig, investigated what happened when the vagus nerve was stimulated with a continuous battery-generated electrical current. The twin vagus nerves go from the cerebellum, at the back of the brain, deep into the thorax, innervating all the major viscera, including the heart. To the Webers' surprise, continuous electrical stimulation of the vagus nerve led the heart rate to drop. The vagus seemed to inhibit the heart, and sufficient stimulation could even make it stop altogether.

The Webers immediately linked their discovery to the way in which the mind can sometimes stop the body from moving or responding: 'The experience that the will restricts convulsions, if they do not occur too strongly, and can inhibit the origin of many reflex movements … also demonstrates that the brain can have an inhibitory action on movements.'[2]

Their findings chimed with the views of Johannes Müller and Marshall Hall, both of whom had recently shown that destruction of the cerebral hemispheres led to unrestrained reflex actions, although they disagreed about the basis of this effect and were engaged in an unseemly squabble over priority. The results also fitted with the ideas of A. W. Volkmann, who in 1838 showed that if you removed a frog's head, the body would show reflex actions that did not occur in the intact animal. As Volkmann explained: 'It becomes clear from this that the brain contains the cause for the hindrance in the activation of the nervous principles ... the influence of the mind possibly hinders this activation.'[3]

A series of studies on other peripheral nerves revealed inhibition of basic physiological processes, and in 1863 the Russian physiologist Ivan Sechenov generalised these insights into a theory about brain function. Sechenov had previously worked with some of the greats of European physiology, such as du Bois-Reymond, Helmholtz and Claude Bernard, and built upon the ideas of the Webers and Volkmann, arguing that the brain must contain two complementary centres: 'one suppresses movements, while the other, on the contrary, intensifies them'.[4] This seemed to explain most aspects of behaviour: 'man not only learns to group his movements through the frequent repetition of associated reflexes, he, at the same time, acquires (also by means of reflexes) capacity to inhibit them'.

This idea enabled Sechenov to sketch out a theory of how the brain works. His starting point was the reflex pathway:

stimulus → central inhibition or intensification → muscular response

This simple chain of reactions was, he claimed, all you needed to understand even the most complex of brain functions. 'A thought,' he stated, 'is the first two-thirds of a psychical reflex.' In other words, a thought corresponds to the external stimulus that induced it and an appropriate central activity; whether that thought is acted upon and the final third of the reflex – a muscular response – is invoked would depend upon circumstances. Sechenov was not alone in this view. For the British neurologist Hughlings Jackson it was self-evident. As he wrote in 1870:

What can an 'idea', say of a ball, be, except a process representing certain impressions of surface and particular muscular
adjustments? What is recollection, but a revivification of such
processes which, in the past, have become part of the organism itself?[5]

Writing for the general public, Sechenov addressed the obvious
criticism that thinking does not feel like 'two-thirds of a reflex', but
rather like an internal process, full of voluntary actions and often
independent of external factors. His answer was rigorous but bleak:

When the external influence, i.e. the sensory stimulus, remains
unnoticed – which occurs very often – thought is even
accepted as the initial cause of action. Add to this the strongly
pronounced subjective nature of thought, and you will realise
how firmly man must believe in the voice of self-consciousness
when it tells him such things. But actually this is the greatest
of all falsehoods; the initial cause of any action always lies in
external sensory stimulation, because without this thought is
inconceivable.[6]

Sechenov was trying to provide a physiological explanation of
the nature of thought, and also to show how patterns of inhibition
and activation of reflexes could produce complex behaviours. As
Henry Maudsley wrote in 1867: 'One of the most necessary functions
of the brain is to exert an inhibitory power over the nervous centres
that lie below it.'[7]

Ferrier knew of Sechenov's ideas and accepted that inhibition was at the heart of how the brain worked. Inhibition, argued
Ferrier, was 'the essential factor of attention' – the organism has to
inhibit responses to extraneous events in order to focus on one particular stimulus. Because of this, he claimed, centres of inhibition
in the brain 'constitute the organic basis of all the higher intellectual faculties', and 'the more developed those centres, the greater
the intellectual power of the organism'.[8] Inhibition seemed to be a
key factor in intelligence. A few years later, the early psychologist
William James (brother of writer Henry James) wrote that 'the entire

drift of recent physiological and pathological speculation is towards enthroning inhibition as an ever-present and indispensable condition of orderly activity'.[9]

Despite all this interest it was not at all clear how inhibition actually worked. There were various theories, all involving some kind of physical metaphor. The Victorian polymath Herbert Spencer argued that there was a limited quantity of nervous force and that when this was exhausted, reflexes were inhibited.[10] The German physiologist Wilhelm Wundt suggested that inhibition and excitation took place simultaneously, and that 'the entire process of excitation is therefore dependent at every movement on a mutual interaction of excitation and inhibition'.[11] The British psychologist William McDougall had a similar idea, according to which the nervous system had a built-in balance, with the activity of one 'neural system' inhibiting the activity of another, such that 'inhibition appears always as the negative or complementary result of a process of increased excitation in some other part'.[12] McDougall described the force contained in a nerve as 'neurin', and, thinking in terms of fluids, suggested that inhibition involved the 'drainage of the free nervous energy from the inhibited to the inhibiting system'.[13] Descartes would have approved.

Other thinkers used more complex hydraulic metaphors, suggesting that inhibition was perhaps produced when the actions of two parts of the system interfered with each other, like two sets of waves meeting, thereby cancelling out or altering their activity.[14] David Ferrier was more straightforward when he admitted 'the nature of the inhibitory mechanism is exceedingly obscure'.[15] No existing model of nervous function, be it based on spirits, fluids, irritation, vibration or electricity, could account for it.

*

Meanwhile, scientists were beginning to explore what the *absence* of inhibition revealed about how the brain works. In 1865 Francis Anstie, a young English physician, suggested that narcotics and anaesthetics induced 'a partial and highly peculiar kind of paralysis of the brain' and that in the cases of hashish and alcohol 'the apparent exaltation of certain faculties should be ascribed rather to the removal of controlling influences, than to positive stimulation of the faculties themselves'.[16] Psychoactive drugs suppress the brain's ability to control, including through inhibition. This could be seen every time an anaesthetic was used in the operating theatre – the highest mental functions were the first to go, leading to a loss of control just before the patient became unconscious.

Control is now a central part of our understanding of brain function, but up until this point it had not been used as a way of thinking about what the brain does.[17] Anstie's view was part of the realisation that one of the overall functions of the brain is to control the body, with the ideas of inhibition and control being tightly linked. With this fundamental insight, new ways of understanding the role of the brain in health and in disease became possible. For example, Hughlings Jackson argued that epilepsy could be understood as a loss of control in the brain, through the absence of inhibition.[18] For the psychologist Conwy Lloyd Morgan, inhibition was an essential feature of how an organism learned to control its behaviour:

> What we term the control over our activities is gained in and through the conscious reinforcement of those modes of response which are successful, and the inhibition of those modes of response which are unsuccessful. The successful response is repeated because of the satisfaction it gives; the unsuccessful response fails to give satisfaction, and is not repeated.[19]

Lloyd Morgan extended this view to link control and consciousness, paralleling the ideas of Sechenov, in which the importance of control in higher organisms such as humans was associated with increased degrees of behavioural flexibility: 'the primary aim, object, and purpose of consciousness is control. Consciousness in a mere automaton is a useless and unnecessary epiphenomenon'.[20]

Distancing himself from Huxley's paradoxical suggestion that humans were conscious automata, Morgan was presenting a sophisticated evolutionary view of the role of consciousness as control, which could only have a function in an organism that was not composed of mere reflexes.

Soon a wide range of disorders were seen in terms of loss of control – somnambulism, insanity, hysterical sexual fits (only in women, obviously), and even asthma. One of the places that in the 1870s and 1880s had the greatest influence on our understanding of the way that the brain controls both mind and body was La Salpêtrière hospital in Paris, where the neurologist Jean-Martin Charcot had his practice. Charcot and his colleagues revealed that a series of disorders with major behavioural symptoms all had their basis in damage to the brain's ability to inhibit and control – these included multiple sclerosis, Parkinson's disease, motor neurone disease and Tourette's syndrome (de la Tourette was one of Charcot's colleagues).

To treat his patients, Charcot used a variety of therapies, including a version of Hitzig's electrotherapy and a vibratory chair (there was a portable version in the shape of a vibratory helmet). But his most innovative approach involved hypnotism, by which he was able to reproduce hysterical symptoms such as sleepwalking by apparently inducing a loss of conscious control. In 1880 a report in *Scientific American* described how Charcot held his finger in front of his star patient, Marie 'Blanche' Wittmann, asked her to concentrate on it, and within ten seconds, 'her head fell heavily to one side … her body was in a state of complete resolution; if an arm was raised by the observer, it fell again heavily'.*[21] With Wittmann in this state, Charcot was able to induce all kinds of hallucinations and symptoms, similar to those reported by his patients. For Charcot, the significance of hypnotism was that by recreating symptoms it could provide insight into the workings of the mind; this greatly impressed an Austrian visitor to the Salpêtrière, Sigmund Freud.[22]

Charcot admitted that he did not understand how hypnosis worked, nor was he greatly concerned by the question – 'facts first, theories afterwards', he said. But in 1881 the Polish physiologist Rudolf Heidenhain argued that 'the cause of the phenomena

---

* There is a well-known, immense painting from 1887 by André Brouillet, entitled *Une leçon clinique à la Salpêtrière* (A Clinical Lesson at the Salpêtrière), that captures the underlying sexual politics of these events. It shows Charcot in front of two dozen male medics wearing dark suits, hypnotising Wittmann, whose white blouse has fallen off her shoulders. Freud owned a print of Brouillet's work. Morlock, F. (2007) *Visual Resources* 23:129–46.

of hypnotism lies in the inhibition of the activity of the ganglion-cells of the cerebral cortex', which was produced simply by 'gentle prolonged stimulation of the sensory nerves of the face, or of the auditory or optic nerve'.[23] There was no direct evidence for this interpretation, in particular in terms of the activity of 'the ganglion-cells of the cerebral cortex', which sounded very scientific but which was basically guesswork. However, together with a Russian colleague, Nikolai Bubnov, Heidenhain drew parallels between the effects of hypnotism and those of morphine, both of which, they claimed, lowered the individual's ability to maintain 'inhibitory processes'.[24] They showed that stimulation of motor areas of the cortex could inhibit excitation in these regions, suggesting that nervous centres in the brain were interacting in ways that resembled inhibition, producing the effect of control.

Both Freud and the Russian psychologist Ivan Pavlov later used the concept of inhibition in their writings on behaviour, but neither was particularly interested in the brain. Once Freud embarked on the path that would lead him to create his spurious but highly influential psychoanalytic framework, he gradually lost interest in the material basis of psychology. In 1893 Freud distanced himself from Charcot's attempt to link hysteria with brain anatomy:

> I, on the contrary, assert that the lesion in hysterical paralyses must be completely independent of the anatomy of the nervous system, since *in its paralyses and other manifestations hysteria behaves as though anatomy does not exist or as though it had no knowledge of it.*[25]

For Freud, brain function could not explain psychology. In 1915 he recognised that there was 'irrefutable proof that mental activity is bound up with the function of the brain as with that of no other organ', but he insisted that his psychological theory had 'nothing to do with anatomy; it is concerned not with anatomical locations, but with regions in the mental apparatus, irrespective of their possible situation in the body'.[26] As he explained in 1916, 'I know nothing that could be of less interest to me for the psychological understanding of anxiety than a knowledge of the path of the nerves along which its excitations pass.'[27] Although in *The Ego and the Id*, written in 1923, Freud suggested in passing that there was an 'anatomical analogy' between his psychological construct the Ego and the representation of the body in the cortex, this had no consequence for his psychological theory, nor, conversely, did his theory make any predictions about potential cerebral lesions corresponding to particular mental disturbances.

There was one brief exception to this general tendency. In 1895 Freud furiously wrote a long manuscript that eventually became known as the 'Project for a Scientific Psychology'. Not only did Freud not publish this work, he soon disowned it, explaining that the whole thing had been 'a kind of madness'.[28] In this strange document, Freud imagined that the brain contained three kinds of nerves, some of which acted like connecting pipes, each type showing different degrees of permeability, thereby enabling these structures to achieve what Freud argued was their aim, which was to reach a state of rest. The fundamental metaphor underlying his speculative theoretical framework was hydraulic – the document repeatedly refers to 'flow' and even 'pressure' in nerves. Whatever the intellectual link between this brief theoretical speculation and his full-blown psychoanalytic theory – followers and opponents disagree – in reality, Freud had nothing novel or insightful to say about how the brain worked.

Pavlov's initial interest was in digestive physiology, and when in the 1890s he extended this to the study of what are known in English as 'conditioned reflexes' (epitomised by the dog salivating to the sound of a bell*), he saw inhibition simply as a phenomenon that reduced the strength of a reflex response. Pavlov eventually attempted to integrate his studies of conditioned reflexes with investigations of brain function, and even with psychiatry, but he was unable to present any further insight into how the brain actually works.[29]

---

*The opening sentence of Daniel Todes's biography of Pavlov states that the great Russian scientist 'never trained a dog to salivate to the sound of a bell'. Todes, D. P. (2014), *Ivan Pavlov: A Russian Life in Science* (Oxford: Oxford University Press).

These two great figures of early twentieth-century psychology both had a major influence on views of behaviour and the mind, but their ideas had no consequence for how we understand the brain.

*

After inhibition and control, a third unexpected aspect of brain function was discovered in the 1860s. It was explored by Hermann von Helmholtz in his *Handbook of Physiological Optics*, which appeared in 1867. For centuries, many of the philosophical discussions about the mind had focused on what happens when we perceive an object. A common-sense explanation was that perception is merely the consequence of the physical stimulation of the sense organs – we see what is in front of us, as though through a window. But Helmholtz realised that things are not so simple. In reality the nervous system, and in particular the brain, plays a highly active part in constructing our perception of even quite straightforward things. The brain does not simply register the outside world, it selects and represents aspects of it. Even our simplest perceptions involve the brain making inferences about what is going on, rather than straightforwardly observing them.

Helmholtz's starting point was the existence of illusions such as when you press your eyeballs and perceive coloured patterns, or the distressing 'phantom limb' illusion when an amputee can still feel the limb even though it is no longer present. Helmholtz argued that these effects, which had led Müller to believe that each nerve had its own kind of energy, were in fact 'an illusion in the judgement of the material presented to the senses, resulting in a false idea of it'. Helmholtz understood that in such cases stimulation of the nerves was being perceived either as though the usual sensory modality was involved (in the case of pressure on the eyeballs) or as though the absent limb was in fact present. His explanation was that the brain did not simply register a stimulus, but rather 'drew a conclusion' about the nature of the stimulation it was receiving. This was something like an inference that flows from a logical syllogism – the eye's function is to detect light, the eyeball is stimulated, therefore the stimulus must consist of light. The phantom limb could be explained in the same way: 'all stimulations of cutaneous nerves, even when they affect the stem or the nerve-centre itself, are perceived as occurring in the corresponding peripheral surface of the skin', Helmholtz surmised.[30]

Helmholtz applied this insight to normal perception and argued that when we perceive, the nervous system draws what he called 'unconscious conclusions' about the nature of what is being perceived. Perceptions are not simple impressions produced by the environment, but rather 'inductive conclusions, unconsciously formed', he declared.[31] Helmholtz's explanation hinted at some kind of process in the nervous system that would be able to draw conclusions without the mind being aware of it. With sufficient repetition, the process became completely unconscious, he argued. We learn to perceive.

Another unconscious conclusion that Helmholtz described was the way that we construct our stereoscopic 3-D view of the world out of the slightly different images that come from each eye (try alternately opening and closing each eye, and you will see how they differ). As his colleague Wilhelm Wundt showed, somewhere in the visual system of the brain, before we become conscious of it, these two images are assembled into a coherent image that enables us to perceive in depth. Our impression of a 3-D world is constructed by our brains out of two 2-D images without us being aware of it.

Two other members of the German school of physiology, Ernst Weber and his student Gustav Fechner, showed that our ability to perceive the differences between two stimuli changes with their amplitude – for example, the heavier two objects are, the larger the difference between them has to be before we can detect it. The same thing applies to other sensory modalities, and the effect follows a near-constant logarithmic relationship between the difference that can be detected and the amplitude of the two stimuli. Another way of putting this is that we are very good at detecting small differences between low-amplitude stimuli. Our brains and our sensory systems obey specific laws and make unconscious conclusions about the world even before we know it.

In an even more dramatic challenge to how we believe that we perceive, Helmholtz argued that perception involves a kind of filter

– the brain does not pay equal attention to all the stimuli it is presented with. For a start, our bodies react to the environment and can often alter our perception accordingly, for example by enlarging the pupil in the dark. As Helmholtz put it, 'We are not simply passive to the impressions that are urged on us, but we observe, that is we adjust our organs in those conditions that enable them to distinguish the impressions most accurately.'[32]

Even more disturbingly, there is a 'blind spot' in the visual field where we literally cannot see – this is because there are no light receptors in the part of the retina where your optic nerve leaves the eyeball. This corresponds to a point slightly to the right or left of the centre of the visual field in the corresponding eye. But we do not perceive a gap in our visual world, and unless we concentrate, we are completely unaware that it exists. One reason for this is that our eyes are in constant motion, even if only slightly, so the empty bit of the visual field is continually being filled in.

Another reason, which particularly intrigued Helmholtz, reveals a general principle about how the brain processes stimulation: 'we are wont to disregard all those parts of the sensations that are of no importance so far as external objects are concerned', he wrote.[33] The brain simply ignores the absent stimulus and fills in the gap, making up the space with a perceptual blur that is based on surrounding shapes and colour, and which we do not notice. Helmholtz suggested that even when dealing with relatively simple stimuli, the brain was continually drawing unconscious conclusions about the nature of the objects that were stimulating the nervous system. The implication was that the complex structures of the brain were somehow able to perform logical operations not only without conscious thought being involved but apparently as a prerequisite to that conscious thought.[34]

Helmholtz's view of the brain as an active organ, and of perception as an imperfect and selective process leading to a view of the world, represented a major breakthrough in our understanding of what the brain does, one that still dominates today. This insight was the pure product of scientific discovery, without the application of a metaphor from technology. On the other hand, in one respect, philosophy had got there first. The eighteenth-century philosophers of perception such as Hume and Kant had argued over whether ideas came from the world (Hume), or whether we use innate concepts in our perception (Kant). Philosophers and historians have argued over whether Helmholtz was truly a Kantian, but his view of perception and brain function chimed with one aspect of Kant's philosophy, which resonates down to the present day.[35] In his *Critique of Pure Reason*, published in 1787, Kant argued that some features of how we perceive are given a priori, that is, without experience. Although Kant was primarily interested in things such as space, time and moral judgements, he put his finger on a key feature of what is happening when we interact with the environment. Our senses are not open valves that simply allow all stimuli into our brain; instead we perceive only certain parts of our environment. A trivial example would be our inability to see ultraviolet light; other animals, such as insects and birds, can do so. More complex filters also exist within our brains.

Many scientists have subsequently referred to what are called in the jargon 'the Kantian synthetic a priori' – our nervous system involves an innate cognitive and neurobiological framework that filters and processes raw sensory stimuli to turn them into a picture of the world.[36] For Helmholtz, the brain does not simply register impressions, it changes and interprets them, making unconscious inferences.[37]

*

Given the difficulties of understanding the brain as a whole, many physiologists preferred to investigate the new ideas of control and inhibition by focusing on the elementary components of the nervous system. This was the approach taken by C. S. Sherrington at the University of Liverpool, who wanted to understand how sets of nerves and muscles or 'reflex arcs' interact to produce reflex behaviour.[38] For Sherrington, the reflex arc was the basic unit of the nervous system and all complex behaviours were composed of combinations of reflexes. For example, these combinations could explain a frog noticing the movement of a fly, jumping at it, catching it in its mouth and then swallowing it.[39]

According to Sherrington, the threshold required to produce a reflex behaviour is lowered by the activity of the reflex action that took place before it, thereby ensuring the rapid passage from one reflex to the next in the chain and producing a single coordinated complex behavioural response. Like Steno 250 years earlier, Sherrington viewed the animal as a complex machine that could be understood by investigating its component parts. As he wrote in *The Integrative Action of the Nervous System*, a landmark book summarising his ideas that was published in 1906 and is still in print: 'In the analysis of the animal's life as a machine in action there can be split off from its total behaviour fractional pieces which may be treated conveniently, though artificially, apart.'[40]

Sherrington provided a precise account of reflexes in dogs, in particular the scratch reflex, where stimulation of skin on the animal's flank leads to a rhythmic scratching movement in the leg (try this by scratching the side of a friendly dog – it does not work so well on a cat). Sherrington showed how each sensory nerve was linked to a particular area of skin, which he called a receptive field, that if stimulated could induce the nerve to respond. Activation of any of these nerves led to the same behaviour – scratching – a muscular response that Sherrington called the 'final common path' of the scratch reflex arcs.[41]

Inhibition was at work here, too – after the dog scratched, the reflex was inhibited for some time, apparently by some brain process, which could be overcome by the application of drugs such as strychnine. Sherrington was convinced that the significance of inhibition could be felt at the highest levels of brain function: 'nerve-inhibition must be a large factor in the working of the mind'.[42]

Starting from an exploration of the comparative anatomy of animals, Sherrington argued that the brain is really just another bundle of nerves. The challenge, he said, was to explain 'the dominance attained by one limited set of neural segments, the brain, over all the rest'.[43] His answer was that the brain evolved as 'an organ of coordination in which from a concourse of multitudinous excitations there result orderly acts, reactions adapted to the needs of the organism'. From an evolutionary perspective, the function of the cerebral hemispheres, in particular in humans, was to ensure control over the body and its interactions with the environment by enabling the organism to provide a wide range of flexible responses. Understanding how that occurred was the challenge of the coming century, he said: 'It is then around the cerebrum, its physiological and psychological attributes, that the main interest of biology must ultimately turn.'

*

Although no one claimed to be able to properly explain how the brain did what it did, everyone who wrote about the subject inevitably revealed something about what they thought, through the words they used, the metaphors they employed and the diagrams they produced. In 1880 the British neurologist Henry Charlton Bastian summarised current knowledge of brain structure and function in *The Brain as an Organ of Mind*.[44] The words he used reveal a mixture of the old and the new. His starting point was the 'impressions' carried by nerves – this old metaphor suggests the physical imprint of a sensory stimulus upon the nerve. These impressions were 'transmitted' along a 'route' of 'ingoing fibres' towards a 'centre' in the brain, where they were 'registered'. Then, though 'structural connections' those impressions found themselves in 'outgoing currents', the role of which was to 'throw the automatic apparatus of movement into action', as Ferrier had put it.

None of this amounted to an explanatory model or a hypothesis about how the brain worked, but the terms that Bastian used were all related to pressure or water (there is no indication that Bastian had an electric current in mind). The vague term 'centre', which was widely used throughout the second half of the nineteenth century, had no real implication beyond that of a site at which nerves were particularly concentrated, with perhaps a sense of localisation of function. But the idea of impressions being 'registered' implied some kind of physical inscription, while the image Bastian adopted from Ferrier was purely mechanical, conjuring up the notion of a steam engine or some other machine lurching into motion following the release of a control lever.

Writing over twenty-five years later, Sherrington's views were not much further advanced. Despite describing nerve function in terms derived from electricity – like others, he spoke of 'conduction' – Sherrington primarily used physical metaphors to explain what was happening. Reflex arc conduction, he suggested, 'may be figuratively described as indicating inertia and momentum', so the action of the reflex arc was like pulling on an elastic band, rather than on a rigid rod.[45] Sherrington thought that an animal was a machine that could be understood by investigating its component parts, and he inevitably pursued that mechanical analogy into the brain. Like all scientific metaphors, Sherrington's view was bounded by the technology of his time. Living in the age of steam, he found it hard to see further than pistons and cylinders, even if they were made of muscle and cartilage rather than iron and steel.

In order to clarify their views to their readers, and perhaps to themselves, many researchers produced schematic diagrams of the anatomy of the nervous system, in particular of the reflex arcs of the spinal cord. There were no metaphors attached to these schemas – these were not 'wiring diagrams' (that analogy lay decades in the future) – but arrows were added to indicate the way in which different nervous centres influenced each other.

For example, in 1886 Charcot presented a figure showing the various centres that are involved when we hear, pronounce, see or write the word *cloche* (bell). The links between the various 'centres', including the part at the top of the figure labelled 'IC', for 'intellectual centres', were largely imaginary. However, because this kind of diagram suggested at what level a particular defect might exist, it could also constitute a guide for brave or foolhardy surgeons venturing into the brains of their patients, indicating where to search for a growth, and where – or where not – to cut. A decade earlier, Ferrier had also used arrows to indicate the 'centripetal or centrifugal direction', that is, whether the nerve fibres flowed outward from the centre, or from the periphery.[46] Ultimately, however, this went little further than a highly simplified anatomical diagram. There was nothing here that could be used to construct a model or a hypothesis about what was actually happening in these centres, or what was travelling down the centripetal and centrifugal nerves.

Sherrington's schema, drawn up over thirty years after Ferrier's, added inhibition to the picture by including plus and minus

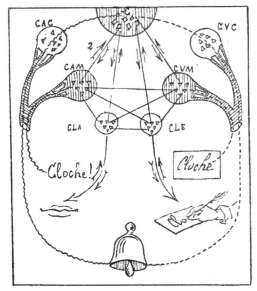

Schema of Charcot's idea for how different parts of the brain are connected. The various abbreviations refer to different 'centres' that Charcot hypothesised, each with a function (vision, hearing, auditory memory, etc.).

signs and trying to describe reflex function (in this case, the knee-jerk reflex) in almost algebraic terms:

> If we denote excitation as an end-effect by the sign *plus* (+), and inhibition as end-effect by the sign *minus* (−), such a reflex as the scratch-reflex can be termed a reflex of double-sign, for it develops excitatory end-effect and then inhibitory end-effect even during the duration of the exciting stimulus.[47]

Translating this into actual nervous activity, turning the diagram into a grounded model of brain function, was impossible at this time. Despite the fact that electrical stimulation was at the heart of so many of the discoveries in the final decades of the nineteenth century, it was often seen primarily as a more subtle and precise form of irritation

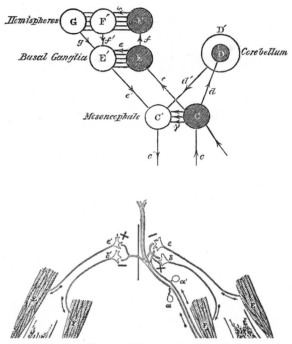

Top: *Ferrier's diagram of the organisation of the brain.*
Bottom: *Sherrington's diagram of spinal cord reflexes. Note plus and minus signs.*

that could reveal function. For the implicit to become explicit, for nervous action to be understood properly and for the basis of brain activity to become part of visions of how it worked, scientists first had to realise what the brain was actually made of.

SEVEN

NEURONS

19TH TO 20TH CENTURY

O ne of the greatest scientific achievements of the nineteenth
century was cell theory – the realisation that all organisms are
made up of cells, and that cells can only come from other cells, showing
that life does not generate spontaneously. Biology had found its fundamental particle. One of the pieces of evidence that led to the rapid
acceptance of this theory was obtained in the 1830s, when the Czech
anatomist Jan Purkyně (generally written as Purkinje) used one of the
latest microscopes to look at thin slices of the human cerebellum.[1]

Together with one of his students, Gabriel Valentin, Purkinje
saw that the cerebellum was made up of globules – vase-shaped
structures full of tiny spots. These globules were gathered together
in a layer, which was situated just above a series of long fibres. In
1838 one of Johannes Müller's students, Robert Remak, showed that
each of those fibres was connected to one of the globules. There were
cells in the brain.

The realisation that the globules and fibres were parts of nerve
cells, and that the brain, like every other part of the body, was made
of cells, was popularised over a decade later (with no recognition of Remak*) by the Swiss anatomist Albert von Kölliker, in his

---

* Remak was a Jew and was unable to become a professor in a German

widely read *Handbook of Human Histology*. Nerve cells seemed to be composed of three regions: a set of branches termed protoplasmic expansions, a cell body or soma and, finally, a long tubular fibre or axis cylinder.

Despite this advance, a major dispute emerged over how the nerve cells were organised. Everywhere else in the body cells were discrete units, each bounded by a membrane. But von Kölliker's beautifully precise drawings suggested that Purkinje's globules and fibres constituted a single organic network, as the fibres branched ever-finer and appeared to fuse, forming a single net-like, or reticular, structure. Furthermore, studies of the first full nervous systems to be investigated – those of jellyfish, which have no brain – showed that their nerves were organised in a kind of net. Von Kölliker did not agree with this view. He was convinced that each nerve cell was an independent structure, but he recognised that he had no direct proof that the reticular theory was wrong. With the techniques of the time, it was simply not possible to be certain if the branches of different cells were separate, and von Kölliker doubted it would ever be possible to resolve the issue.

The answer began to come into focus in 1873, when the Italian anatomist Camillo Golgi had a minor accident in his laboratory. He spilt some silver nitrate onto a slice of tissue that he had previously hardened with potassium dichromate. To his annoyance, the reaction of the two chemicals made the tissue turn black, apparently ruining it. But when Golgi looked at the sample under the microscope, he discovered that only a tiny proportion of the nerve cells had been stained, and they could now be distinguished down to their finest details, showing up as black silhouettes against the light background. Paradoxically, the fact that very few cells were coloured meant that it was possible to precisely describe the structure of single nerve cells. Had all the cells been affected, the result would have been a dense, uninterpretable mess.[2]

---

university under that country's repressive legislation. Remak, together with Virchow, later observed that in vertebrates that some nerve cells are covered with a white substance – what is now called myelin – while others are not. This difference produces the distribution of grey and white matter in the brain – grey matter includes most of the neuronal cell bodies and dominates your cortex.

Over the next few years, Golgi used this difficult technique – initially called the black reaction, but soon simply called the Golgi method or the Golgi stain – to explore parts of the vertebrate brain, such as the cerebellum, the olfactory bulb, the hippocampus and the spinal cord. The world Golgi saw down his microscope was unimaginably complex – the nerve branching that had been revealed with earlier methods turned out to be just the beginning. It could now be seen that the branches branched, and those branches branched again.

Despite the increased resolution provided by the new technique, it was still not possible to see whether the finest intertwined branches of two neighbouring nerve cells were truly independent. Golgi was convinced that the branches were indeed separate, but he clung to the reticular view of the nervous system, arguing that nerve cells were instead fused at the level of the axis cylinders. Although Golgi recognised that there might be chemical or other differences that corresponded to functional differences between brain cells, he was sure that any activity in a nerve cell would be shared across the hypothetical network.[3] As he put it, there was 'certainly not an isolated action of cellular individualities, but a simultaneous action of extensive groups'. Golgi was so confident of this that in 1883 he drew the obvious conclusion about how the brain worked and rejected any localisation of function. Although he praised the 'ever celebrated' results of von Fritsch and Hitzig and accepted that he could not deny 'the physiological doctrine which ascribes to various convolutions different functions', he nevertheless concluded:

> the concept of the so-called localisation of the cerebral functions, taken in a rigorous sense (i.e. that certain determinate functions may be referred to one or another zone, exactly limited), cannot be said to be in any manner supported by the results of minute anatomical researches.

Golgi was definitely on the side of those who were opposed to localisation.

*

Golgi's method was notoriously difficult to master and it took some years for it to become widespread. When other researchers eventually published their observations, they did not agree with Golgi in one crucial respect. In the mid-1880s, Wilhelm His of the University of Leipzig reported he could see no fusion between nerve cells and concluded that they were indeed each an independent structure, like other cells. His also coined a new term to describe the complex tree-like part of a nerve cell – he called them dendrites after the Greek for tree (*dendron*). At around the same time a Swiss scientist, August Forel, cut the nerve fibre leading to the tongue and then, after a few days, studied which tissues in the brain died because they had been separated from the main part of the cells which supplied them with nourishment. To Forel's surprise, only a tiny area of the brain was affected, indicating that nerve cells were not interconnected. The very specific and limited degeneration observed by Forel suggested that each cell body and its dendrites formed a single unit.

The final blow to the reticular view of the nervous system came from the work of Santiago Ramón y Cajal, a Spanish neuroanatomist whose contribution to science was on a par with that of Vesalius. As well as being a skilled anatomist, Cajal, as he is generally known, was also a talented artist and photographer, even inventing his own way of producing colour photographs. There is a famous self-portrait of him in his laboratory in 1885; wearing a stained smock and a stylish cap, Cajal is seated at a table with three microscopes, his head leaning on his hand. Behind him are shelves of bottles and vials containing chemicals – the key to unlocking the hidden structure and function of the brain. As he later put it: 'We saw that an exact knowledge of the structure of the brain was of supreme interest for the building up of a rational psychology. To know the brain, we said, is equivalent to ascertaining the material course of thought and will.'[4]

Cajal's world changed in 1888, in what he called 'my greatest year, my year of fortune'.[5] A colleague from Madrid showed him some nerve cells stained using Golgi's method; Cajal's description of what he saw gives a vivid impression of the power of the technique:

What an unexpected sight! Against a perfectly clear yellow background, I could see thinly-distributed black threadlets,

some slender and smooth, others thick and thorny, and dark
structures – triangular, star-shaped or spindle-like. It was as
though they had been drawn with Chinese ink on transparent
Japan-paper. For anyone used to the impenetrably tangled
thicket that appears when tissue is stained with carmine or
logwood, where you are doomed to guesswork, this was profoundly disconcerting. Here, everything was clear and simple.
There was no need to guess; all you had to do was look …
Amazed, I could not take my eye from the microscope.[6]

Stunning as this was, Cajal soon found ways of improving the
Golgi stain. By studying immature brains from a variety of animals
including birds and fish, and using a number of technical tweaks,
such as thicker sections and a second staining, Cajal rendered the
method more reliable and informative. His illustrations, some of
which are still unsurpassed, showed the structure of the brain in
unprecedented detail and are much loved by modern neuroscientists
for their clarity, impact and beauty. But these figures were also artificial constructions – as Cajal cheerfully admitted, each figure involved
observations of many different microscopic slices of the brain, which
he then painstakingly assembled into a single, highly informative
image. The images were accurate, but despite their natural appearance, they were also artificial.

Cajal's observations revealed that the brain and the peripheral
sense organs such as the retina had a clear but mysterious organisation. The dendrites of cells were oriented towards the external
environment, while their axis cylinders were closer to the centre of
the brain. Using his version of the Golgi stain, Cajal could see that
nerve cells had many different shapes, and that similar-shaped cells
were grouped together in layers. The tempting conclusion was that
this organisation was somehow linked to how the brain worked,
although Cajal could not imagine what that link might be. But he
could use his exquisitely precise observations to settle the thorny
question of whether nerve cells were connected in a net.

First, he showed that the axis cylinders were not fused, as Golgi
claimed; then he suggested that the dendrites were neither fused nor
nutritional, but instead had a vital functional role. His explanation

invoked the most complex technological metaphor of the time, the telegraph. Cajal suggested that the Purkinje cells in the cerebellum were linked to another type of cell, the granule cell, 'somewhat as a telegraph pole supports the conducting wire' while the dendrites in these cells established 'contacts of transmission' with nearby cells.

Cajal's use of a telegraph image coincided with an analogy that had been made by the French anatomist Louis-Antoine Ranvier in 1878, when he speculated that the myelin seen on the outside of motor and sensory nerves in vertebrates acted as a kind of insulation, much like the way that undersea telegraph cables were constructed.[7] Cajal revealed that the structure of the olfactory bulb provided an example of how dendrites took 'currents from nerve fibres' – the sensory cells in the nose converge in the brain, forming a series of round masses known as glomeruli, while the dendrites of another class of cells are connected to these masses, and their axon cylinders go on deeper into the brain. An equally precise, but very different anatomical organisation could be found in the retina, Cajal showed.[8]

In October 1889, Cajal went to the Berlin Congress of the German Anatomical Society, where he showed his amazing microscope slides. He later recalled:

> I began to explain to the curious in bad French what my preparations contained. Some histologists surrounded me, but only a few ... Undoubtedly, they expected a fiasco. However, when there had been paraded before their eyes a procession of irreproachable images of the utmost clearness ... the supercilious frowns disappeared. Finally, the prejudice against the humble Spanish anatomist vanished and warm and sincere congratulations burst forth.[9]

One of those who was so impressed by what they could see down the microscope – the cells were stained dark red or black, standing out against a yellow background – was the doyen of neuroanatomy, von Kölliker. Von Kölliker soon replicated Cajal's findings and brought the Spaniard's work to the attention of the international scientific community. As Cajal later recalled, 'it was due to the great authority of Kölliker that my ideas were rapidly disseminated and appreciated by the scientific world'.[10]

The work of Cajal, von Kölliker and others was in turn summarised in 1891 by the German anatomist Wilhelm von Waldeyer, who reported that the work of a Norwegian student, the future polar explorer Fridtjof Nansen, showed there was no fusion of nerve cells.[11] On the basis of all this evidence, von Waldeyer argued that nerve cells were separate, discrete entities, which he called neurons (sometimes written as neurones and from the Greek for fibre).[12] Another significant step in the creation of the modern anatomical vocabulary of the nerve cell came in 1896, when von Kölliker, now aged eighty, came up with the term axon to refer to the axis cylinder.[13]* Everything was now in place, and this view – which soon became known as the neuron doctrine or neuron theory – was rapidly accepted and formed the basis of all future studies of the nervous system.[14]

Nevertheless, Golgi continued to deny that neurons were independent cells. The dispute carried on through to 1906, when Golgi and Cajal were jointly awarded the Nobel Prize (the two men met for the first time at the award ceremony in Stockholm). Golgi's acceptance speech was grudging and rather cranky, focused entirely on his opposition to the neuron theory, and emphasising that, for him, the nervous system, and in particular the brain, had a 'unitary action'. He was convinced that the organisation of different areas could reveal nothing about function. As he put it: 'specific function is not associated with the characteristics of the organisation of centres, but rather with the specificity of peripheral organs destined to receive and transmit impulses'.[15]

Golgi thought that different sensory organs produced different kinds of sensory activity, just as Müller had argued half a century earlier with his 'law of specific nerve energies'. Despite Golgi's enormous contribution to science, his views were distinctly behind the times.

---

*Von Waldeyer had a knack for neologisms – three years earlier he had coined the term chromosome to describe the enigmatic string-like structures in the cell that became coloured when stained (*chromo some* = coloured body).

*

In February 1894 Cajal gave a prestigious lecture to the Royal Society of London. He surveyed over half a century of microscopic studies of brain structure, describing his own unique contribution and exploring various ways of thinking about how the brain works.[16] His starting point was the widely held view that the mammalian brain constitutes 'the most subtly complicated machine to be found in all of nature'.[17] But unlike previous thinkers, Cajal was able to describe what he called the units of this structure, and suggested they functioned something like the components of the telegraph networks that now covered much of Europe and North America:

> The nerve cell consists of an apparatus for the *reception* of currents, as seen in the dendritic expansions and the cell body, an apparatus for *transmission*, represented by the prolonged axis cylinder, and an apparatus for *division* or *distribution*, represented by the nerve terminal arborisation.[18]

These three functions of the different parts of the neuron – reception, transmission and distribution – were highlighted by the illustrations that accompanied the lecture, which included a key interpretative feature that Cajal had begun to use in 1891: arrows to indicate 'the probable direction of nervous currents and the dynamic relationships of cells'.[19] Rather clumsily, Cajal described this as the dynamic polarisation of neurons:

> In organs where the origin of excitation is well-established we see that the cells are polarised, such that the nervous current always enters through the protoplasmic apparatus or the cell body, and that it leaves via the axis cylinder which transmits it to a new protoplasmic apparatus.[20]

Cajal was not the only person to come up with this idea – at around the same time, the Belgian neuroanatomist Arthur van Gehuchten was making similar suggestions.[21] This principle that the nervous current could go in only one direction was obvious when it came to

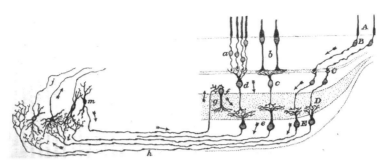

Cajal's drawing of the retina. Light is detected by the retinal cell marked A.

the microscopic organisation of sensory systems such as the retina – sensory impressions moved from the periphery to the centre. It had also been known for decades at the large-scale level of nerve fibres – by the 1830s, following the work of the British anatomist Sir Charles Bell and the French physiologist François Magendie, it was accepted that reflex arcs in the spinal cord worked in only one direction. Tapping the tendon below your knee causes your thigh muscle to contract, but you cannot stimulate your thigh and make the tendon respond.

At the same time as Cajal and van Gehuchten were developing their ideas of unidirectional function at the microscopic level, the early psychologist William James generalised the conclusions of the gross anatomical and functional studies of nerves and muscles and the pathways they followed to create reflex arcs. As he put it in his 1890 book *The Principles of Psychology*:

> paths all run one way, that is from 'sensory' cells into 'motor' cells, and from motor cells into muscles, without ever taking the reverse direction. A motor cell, for example, never awakens a sensory cell directly, but only through the incoming current caused by the bodily movements to which its discharge gives rise. And a sensory cell always discharges or normally tends to discharge towards the motor region. Let this direction be called the 'forward' direction. I call the law an hypothesis, but really it is an indubitable truth.[22]

To underline his point, James accompanied his argument with a number of diagrams showing the organisation of different cell types. In these figures the cells were all connected, as if in a network, and, like Cajal – but a year earlier – arrows were used to denote the direction of the hypothetical nerve currents.

Despite the highly organised structure of the nervous system, Cajal's view of how the whole thing might work was far from mechanical. The complex branching patterns of the dendrites suggested that function could involve alternative pathways, depending on the strength of what Cajal called the sensitive impression. Weak excitation would pass directly down the network, while stronger excitation, Cajal suggested, could propagate through the branches to neighbouring cells and as a result 'the whole system of short contralateral branches would be influenced'.[23]

Although he highlighted the apparent similarity between the functions of the different parts of the neuron and the workings of the telegraph system (reception → transmission → distribution), Cajal felt that the telegraph was not a good model for how the brain worked.[24] His study of embryonic development told him that the complexity of a nervous system came not only from the number of units that it contained but also from the connections between those units, connections that changed with experience. Cajal argued that experience would lead to 'a greater development of the protoplasmic apparatus and of the system of collateral nervous branches'. This applied not only to the strengthening of existing associations but also to 'the creation of completely novel intercellular connections'.[25] Learning, Cajal claimed, led to increased connectivity and revealed what the Belgian scientist Jean Demoor called the plasticity of cerebral neurons.[26] Cajal realised that this plasticity meant that only limited understanding could be gained by seeing the brain as a kind of telegraph system:

> A continuous pre-established network – a kind of grid composed of telegraph wires in which neither new nodes nor new lines can be created – is something rigid, immutable, incapable of being changed, which clashes with the widespread impression that the organ of thought is, within certain limits,

malleable and capable of perfection, above all during its development, by means of well-directed mental exercise.

Unable to point to any more sophisticated technological metaphor, Cajal retreated into describing the brain in terms of other forms of living matter:

> At the risk of making a far-fetched comparison, I would defend this idea by saying that the cerebral cortex is like a garden full of an infinite number of trees – pyramidal cells – which, by careful cultivation, can produce more branches, push their roots deeper, and produce ever more varied and exquisite flowers and fruits.[27]

Other thinkers were not afraid to use more modern technological metaphors to explain what the brain might do. In his Foreword to Cajal's 1894 book *Les Nouvelles idées sur la structure du système nerveux*, the French anatomist Matthias Duval explained that the independence of nerve cells implied that the nervous system and the functions it embodies were not fixed but, as he put it, malleable:

> On their journey, the nerve pathways of conduction and association appear to be endowed with an infinite series of switches, and thus we see that exercise may accentuate transmission along certain more specific routes, in accordance with skills that have been learned.[28]

Duval's idea was that organic structures that behaved like switches would enable even an anatomically rigid structure to be functionally plastic – different routes could be taken by nerve impulses, switched down different pathways, depending on experience. This is the earliest suggestion I have found that the organisation of the nervous system involves switches, although the word had been used in relation to electricity for over thirty years.

Two years later, in his essay 'Matter and Memory', the French philosopher Henri Bergson used a similarly modern metaphor to explain what the brain might do. Ironically, this was primarily in order to play down the significance of the brain – Bergson had an idealist position about the nature of mind and rejected the idea that thought and brain activity were the same thing. But his insight into the potential parallel between brain function and the highest form of contemporary technology was telling. As Bergson put it:

> the brain is no more than a kind of central telephonic exchange: its office is to allow communication, or to delay it … it really constitutes a centre, where the peripheral excitation gets into relation with this or that motor mechanism, chosen and no longer prescribed.[29]

Telephone exchanges had been in operation for about twenty years when Bergson used the analogy. They worked like this: when a caller picked up the phone, a light came on in the exchange over a slot representing the caller's number; the operator manually connected a cable to that slot, asked the caller for the number they wanted to connect to, and would then put the other end of the cable into a slot corresponding to the desired receiver – either the precise location if it was in the exchange area, or, if further afield, to a slot corresponding to a remote exchange, where the process would be repeated.

The parallel between the telephone exchange and the brain was popularised in the United Kingdom's most well-established forum for public engagement with science, the Royal Institution Christmas Lectures. In 1916–17, in the middle of the First World War, a series of talks on 'The Engines of the Human Body' were given by the physiologist and surgeon Professor Arthur Keith. The audience at these lectures consisted mainly of children, so Keith's explanation was fairly simple. In his lecture on the nervous system, he drew a parallel between the cells in the brain and the human operators in a telephone exchange, both of which he considered as 'relay units'.[30] In his extended comparison Keith focused on reflex actions or responses that are not under conscious control – from the power of tickling to the sequence of responses involved in the eyes watering to remove a dust particle. When he did attempt to untangle the mystery of voluntary behaviours, using the example of someone with a painful stone

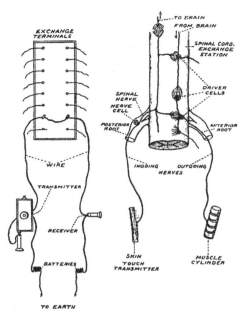

Keith's comparison of the spinal cord (right) and a telephone exchange (left).

in their shoe, Keith did not really explain anything. After describing how the pain message got to the brain, he continued:

> To obtain relief the 'driver cells' of the cortex have to be set in motion; … they control the driver units in the local exchanges, and combine their actions so that the muscular engines carry out the movements which are determined on by operations effected within the exchange systems of the cortex.[31]

This does not explain how the driver cells know how to reduce the sensation of pain, how that outcome is selected from many other possible patterns of activity, or how the cells know when to stop producing their pain-reducing behaviour. Furthermore, although the idea that the brain routes messages to an appropriate destination is a powerful one, if taken literally it meant that each cell was only connected to one other cell, and that neuronal transmission was linear. Neuroanatomy showed that this was completely naive.

Nevertheless, the significance of Keith's extended metaphor was that he identified the functions of the components of the nervous system – transmitters, ingoing and outgoing messages, relay units or switches – through a technological comparison. This example of popular science communication shows how the view of brain function was shifting as a result of both increased anatomical knowledge and more complex contemporary technology.

Despite not using any technological metaphor, in 1899 Cajal was bold enough to suggest how the indescribably complex network of cells in the human brain might produce awareness:

> the impulse provokes a chemical change in the neuronal arborisations which, working in turn as a physico-chemical stimulus on the protoplasm of other neurons, would create new currents in these. The state of consciousness would be precisely tied to these chemical changes brought about in the neurons by these nerve endings.[32]

Although this claim is generally accepted today, for Cajal it was more an article of faith than a scientific explanation. He could propose no mechanism, no analogy to help suggest how chemical changes could produce consciousness. To be honest, over a century later we have got no further.

*

At the beginning of the twentieth century, scientists studying how nervous systems work were faced with a major problem. Cajal and others had shown that neurons were all independent structures, and it was known that some kind of electrical charge passed down neurons, from the dendrite to the axon, much like a telegraph or telephone message down a wire. It was less clear what happened next. If neurons had all been part of some great neuronal net, as Golgi and others had suggested, then that charge would simply have found its way through the network. But that was not how nerves were organised in most animals. Somehow, the nervous impulse passed from one cell to another, even though they were separate. For Cajal, the best analogy for what was happening was to be found in the world of technology: 'current must be transmitted from one cell to another by way of *contiguity* or *contact*, as in the splicing of two telegraph wires'.[33] But this was at best a hypothesis and, in reality, merely an assumption. Despite his skills, Cajal had no proof of what happened when two neurons met, nor how the current was transmitted.

Sometimes, a problem has to be named before it can be fully understood. In this case, the breakthrough in understanding the transmission of nerve impulses began with naming the place where two neurons meet. In 1897 Sherrington was asked to contribute to a new edition of the *Text Book of Physiology*, edited by Michael Foster, Professor of Physiology at Cambridge. In his chapter, Sherrington introduced a term to describe how two cells interact:

> So far as our present knowledge goes we are led to think that the tip of a twig of the arborescence is not continuous with but merely in contact with the substance of the dendrite or cell-body on which it impinges. Such a special connection of one nerve-cell with another might be called a synapsis.[34]

'Synapsis' was taken from the Greek for 'clasp', because it seemed that the arborescence of the axon of the incoming cell clasped the dendrites of the next cell. Within two years, synapsis, which was already in use in cell biology, had become synapse, the term we use today.

Sherrington went further than merely naming this neuroanatomical space: he speculated that it was not simply a passive gap between the two cells, but that it might actually alter the nature of the nerve impulse as it passed from one cell to another:

> we seem entitled to assume that each synapse offers an opportunity for a change in the character of nervous impulses, that the impulse as it passes over from the terminal arborescence of an axon into the dendrite of another cell, starts in that dendrite an impulse having characters different from its own.[35]

In 1906 Sherrington developed his ideas about the synapse in *The Integrative Action of the Nervous System*, trying to link the new neuroanatomy with what was known about nerve function. For Sherrington, nerve cells 'have in exceptional measure the power to spatially transmit (conduct) states of excitement (nerve-impulses) generated within them' and these nerve impulses were then integrated by the nervous system, leading to appropriate behaviour.[36] This integration worked 'through living lines of stationary cells along which it despatches waves of physico-chemical disturbance, and these act as releasing forces in distant organs'. The exact nature of that disturbance was still unclear, but Sherrington's description of what nervous systems do, using a mixture of physico-chemical and telegraphic metaphors, marked a change from nineteenth-century views. Furthermore, by describing synapses as 'surfaces of separation', Sherrington focused in on hitherto unsuspected microscopic locations – the surfaces of an axon and of a dendrite – with the implication that the behaviour of these surfaces might hold the secret to what happens when a nerve impulse moves across them.

His starting point was the physical structure of the neuron, which he described as 'a conductive unit whereby a number of branches (dendrites) converge toward, meet at, and coalesce in a single outgoing stem (axone). Through this tree-shaped structure the nervous impulses flow, like the water in a tree, from roots to stem.'[37] While this might seem like an old-fashioned metaphor, thinking about nerve action in terms of water flow was no different from the way that the movement of electricity was, and still is, described – as a current. But when it came to the synapse, the water analogy broke down, because something had to pass across that gap. When Sherrington examined the available data showing what happened to the nerve impulse on either side of a synapse, he realised there was another kind of funnelling taking place which suggested the synapse worked something like a row of falling dominoes:

> At each synapse a small quantity of energy, freed in transmission, acts as a releasing force to a fresh store of energy not along a homogenous train of conducting material as in a nerve-fibre

pure and simple, but across a barrier which whether lower or higher is always to some extent a barrier.

That barrier was the synapse, which produced what Sherrington termed 'resistance' in the 'conductive chain' of neurons. The consequence was what he called 'the valved condition of the reflex circuit' – reflexes only work one way.

Cajal's dynamic polarisation of neurons was a name for the one-way activity of neurons; when this was coupled with the activity of the synaptic surfaces, the reflex circuit behaved as though it contained valves (thinking in terms of valves and circuits was also inspired by water-supply systems). Sherrington suggested that the explanation of the valve-like behaviour of the surfaces of separation 'may lie in a synaptic membrane more permeable in one direction than in the other' – something similar had recently been found with regard to salt moving across the intestinal wall.[38] The explanation of synaptic function apparently lay in the structure of the membranes of the two cells involved. Exactly what was happening at the synapse would take decades of hard work to establish, as the supporters of two contradictory views slugged it out in one of the longest scientific disputes of the twentieth century, which became known as the war of the soups and the sparks.[39]

*

In 1877 du Bois-Reymond had tried to understand how excitation in a nerve might produce contraction in a muscle. He came up with two alternatives, which were to dominate over seventy years of thinking about the question:

> Of known natural processes that might pass on excitation, only two are, in my opinion, worth talking about: either there exists at the boundary of the contractile substance a stimulatory substance in the form of a thin layer of ammonia, lactic acid, or some other powerful stimulatory substance, or the phenomenon is electrical in nature.[40]

In other words, either nerve cells affected the muscle by some chemical process or electricity jumped from the nerve to the muscle, directly inducing contraction.

Up to the end of the nineteenth century, most of the research into what nerves do had been based on the control of movement, including in the much-studied reflex arcs of the spinal cord. But there are other parts of the nervous system that are not involved in movement, such as the vagus nerve that controls heart rate and which had provided proof of inhibition in the nervous system. These make up what is called the autonomic nervous system – the term was coined by the Cambridge physiologist John Langley, who, like Sherrington, was a student of Foster.[41]

At the turn of the twentieth century Langley began to study autonomic control of the viscera (including the salivary glands, stomach, pancreas, liver, bladder, intestines and penis). It had long been known that drugs such as curare could alter or even completely block autonomic function; by the end of the nineteenth century it was realised that they acted on the neuromuscular junction – the point where the autonomic nerve comes into contact with the muscle. Langley studied the effects of adrenaline, a substance extracted from a small gland found just above the kidneys (hence its name), a gland that was known to be essential for life. He discovered that adrenaline basically had the same effect as that produced by the activation of the autonomic nervous system – it inhibited the action of the intestines and bladder, dilated the pupils and increased blood pressure. A few years later, Langley's colleague Thomas Elliott concluded that 'adrenaline might then be the chemical stimulant liberated on each occasion when the impulse arrives at the periphery'.[42] However, Elliott thought that the nervous impulse led the organ to secrete adrenaline, not that it was produced by the nerve itself. Indeed, in 1921 Langley – a stickler for the facts, and hostile to speculation – dismissed the idea that adrenaline might act in the synapse 'for it would involve the secretion of a substance from the nerve endings', and that was impossible.[43]

One of the key figures who realised that nerves do indeed secrete substances was another British scientist, Henry Dale. In the years before the First World War Dale studied the physiological effects of extracts of the ergot fungus, including some that could reproduce the effects of adrenaline and autonomic nerve stimulation. One of his earliest discoveries was that substances such as nicotine could alter nerve function in the autonomic system. Dale found that one of the extracts from ergot, acetylcholine, basically shut the heart down (when Dale first administered it to a cat, he thought the substance had killed the animal, for he could detect no heartbeat).[44] Dale was initially convinced that acetylcholine was nothing more than a powerful drug – he could find no evidence that it or anything similar existed in the body.[45] Even though he slowly accumulated evidence that various compounds could either mimic or block the activity of the autonomic nervous system, Dale steadfastly avoided suggesting that these substances were normally present in the body, for lack of evidence.

The breakthrough came in 1920, when the German physiologist Otto Loewi had a dream that revolved around an idea he had discussed with Elliot many years previously, about the possibility that chemical substances might be released during muscle stimulation by a nerve. The story of Loewi's dream, as he recounted it forty years later, was as follows:

> The night before Easter Sunday of that year I awoke, turned on the light, and jotted down a few notes on a tiny slip of thin paper. Then I fell asleep again. It occurred to me at six o'clock in the morning that during the night I had written down something most important, but I was unable to decipher the scrawl. The next night, at three o'clock, the idea returned. It was the design of an experiment to determine whether or not the hypothesis of chemical transmission that I had uttered seventeen years ago was correct. I got up immediately, went to the laboratory, and performed a simple experiment on a frog heart according to the nocturnal design.[46]

Whatever the exact truth of Loewi's tale – the precise details changed each time he recounted the story – the experiment was a success.[47] Or so he claimed. He studied two frog hearts, inhibiting the activity of one by stimulating its vagus nerve and then removing some salt solution that had previously been introduced into it and injecting this into the second organ, which then slowed down. Despite Loewi's confident conclusion that this showed that a substance secreted by the vagus nerve inhibited heart movement, most scientists did not accept his results. Either they were unable to replicate his findings, or they were simply unconvinced by the rather vague figures that accompanied his article.[48] Loewi piled up the evidence – he published seventeen articles on the topic in the space of a few years – but many researchers were still dubious, because of the replication problems. We now know these difficulties occurred partly because Loewi had been extraordinarily lucky – the substance he was in fact studying, acetylcholine, is very fragile, but was less likely to degrade if the amphibians were still in their cold winter state, as they were when he did his initial experiments.[49] Researchers who tried to replicate his work during the summer months generally failed.

By the beginning of the 1930s, improvements in the experimental apparatus and a greater understanding of how acetylcholine might be broken down by naturally occurring enzymes led to growing confidence that the effect was real. Surprisingly, even Loewi did not consider that acetylcholine was an example of a more general phenomenon – like most scientists, he did not think that the synapses involved in movement could work by chemical transmission.

At around this time, Henry Dale turned his attention to the problem of what precisely was happening in the synapse. His work was soon buoyed by the arrival in his laboratory of Wilhelm Feldberg, a Jewish scientist who fled Germany shortly after the Nazis came to power. Feldberg brought with him a complex technique for detecting minute amounts of acetylcholine, which involved passing extracts from a nerve over a particular muscle dissected out from a certain strain of leech; a gauge was attached to the muscle and would give a read-out of how much contraction occurred. Despite the complexities of the experiment, within three years of arriving in Dale's laboratory Feldberg had published twenty-five papers showing that a wide variety of autonomic nerves secreted acetylcholine, including all the branches of the vagus nerve. The technique was precise enough to be able to show that the substance was secreted into the synapses of the autonomic nervous system, and also that the same stuff was present in the synapses of nerves involved in voluntary movement, although Feldberg and Dale could not demonstrate its function. In an unusually spritely move by the Nobel Prize Committee, their 1936 award went to Loewi and Dale for demonstrating what was called neurohumoral transmission. Loewi had dreamed the experiment; Dale – with the essential help of Feldberg, who was not rewarded – had shown it to be true.

The disagreement between those who advocated electrical synaptic transmission (sparks) and those who argued in favour of a chemical effect (soups) had been rumbling ever since du Bois-Reymond first highlighted the two possibilities sixty years earlier. Now the dispute became much more heated. Probably the most ardent advocate of the sparks view of synaptic activity was the opinionated Australian physiologist, John 'Jack' Eccles, who had been Sherrington's student. Eccles was convinced that all synapses in the central nervous system were electrical, but faced with the growing evidence from Dale and others, he gradually accepted that chemical action at the synapse might play a minor role in nerve transmission.

This inflexion in the sparks position did little to calm matters, and arguments over the question sometimes got out of hand – in 1935 the future Nobel Prize winner Bernard Katz attended his first meeting of the Physiological Society in Cambridge, and was astonished to see 'what seemed almost a stand-up fight' between Eccles and Dale (Eccles himself described it as 'a very tense encounter').[50] These disputes led to no long-lasting ill feeling, and Eccles and Dale were actually on good personal terms, no matter how aggressive their arguments appeared to outsiders.

Eccles persisted in his vigorous opposition to the soup hypothesis until the early 1950s, when data from his own laboratory finally convinced him that he was wrong. In 1947 he had put forward a theory to account for inhibition from an electrical point of view, suggesting that a small cell near the synapse, now called the Renshaw cell, could alter the polarity of the postsynaptic neuron, thereby effectively countering the transmission of the electrical signal. (As with Loewi, this idea came to Eccles in a dream.[51]) But within four years, Eccles's dream had evaporated, destroyed by a cruel fact:

the Renshaw cell did indeed affect the postsynaptic cell, but in the opposite way to that predicted by Eccles and could not therefore explain inhibition. Reporting the experiment in 1952, Eccles and his colleagues wrote: 'It may therefore be concluded that the inhibitory synaptic action is mediated by a specific transmitter substance that is liberated from the inhibitory synaptic knobs'; they went further, accepting that it was probable that 'excitatory synaptic action is also mediated by a chemical transmitter'.[52]

The victory of the soups was assured, and the role of what eventually became known as neurotransmitters in nervous function was slowly accepted. But these elegant and game-changing studies had little immediate impact on understanding how the brain worked, for the simple reason that almost all of them focused on the autonomic nervous system and the relatively slow movements of the visceral muscles. Many scientists were convinced that the more rapid movement that was the focus of the central nervous system would preclude any kind of chemical stimulation in the synapse – few were prepared to consider that neurotransmitters might also function in the brain. Although in the mid-1920s Sherrington and others had accepted the possibility that inhibition in the brain might have a similar chemical basis to that in the autonomic nervous system, testing this hypothesis was technically challenging. It required isolating nerves from the brain and eliminating other influences on their activity. For decades, these difficulties were insurmountable.

As well as enriching our understanding of how nerves function, the discovery of synaptic transmission highlighted a major problem with the dominant metaphors being used to understand how the brain worked. In the nineteenth century, the discovery of electrical activity in nerves, paralleled by the invention first of the telegraph and then of the telephone, had helped frame attempts to conceptualise brain function. But by the 1930s it was apparent that this analogy, no matter how seductive, was imprecise at the most basic level. The nervous system might be composed of an infinite series of switches, but these switches did not work in the same way as those in a piece of electrical equipment. Biological discovery was outstripping the dominant technological metaphor and revealing that, however convincing Professor Keith may have been to his young audience at the

Royal Institution, the brain is not a telephone exchange. Other metaphors were going to be necessary to understand what the brain does, and how it does it.

EIGHT

MACHINES

1900 TO 1930

In October 1922, a play opened in New York that would change the world with a single word. Entitled *RUR* and written by Czech playwright Karel Čapek, it had first been performed eighteen months earlier in Czechoslovakia; by the time the play opened in London, in 1923, it had been translated into thirty languages. The play's global influence lies in the fact that the title stood for Rossum's Universal Robots – this was the origin of the now universal word 'robot', which Čapek adapted from the old Czech word for servitude. In the play, society relies upon the work of docile robots, developed by a scientist called Rossum; when the robots are given an element of humanity, they kill their overlords. But despite being made of some weird mechanised flesh (technically, therefore, they are in fact cyborgs), the robots cannot reproduce. In the final act, two robots overcome their sterility. They are the new Adam and Eve.

Part reworking of *Frankenstein*, part expression of the fear of automation, part satire on twentieth-century capitalism, *RUR* expressed the growing worldwide fascination and anxiety that machines might one day mimic human abilities. The speed with which Čapek's new word spread into all major languages showed the global lexical void that existed – we knew about robots, we just did not have a word for them. Both the word and the concept spread like wildfire. In 1927 one of the greatest films ever made, Fritz Lang's *Metropolis*, featured a robot that was built to discredit the female leader of a workers' rebellion. Fantasies of domestic robots appeared in magazine features on the home of the future.[1] Science-fiction writers began to play with the new concept, forecasting both heaven and hell. La Mettrie's shocking eighteenth-century suggestion that humans were machines had been turned on its head – in the twentieth century, it seemed, machines would become human.

Although the idea of creating an independent automaton went deep in culture – at least as far back as the ancient Greeks – fascination at the link between humans and machines had been growing in the early years of the century.[2] Increased mechanisation of manufacturing, followed by the development of Ford's production line and the imposition of a limited set of repetitive actions by those using it, seemed to be making factory workers part of the machines they tended. With the outbreak of war in 1914, and the developments in killing technology that accompanied it, fascination became fear. A single work of art summarised that shift. In 1913 the British sculptor Jacob Epstein created a triumphant work in which a humanoid figure – all sharp edges and with a beak-like face – stood astride a tripod made out of an industrial rock drill. Intended as a celebration of modernity and machinery, *The Rock Drill* was exhibited only once in this form, shortly after its creation. When it was displayed again, in 1916, it was radically transformed, becoming simultaneously menacing and pathetic. Only the torso, head and one arm were retained, and the whole thing was cast in gunmetal. Epstein had amputated key parts of his man-machine, leaving it immobile and powerless, just as millions of men were being mangled and killed by terrifying mechanical weapons that were produced on massive production lines, and then deployed to destroy human bodies.

Despite this widespread cultural ambiguity about the link between human and machine, most scientists enthusiastically embraced the machine metaphor to explain our bodies – A. V. Hill's 1926 Royal Institution Christmas lectures on physiology appeared under the title *Living Machinery*, while in 1929 the physiologist C. Judson Herrick wrote a long book about the nature of life entitled *The Thinking Machine*.[3] This use of machine imagery by scientists was partly a response to an attempt by some philosophers to push back against the materialist implications of recent scientific discoveries, in particular those involving behaviour, heredity and development. These philosophical positions involved a revival of vitalism – explaining biology not through materialist mechanisms, but instead by some unique spiritual attribute shared by all living things.

One of the main targets of the revivalist vitalists was a new framework for understanding behaviour in animals and humans. In the early years of the century the physiologist Jacques Loeb, followed by his student, the psychologist J. B. Watson, argued that scientists should concentrate on merely observing human or animal behaviour, rather than seeking an explanation in some inner mental life.[4] Loeb explained most movements in terms of simple underlying processes called taxes and tropisms. For example, according to Loeb, an animal moves away from the light because it shows negative phototaxism. While this produced a neat classification of behaviours, it assumed that there was a common force driving, say, all movements away from light. It turned out there was no such thing. Far from providing an explanatory framework that could be tested by examining the role of nervous systems and brains, Loeb's taxes and tropisms were ultimately circular definitions that explained little. Watson proclaimed the need for a behaviourist psychology and built upon the work of Sechenov and Pavlov, who had explained behaviour in terms of conditioned reflexes. Although Watson soon abandoned science and went into advertising, the behaviourism he helped create proved enormously influential, in particular in the USA. But by concentrating purely on behaviour, and becoming increasingly distant from the origin of that behaviour in the brain, behaviourism was unable to develop any real insight into how the brain worked. Indeed, Watson's followers, such as B. F. Skinner, who dominated US psychology for decades, were not interested in the question.

The vitalists who opposed these developments were primarily driven by two considerations.[5] Coupled with a deep-rooted opposition to the materialist view of life and mind, there was a new criticism, based around the idea of teleology, according to which there was some inner purpose in life that expressed itself in development, physiology and behaviour. The materialist view, it was claimed, could not explain goal-directed behaviour, which was unique to living matter. The only explanation of such phenomena must be some kind of spiritual inner urge that was common to all life. The scientists who opposed this vitalist view had a problem: there was still no good explanation of apparent goal-directed phenomena in physiology and behaviour. The answers, however, were on their way.

*

In the years before the outbreak of the First World War, some scientists and engineers began to develop models of the nervous system using machines, either real or imagined. Their idea was not simply to copy behaviour, as in an automaton, but to gain some insight into the processes and structures that were involved in producing behaviour in a living system.

In 1911 Max Meyer of the University of Missouri described how a machine could perform some of the basic functions of a nervous system. Meyer followed the new conventions of electrical wiring diagrams to present his models, but everything about his view of how the nervous system functioned was hydraulic in its conception.[6] The limits of this pressure-based model were made clear two years later, when S. Bent Russell, an engineer from St Louis, published the plans of a device that would 'simulate the working of nervous discharges by purely mechanical means'. Russell claimed that his proposed apparatus – a steam-punk concoction of spur valves, cylinders and connecting rods – functioned according to a logic that was 'not altogether unlike' that of Meyer.[7] Russell's description of his device was confident, even though it does not appear he ever built a prototype: 'We have shown a practical arrangement of mechanical transmitters and receivers that will respond to signals and control movements like a nervous system and that possesses associative memory as it can learn by experience.'[8]

Meyer was irritated that Russell had ignored his system for drawing diagrams of the nervous system and poured scorn on the whole enterprise, demanding to know how each of the dozens of components corresponded to anatomical structures. Even if the device were to work, without a link to anatomy its scientific value would be limited. These criticisms applied equally to Meyer's own ideas, which contained no way for the system to recognise that it had completed a task, or to refine its performance if it had completed it inadequately. The anatomical basis of one of the basic features of learning was absent.

Not all the technological toying with behaviour was so benign. In the 1910s a US radio engineer called John Hays Hammond was working on plans for a self-directing torpedo; he was particularly interested in Loeb's ideas about taxes – how animals move towards, or away from, a stimulus. In 1912, in collaboration with Hammond, Benjamin Miessner built what they called an electric dog (in reality a box on three wheels). The dog – which a few years later was exhibited under the name Seleno – had two light detectors made of selenium (hence the name) at the front and used the signals from these receivers to navigate towards a torch, at a rate of about a metre a second.[9]

Loeb cited Hammond and Miessner's creation of the dog as 'proof of the correctness of our view', drawing the unjustified conclusion that because a machine could reproduce the behaviour of an animal, that meant an animal was simply a machine:

> We may feel safe in stating that there is no more reason to ascribe the heliotropic reactions of lower animals to any form of sensation, e.g., of brightness or color or pleasure or curiosity, than it is to ascribe the heliotropic reactions of Mr. Hammond's machine to such sensations.[10]

The aim of Hammond and Miessner's device was not primarily scientific and was certainly not related to concerns about what animals – or machines – might feel. In 1916, just as the US was preparing to enter the war, Miessner explained that the same principle that functioned in Seleno the dog enabled the Hammond torpedo to home in on the sound of a ship's engines and destroy it. Amidst his pride, Miessner expressed an early streak of technofear as he thought about the potential implications:

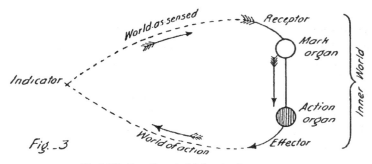

Uexküll's 'function circle' showing how nervous
systems sense and act upon the world.

> The electric dog which is now but an uncanny scientific curi-
> osity may within the very near future become in truth a real
> 'dog of war', without fear, without heart, without the human
> element so often susceptible to trickery, but with one purpose:
> to overtake and slay whatever comes within range of its senses
> at the will of its master.[11]

None of these attempts to describe how nervous systems might
function by creating mechanical imitations had any immediate
scientific consequence. But after the war, scientists began to think
in more abstract ways about how animals – including humans –
interact with the world. The Estonian biologist Jacob von Uexküll
came up with two key insights.[12] At the beginning of the century,
he had highlighted the existence of what in German was called the
Umwelt, or inner sensory world, of each species, which was rooted
in its ecology. Uexküll explored this idea in terms of Kant's a priori
hypothesis about sensation, in a similar way to the Dutch pharmacologist Rudolf Magnus, who wrote: 'The nature of our sensory
impressions is thus determined *a priori*, i.e. before any experience,
by this physiological apparatus of our senses, sensory nerves, and
sensory nerve centres.'[13] This approach is now part of our way of
understanding how natural selection has shaped brains and nervous
systems, and, when it comes to other animals, how we think of what
it is like to be, say, a bat. Uexküll's second innovation came in the

form of some intriguing diagrams that he called 'function circles'. These showed how a nervous system or brain could sense the world and act upon it to effect a particular aim. Uexküll was not concerned with trying to turn this schema into a device, but rather with understanding how in principle behaviour might emerge from it. The key feature that Uexküll included was the element that had been absent from Mayer's schemes – the system could sense how its output had changed the world and alter its functioning accordingly.[14]

This insight was also found in the work of Alfred Lotka, the US mathematician and founder of theoretical population ecology. In his 1925 book *Elements of Physical Biology* Lotka described a toy clockwork beetle that apparently showed purposive behaviour, in that it was able to sense when it was about to fall off a table and take evasive action. The underlying mechanism was trivial. The beetle was driven by a pair of wheels, with a freely moving wheel set at right angles just in front of the driving wheels. On the beetle's head were two metal antennae that touched the ground, lifting the free wheel slightly so the toy moved unhindered in a straight line. If the beetle came to the edge of a surface, the end of the antennae would fall down, lowering the front of the beetle's body, making the free wheel touch the ground. The forward motion of the driving wheels was now transformed into a circular motion by the movement of the free wheel, making the beetle turn away. The beetle would carry on rotating until the antennae were back on the surface, whereupon the free wheel would be raised off the ground and the device would go straight ahead again.

Lotka interpreted this simple toy in terms of three kinds of organ – an effector (the driving wheels), an adjustor (the transverse wheel) and a receptor (the antennae). As he put it, the adjustor '"construes" the information furnished by the receptor antenna, and modifies in accordance with this information the law of motion of the toy, in such manner as to preserve the beetle from a fall'.[15] Lotka had given a clear example of a system that could show apparently purposive, goal-directed behaviour on the basis of a simple reflex arc. By thinking abstractly about the kinds of organs that were involved – receptors, adjustors and effectors, effectively the same three components that Uexküll identified in his function circles – Lotka showed how those

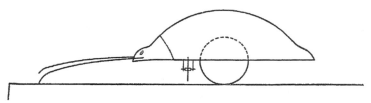

A diagram of a toy beetle used by Lotka to show apparent purposive behaviour.

concepts could be applied to a wide variety of situations in which animals responded in an adaptive, apparently purposeful way.

Post-war technological developments began to shape how scientists saw the nervous system and the brain. In 1929 the Yale psychologist Clark L. Hull described a model of the conditioned reflex that used electronic components; it was soon followed by two improved versions.[16] Consisting of a series of resistors and storage cells connected in parallel, complete with buttons and lights, the apparatus changed its behaviour with repeated use. The aim, claimed Hull, was to 'aid in freeing the science of complex adaptive mammalian behaviour from the mysticism which ever haunts it'.[17] Hull wanted to show how complex forms of adaptive behaviour could emerge from simple structures and functions without any recourse to vitalist notions, even if he was unsure about the link between his model and real anatomy or physiology. Hull explicitly stated that 'no claim is made that these mechanisms are duplicates of the corresponding organic processes', but he thought the approach could provide insight into the otherwise mysterious process of learning.[18]

In 1933 a University of Washington student called Thomas Ross made a further development, which he outlined in a *Scientific American* article with the provocative title 'Machines That Think'.[19] The article contained the plans for an electrical device that could learn to find its way through a short maze. Ross described his project as follows: 'to test the various psychological hypotheses as to the nature of thought by constructing machines in accord with the principles that these hypotheses involve and comparing the behaviour of the machines with that of intelligent creatures'.[20] Three years later, he put it more pithily: 'One way to be relatively sure of understanding a mechanism is to make that mechanism.'[21]

With the aid of Stevenson Smith, a psychology professor, the device was made mobile and became a three-wheeled 'robot rat', looking somewhat like a skateboard with an alarm clock on top of it. This device was able to negotiate a simple maze consisting of twelve Y-branches and to learn the path it had taken, using a crude mechanical analogue memory. At each Y, one of the branches led to a dead end; when the device encountered the wall in the dead end, a lever on the front of the machine was depressed, sending the device into reverse until it came back to the Y, whereupon it took the other branch. By proceeding in this way, the robot could eventually get to the end of the maze. The machine also contained a physical 'memory disc' – when it encountered a dead end and the reversing lever was pushed, a tab on the disc was raised, such that once the maze had been successfully completed, if the machine was put back at the beginning of the track, it could find its way without making a mistake. It had apparently learned the correct path.

In an interview with *Time* magazine, Smith said, 'This machine remembers what it has learned far better than any man or animal. No living organism can be depended upon to make no errors of this type after one trial.'[22] Precisely. While impressive for the general public, the device shed no light on learning as a process – it could not generalise what it had learned to any other maze, nor could it cope with the slightest change to the maze upon which it had been trained. Finally, the mixture of trial-and-error learning with immediate, immutable memorisation of the correct response did not correspond to any form of learning seen in the natural world.

All these attempts at model-building, from Meyer's diagrams to Ross's robot rat, were limited because none of them was based on the real way that nervous systems functioned. By starting with simple mechanical or electric models, scientists were limited in the kinds of behaviours and nervous system activities that they could model. At the same time as these models were being built out of wires and metal, neurophysiologists were realising that real nervous systems worked in a very different way.

*

The electrical nature of the nervous impulse had been clear from the middle of the nineteenth century, and in 1868 Helmholtz's student Julius Bernstein had discovered that a wave of negative polarisation moved down the nerve with exactly the same dynamics as the nerve impulse.[23] Although it was tempting to conclude that these electrical changes were identical with the nerve impulse, there was no proof and no explanation. In 1902, after nearly four decades of work, Bernstein put forward a theory to explain what the link might be.[24] His idea revolved around the movement of ions – charged particles – that were in solution inside and outside neurons. Moving a positively charged potassium ion from the inside of the cell to the outside meant that the inside of the cell now had a slightly negative charge compared to the outside. According to Bernstein's model, the neuronal membrane was semi-permeable: when the neuron was at rest, the concentrations of ions on the inside and outside of the cell were fixed, but when a nervous impulse passed down the cell, the membrane temporarily and locally changed its nature, and small numbers of ions were moved in or out, creating a wave of depolarisation.[25] As had long been suspected, the electrochemical transmission of a nerve impulse was very different from the movement of electricity down a telegraph cable or a telephone wire. Biology was proving to be more complicated than technology.

If the physical form of the nerve impulse was unexpected, the way that nerves behaved also held surprises. In 1898 Francis Gotch, Professor of Physiology at Oxford, showed that if a nerve fibre – a bundle composed of many neurons – was stimulated twice in rapid succession, the second stimulus did not evoke a response if the two stimuli were separated by less than 0.008 seconds.[26] This interval, the refractory period, is a fundamental characteristic of all neurons. Gotch found that, as expected, the stronger the stimulation of the nerve fibre, the stronger the response, but he also observed that the response always showed the same time-course, irrespective of the strength of the stimulus. Gotch drew a parallel between his results on motor nerves and the well-known effect in the heart whereby the muscle either responded to stimulation or it did not – this was known as an 'all-or-none' response.[27]

To explore if all nerve fibres, both sensory and motor, shared this all-or-none response, Keith Lucas at Cambridge devised sensitive new equipment that enabled him to confirm Gotch's hunch in a motor muscle fibre. If the stimulus was above a threshold, the muscle showed a response, but if it was too weak, there was no response at all.[28] To obtain direct proof of what was happening in a single nerve fibre, Lucas asked Edgar Adrian, a young PhD student, to study the question. For Adrian, this was the turning point in his life, opening a door that eventually led to the greatest of achievements. He remained at Cambridge until he retired, becoming Master of Trinity College and eventually Vice-Chancellor of the university; he was elected President of the Royal Society, made a hereditary peer, won the Nobel Prize at the age of forty-two and saw his son also become an FRS, while two of his protégés, Alan Hodgkin and Andrew Huxley, received the Nobel in 1963. As well as gaining these glittering prizes, Adrian had a lifelong interest in psychoanalysis (he twice nominated Freud for a Nobel Prize[29]) and studied the functioning of the nervous system in a wide-ranging bestiary (including eels, frogs, goldfish, water beetles and himself). Despite this fame and influence, few people except a handful of attentive neuroscientists have now heard of him.[30] And yet Adrian not only changed how we think about what neurons do, he also introduced a new language that helped shape our views about how brains work.

Working with Lucas in the halcyon days of late-Edwardian England, before the horror of mechanised war pushed the world into the future, Adrian was soon able to find evidence that muscular nerve fibres operated on the all-or-none principle. But it was not clear if the same was true of sensory nerves, nor was it evident how the individual neurons within the fibre were responding.[31] In August 1914 war broke out and Adrian and Lucas turned their attention elsewhere – Lucas worked for the Royal Aircraft Factory while Adrian finished his medical studies. In 1916 Lucas was killed in a horrific mid-air collision over Wiltshire.[32] This deprived Adrian of a mentor and a colleague; it also marked him for decades to come – in all of his major writings, Adrian referred to Lucas and his work with a palpable sense of loss.

After the war was over, Adrian returned to Cambridge and took up where he had left off, exploring whether the all-or-none rule also applied to sensory nerve fibres. The war had driven the development of new radio technology, in particular the improvement of valves to amplify faint radio signals. In principle, these devices could also be used to amplify the weak electrical activity in a nerve fibre. During the war, Lucas had discussed this possibility with Adrian at what turned out to be their last meeting; the same idea occurred to a number of scientists, including the Harvard researcher Alexander Forbes. After the war, Forbes, along with his student Catharine Thacher, was able to use these valves to amplify signals in a frog nerve fibre more than fiftyfold.[33] Forbes was a friend of Adrian – he had visited Cambridge in spring 1912, spending three weeks in Lucas's laboratory and becoming beguiled by what he called 'the charm of Lucas's personality'.[34] That visit lasted longer than expected, and as a result Forbes and his wife had to delay their planned return to the USA – they had been due to sail on the maiden voyage of the *Titanic*.

In 1921 Forbes returned to Cambridge, bringing a few of the precious valves for Adrian's laboratory.[35] It took some time for Adrian to fully exploit the new technology – he was very busy in the early 1920s, what with getting married, becoming a Fellow of the Royal Society and spending a large amount of time teaching Cambridge undergraduates. The breakthrough came in 1925, when a Swedish researcher, Yngve Zotterman, came to work in Adrian's laboratory. Initially, things did not go well – Zotterman found that Adrian had 'a very volatile disposition', partly caused by 'a mass of lectures' which left him bad tempered from exhaustion. As Zotterman wrote to a friend in December 1925: 'it has been a little difficult to work with him this last week as he can become beside himself if one merely leaves a tap dripping'.[36]

Despite this friction, Zotterman's visit led to a major discovery.[37] Using a new amplifier, Zotterman and Adrian were able to record the activity of sensory nerve fibres attached to stretch receptors in a frog's leg. They were apparently able to strip back the fibre until there was just one neuron remaining and record how it responded. The activity of the most basic unit of the nervous system could now be studied. Through this work, Adrian and Zotterman made three major discoveries that shaped our view of how nervous systems work.

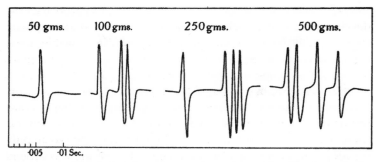

Responses of stretch-detecting neurons to increasing weights. The shape of each spike response remains the same; what changes is the frequency of the spikes.

Firstly, they showed that sensory neurons respond in an all-or-none fashion – if the stimulus is above a threshold, the neuron fires, otherwise it will not. Second, they showed that if a neuron is repeatedly stimulated, for example by a continuous stimulus, the cell will soon stop responding – this was nothing like any of the mechanical models that had been built. Finally, when the neuron fires, the amplitude and shape of the response – these were soon known as spikes because of their shape when visualised on a smoke-covered drum or paper, or, later, on a cathode ray screen – is constant, but the frequency of firing changes with stimulus intensity. Neurons tell the nervous system how intense a stimulus is by changing their firing rate, but each of the component responses from a given cell is identical. These effects can be seen in a figure from one of their papers, showing an increased firing rate of identically shaped spikes as the weight pulling on a fibre was increased.

As a result of this discovery, Adrian received the Nobel Prize in 1932, together with Sherrington. Both men were nominated by A. V. Hill, who wrote of Adrian's work that it was 'of great beauty and of mixed simplicity and subtleties … one of the greatest achievements in physiology of the last quarter century'.[38]

*

Shortly after winning the Nobel Prize, Adrian turned from the simplest component of the nervous system to its most complex form, as he explored the recent discovery by Hans Berger that, astonishingly, the electrical activity of the human brain could be recorded through the bony skull using external electrodes and powerful amplifiers – Adrian described this finding as 'remarkable'.[39] Even more surprisingly, Berger reported that if the subject closed their eyes there was a clear rhythm to the electrical signal, as though the brain was showing coordinated behaviour. In 1934 Adrian and Bryan Matthews explored the nature of what they called the Berger rhythm (now known as alpha waves). Berger had reported that the rhythmic signal could be seen if the subject sat calmly with their eyes closed, but that it disappeared if the eyes were opened or if the subject was required to concentrate very hard, for example on some difficult mental arithmetic. Adrian turned out to be a dab hand at using his own brain to produce the rhythm on demand – he even demonstrated it at a meeting of the Physiological Society. Although Berger claimed that the whole of the brain was involved in this synchronised activity, Adrian and Matthews were able to localise the rhythm's source to the occipital lobe, at the back of the brain, which was thought to be involved in vision. To their great surprise, they found that the brain of a water-beetle produced very similar rhythms when left in the dark, and, as in Adrian's own recordings, the rhythm disappeared if the light was turned on.

Adrian and Matthews showed that, in humans, the key to disrupting the rhythm came from the perception of patterns, or even the attempt to see a pattern in the dark. Like Berger, they concluded that the rhythm was in some way associated with the mechanisms of visual attention – when the subject was not actively using their visual sense, the neurons 'discharge spontaneously at a fixed rate (as in other parts of the central nervous system) and tend to beat in unison'.[40]

As to how the activity of all those neurons might be related to consciousness, Adrian was circumspect:

the whole problem of the connection between the brain and the mind is as puzzling to the physiologist as it is to the

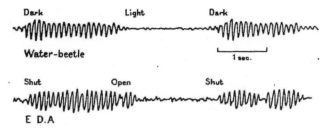

Figure from Adrian and Matthews showing brain activity in
a water-beetle (top) and a human (Adrian – bottom).

philosopher. Perhaps some drastic revision of our systems of knowledge will explain how a pattern of nervous impulses can cause a thought, or show that the two events are really the same thing looked at from a different point of view. If such a revision is made I can only hope that I may be able to understand it.[41]

Despite these difficulties, Adrian's research on nerve function provided evidence for a clear correlation between the activity of neurons and perception. He explained the issue for the general reader in a figure summarising the data he had collected on the effect of continuous pressure on the activity of a sensory nerve:

> The excitatory process in the receptor declines gradually, and as it declines the intervals between the impulses in the sensory fibre become longer and longer. The impulses are integrated by some central process, and the rise and decline of the sensation is a fairly close copy of the rise and decline of the excitatory process in the receptor.

Wary of the difficulty of proving that nervous activity and conscious perception were the same thing, Adrian concluded of his diagram: 'It does not bridge the gap between stimulus and sensation, but at least it shows that the gap is a little narrower than it was before.'

Adrian's summary of the link between stimulus, nervous activity and sensation.

Although he felt that 'undoubtedly the most interesting thing' done by 'the message which travels up the sensory nerve fibre ... is to produce a change in the content of our mind', Adrian could not prove that impulse frequency and sensation were truly the same thing.

*

As with most scientific discoveries, if Adrian had not done this work, someone else would have done so at about the same time. That is the nature of science – with very few exceptions, if researcher X had fallen under a bus, events would have proceeded on something like the same path, through the work of researcher Y. But in one key respect Adrian's contribution to our understanding of how the brain works did not depend on his experimental studies and was much more due to his own unique attributes. Throughout his early career, Adrian produced popular accounts of his work, leading him to think about what nerves do in a rather different way from that expressed in his scientific papers. It was in these writings, searching for ways of explaining what he had discovered to the general public, that Adrian assembled some existing terms in novel ways, with lasting effect. These concepts – messages, codes and information – now form the basis of our fundamental scientific ideas about how the brain works.

Nerve impulses had previously been described as messages that were transmitted – this was integral to the telegraph metaphor that was so popular in the nineteenth century – but virtually nobody thought about what the message might consist of. In his ground-breaking experimental work, Adrian had been able to deconstruct the nerve impulse and had revealed that it was composed of excep-tionally brief pulses. Each of these pulses had the same shape, and yet despite this lack of variability, nerve activity was able to carry a message. To explain this, Adrian made an analogy that now seems obvious, but at the time was utterly novel:

> The message consists merely of a series of brief impulses or waves of activity following one another more or less closely. In any one fibre the waves are all of the same form and the message can only be varied by changes in the frequency and duration of the discharge. In fact the sensory messages are scarcely more complex than a succession of dots in the Morse Code.[42]

Nowadays, the idea that organic structures such as genes or neurons contain some kind of code is relatively banal. Schoolchildren learn about the genetic code, while neuroscience students explore different forms of neural codes. But when Adrian was writing in the early 1930s, this was a completely new way of thinking about what neurons might do, and how brains might function. Furthermore, it pointed the way to a whole new area of research: if the message con-tained a code, then it should be possible to break that code, to reveal what neurons were telling the brain. In the absence of any detailed studies enabling a full answer as to what was in the code, Adrian reached for an abstraction that, while it was not original, took on a different form when allied with codes and messages. His answer was that the nerve message contains information.

Others had used the term before. In the middle of the nineteenth century Dr Spencer Thomson told the readers of a medical dictionary that 'the brain may be likened to a great central telegraph office, to which the wires – nerves – convey the information from all parts of the body that supplies are wanted'.[43] And in describing the workings of the clockwork beetle in 1925, Lotka had written about the device construing information. But Adrian's use was directly linked to the functioning of the nervous system. For example, Adrian argued that 'the central nervous system is able to get every scrap of information out of the message' that was sent from each receptor – indeed, he argued that receptor function enabled the organism 'to extract information about the external world', and that a major challenge for scientists was 'to estimate what sort of information reaches the central nervous system'.[44] As far as Adrian was concerned, the whole point of a nervous system was to transmit encoded information about the world along neurons.

In the mid-1920s, mathematicians such as the great statistician and geneticist R. A. Fisher were also picking up on 'information', using it to describe statistical concepts, although they had yet to settle on a single definition. Whether Adrian knew of these attempts to mathematise information is unclear, but he recognised that work on the nature of the nervous message would inevitably go in this direction. As he wrote to his friend Forbes in April 1929:

> The electric response of nerve is really beginning to show something about itself now – it almost makes me wish I hadn't gone off on to nerve endings and such like – but it will soon get into the realms of physics and chemistry and mathematics and I know my failings, or at least a few of them![45]

That is exactly what would happen in subsequent decades. The significance of Adrian's realisation that there was a neural code, and his intuition that the message contained some kind of information, were part of a transformation of our understanding of how nervous systems and brains work. This transformation took place not in laboratories full of electrodes and impaled frogs, nor in a world of wires and robots, but in front of dusty blackboards as mathematically minded scientists took the most abstract approach possible to modelling what the brain does.

NINE

CONTROL

1930 TO 1950

O nce upon a time, there was a brilliant but rather odd boy called Walter, who lived in Detroit. Walter's working-class family thought he was a freak, as did other children. In 1935, aged twelve, Walter fled into a public library to escape some bullies. Safely inside, he found himself in front of a copy of *Principia Mathematica*, a three-volume work of fearsome mathematical logic written by Alfred North Whitehead and Bertrand Russell. Intrigued and entranced, over the next few weeks Walter returned repeatedly to the library to study the book, poring over the equations and assimilating its arguments.

That story may not be true, but this one is: three years later, in 1938, Walter, now aged fifteen, ran away from home and ended up in Chicago. Somehow, he found his way into the office of Rudolf Carnap, the Professor of Philosophy at the University of Chicago, who had recently published *The Logical Syntax of Language*. According-ing to Carnap, Walter said 'he had read my book and that a certain paragraph on a certain page was not clear to him … So we took down my copy of the book and opened it at the page in question and care-fully read the paragraph … and it was not clear to me either!'[1]

The boy's name was Walter Pitts and the stories about him are legion, and mainly unverifiable. One account of his life begins: 'There are no biographies of Walter Pitts, and any honest discussion of him resists conventional biography.'[2] Pitts seemed so extraordinary and bizarre that his friend Norman Geschwind said outsiders could think he was the product of some kind of collective delusion.[3]* But Pitts was real enough, and his work with the neurologist Warren McCulloch on the logic of nervous system function changed how we think about the brain.

Despite being only fifteen years old, with no academic qualifications (he was entirely self-taught and never got a degree of any kind), Pitts's grasp of mathematics and logic was so profound that he was allowed to attend a weekly seminar on mathematical biophysics organised by Nicolas Rashevsky at the University of Chicago.[4] Rashevsky's interest in fusing mathematics and biology was part of a trend that began in the 1920s and 1930s, whereby mathematically minded scientists began exploring various biological phenomena from population genetics to ecology.[5] But in these cases, mathematical models were generally used to make predictions that could then be tested by observations. Rashevsky's approach was very different. As far as he was concerned, any link between his mathematical models and reality was purely incidental – finding a concrete expression of his ideas was, he said, 'beside the point'.[6]

The discussions in Rashevsky's seminars involved a new way of thinking about biological systems, using a vocabulary that is utterly familiar to us, but which was completely novel at the time – terms such as 'feedback', 'circuits', 'input' and 'output'. 'Feedback' was first used in the early 1920s with regard to electrical circuits, in particular radio signals, but the underlying phenomenon had been known for centuries – negative feedback had been used since antiquity to stop water flowing into a tank once a certain level had been reached, while the nineteenth-century physiologist Claude Bernard had implicitly recognised its existence when he described how the body seeks to maintain a steady internal state (in 1926 Walter Cannon coined the term 'homeostasis' to describe this process). 'Circuit' had been used with regard to the movement of electricity from the middle of the

---

*In 2018 I asked the co-discoverer of the DNA double helix, Jim Watson, then aged ninety, if he had ever met Pitts. His rheumy eyes lit up. 'Oh yes!' he said, 'He was really crazy!'

eighteenth century, while from the turn of the twentieth century 'input' and 'output' were applied to both physiological activity and electrical signals. In the years after the end of the First World War, scientists began to apply these terms to biological phenomena, in particular those related to the nervous system. In 1930 the New York psychiatrist Lawrence Kubie published an article with the title 'A Theoretical Application to Some Neurological Problems of the Properties of Excitation Waves Which Move in Closed Circuits', in which he suggested that some neurological problems, such as the tremors seen in Parkinson's disease, or epileptic fits, might be explained by activity in neuronal circuits moving round and amplifying itself.[7]

By 1940 Pitts, now aged seventeen, was analysing hypothetical patterns of excitation and inhibition in neural circuits, and within two years he published two articles on the topic.[8] That same year, Pitts was introduced to Warren McCulloch by a close friend, Jerome Lettvin. Or perhaps it was in 1941, when McCulloch says he gave a paper to Rashevsky's seminar group.[9] As always with Pitts, the facts are hard to pin down. Whatever the case, Pitts and McCulloch hit it off, and their collaboration led directly to the most common metaphor now used to explain how the brain works: it is a computer.

Except things did not happen that way. In fact, the link between the nervous system and electronic machines was first used the other way round – to suggest that a computer is a brain.

*

McCulloch and Pitts were different in so many respects. McCulloch was an established and cultured academic in his forties, with a family and a large house, while Pitts was an awkward teenage runaway. But they shared a common interest in what was seen as one of the most exciting developments in science: using logic to understand biological phenomena. After obtaining degrees in philosophy, then psychology and finally medicine, in 1934 McCulloch began working with the Yale neurophysiologist Dusser de Barenne, who in turn had worked with Rudolf Magnus and had become interested in his idea that modern sensory physiology provided a materialist basis for understanding Kant's idea of a priori knowledge.[10] All this was transmitted to McCulloch, who in 1959 explicitly explored it in an article about how frogs see.[11]

During this period McCulloch attended a Yale seminar series focused on mathematical approaches to biology; this was run by the psychologist Clark Hull, who in 1929 had proposed that electrical model of the conditioned reflex. In 1936 Hull gave a lecture entitled 'Mind, Mechanism and Adaptive Behavior' in which he presented thirteen logical postulates and their associated theorems that, he claimed, could explain the emergence of adaptive behaviour from simple principles.[12] Hull's aim, as with his electrical model, was to link complex behaviour in an explanatory chain that ran right down to the electron. Little came of Hull's postulates, but they encouraged McCulloch to think more about applying logic to biology.

In 1941 McCulloch moved to the University of Illinois at Chicago. Despite being in a different university, he joined Rashevsky's group and at some point encountered Pitts. McCulloch was nearly forty-two while Pitts was still not an adult but the two men immediately struck up a close friendship. Soon afterwards, the homeless Pitts and his friend Lettvin moved into McCulloch's house. According to the mathematician and neuroscientist Michael Arbib, who worked with McCulloch in the 1960s, McCulloch and Pitts spent 'endless evenings sitting around the McCulloch kitchen table trying to sort out how the brain worked', with McCulloch – who was said to look like an Old Testament prophet painted by El Greco – swilling whisky and puffing endlessly on cigarettes.[13] Pitts's contribution to this partnership should not be underestimated. The brilliant mathematician Norbert Wiener said of him: 'He is without question the strongest young scientist whom I have ever met ... I should be extremely astonished if he does not prove to be one of the two or three most important scientists of his generation, not merely in America but in the world at large.'[14] 'Strongest' could easily have been a typo for 'strangest'.

In December 1943, McCulloch and Pitts published an article entitled 'A Logical Calculus of the Ideas Immanent in Nervous Activity'.[15] As the title suggested, McCulloch and Pitts explored the implications of how neurons fire and how they are connected, and attempted to describe this in terms of logic. Unfortunately, Pitts chose to use Carnap's arcane and rather eccentric notation for his equations – given that most people would already have found the article very hard to understand, it now became, according to Michael Arbib, 'almost impenetrable', while for the historian of science Lily E. Kay the paper was 'a nearly incomprehensible abstraction'.[16] Nevertheless, around the dense stretches of symbolic logic there were patches of clear textual explanation that showed what McCulloch and Pitts were intending.

McCulloch had been thinking about this approach to biology for over fifteen years.[17] His key insight came when he realised that the all-or-none nature of an action potential was the equivalent of a proposition in logic – a statement that is either true or false. The neuron either fires or it does not. This was an example of what McCulloch called a 'psychon' – a basic mental 'atom', which combined with others to form more complex phenomena. He now understood that it should be possible to describe the activity of a series of neurons – what he called a 'nervous net' – in terms of a series of propositions. However, McCulloch found that representing this in strict logical terms was beyond his ability – until he met Pitts. 'It is to him that I am principally indebted for all subsequent success,' McCulloch later wrote.[18]

The ten theorems McCulloch and Pitts described in their paper, each represented by a diagram of interlinked neurons drawn by McCulloch's daughter Taffy, were explicitly cast in terms of the logical algebra developed by George Boole nearly a century earlier.[19] Boolean logic is based on true or false statements, which in conjunction with the basic operations AND, OR and NOT, enable the computation of arithmetic. McCulloch and Pitts showed that these operations could be embodied in the elementary structures of the nervous system. So, for example, the neurons in figure c of the paper represent a Boolean AND function: neuron 3 will fire only if both neuron 1 and neuron 2 are firing. Similarly, figure b represents the OR function: neuron 3 will fire if either neuron 1 or neuron 2 is active; while figure d shows the NOT function: neuron 3 will fire only if neuron 1 fires and neuron 2 does not.

By combining these basic functions, McCulloch and Pitts could explain quite complex phenomena, such as the well-known heat illusion, which they described as follows: 'If a cold object is held to the

LOGICAL CALCULUS FOR NERVOUS ACTIVITY

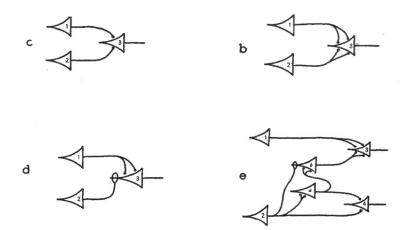

*Figures from McCulloch and Pitts showing how neuronal
organisation might embody aspects of Boolean logic.*

skin for a moment and removed, a sensation of heat will be felt; if it is applied for a longer time, the sensation will be only of cold, with no preliminary warmth, however transient.'[20]

Figure e shows a net that can account for this illusion, involving nothing more than a neuron to detect heat (1), another to detect cold (2), and two neurons representing the sensation of heat (3) and cold (4), connected in a net via two neurons (*a* and *b*). As McCulloch and Pitts put it: 'This illusion makes very clear the dependence of the correspondence between perception and the "external world" upon the specific structural properties of the intervening nervous net.' A basic psychological phenomenon could be modelled by a logical circuit.

McCulloch and Pitts's ambitions went far beyond explaining sensory illusions. They argued that all key aspects of mental activity 'are rigorously deducible from present neurophysiology', and even that psychiatric problems could eventually be understood in terms of some 'disturbed structure'. Their paper showed that nervous systems could be thought of in a highly abstract way; this insight seemed more powerful than any of the physical models that had been proposed in the previous decades. It represented a major departure from the dominant approach in understanding how the brain works, which for over half a century had been based around localising functions in the cortex, but which had done little more than identify vague 'centres' that were involved in various motor functions. How those functions were actually executed remained obscure. The real novelty of McCulloch and Pitts's work was that it focused attention on processes rather than on anatomical regions. Explaining the brain now appeared to involve describing algorithms that could be embodied in networks of neurons, or in interactions between organs. The key issue was the relation between the component parts and the way that function emerged from organisation – what McCulloch and Pitts called the immanent logic of neuronal structures.

Although this approach undoubtedly changed how we view the brain, its influence on studies of actual nervous system function is less clear. This was partly because precise knowledge of neural circuits was so poor at the time, but also because, as McCulloch and Pitts recognised, few actual nerve nets corresponded to their abstract, extremely simplified version. Many neurophysiologists could not get past the fact that the detail did not correspond to the nature of real nervous systems – a model that was so far from biological reality seemed pointless. Although examples of single neurons functioning as AND gates have since been identified, unlike the model suggested by McCulloch and Pitts, they are not simply additive. Instead, as seems to be typical in biological systems, they are nonlinear and multiplicative. Life is more complex than logic.[21]

The paper's greatest influence – which accounts for many of its 4,500 citations – was on the embryonic world of computing. In the USA, the mathematician John von Neumann was already using Boolean logic in his thinking about computers; Norbert Wiener drew von Neumann's attention to the McCulloch and Pitts article, and in particular its implication that although their focus was on the cellular embodiment of AND/OR/NOT operations in the structure of the nervous system, the theory could apply to any substrate – biological, mechanical or electronic.

Von Neumann was a ferociously brilliant man; as well as playing a leading role in the Manhattan Project (he designed the implosion-based detonation device for the bomb that obliterated Nagasaki, and helped select the target[22]) he also developed the basic elements of game theory, now used in economics and ecology, and, most importantly, he began planning for the future development of computers. In June 1945 von Neumann wrote a proposal for a stored-program general purpose computer, which dealt with 'the structure of a very high speed automatic digital computing system, and in particular with its logical control'.[23] The computer I am writing this on, like the phone in your pocket, works on the basis of von Neumann's ideas.

Despite being framed in the language of binary logic and imagined in terms of electrical wiring and glowing valves, at the heart of von Neumann's conception of the structure and logical control of a computing system were McCulloch and Pitts's hypothetical nerve nets. From the opening pages of his document, von Neumann used a biological analogy to justify his proposal of what a computer might be:

> the neurons of the higher animals ... have all-or-none character, that is two states: Quiescent and excited ... Following W. S. McCulloch and W. Pitts we ignore the more complicated aspects of neuron functioning: Thresholds, temporal summation, relative inhibition, changes of the threshold by after-effects of stimulation beyond the synaptic delay, etc. ... It is easily seen that these simplified neuron functions can be imitated by telegraph relays or by vacuum tubes.

Von Neumann continued:

> Since these tube arrangements are to handle numbers by means of their digits, it is natural to use a system of arithmetic in which the digits are also two valued. This suggests the use of the binary system. The analogs of human neurons are equally all-or-none elements. It will appear that they are quite useful for all preliminary, orienting, considerations of vacuum tube systems.

Von Neumann was justifying his choices about how to develop the structure and function of a computer by referring to a biological model. At this moment of its birth, von Neumann's computer was seen as a brain. The direction of the metaphor between machine and brain had switched. Before the metaphor settled into its current form – seeing the brain as a computer – there were a number of years in which studies of brains and computers interacted in the most dynamic fashion possible.

*

McCulloch and Pitts made a unique contribution to our understanding of what brains do, and also, inadvertently, to the development of the computer, but they were not alone in their approach. In the early months of 1942 McCulloch was invited to a select conference, held at a New York Park Avenue hotel, on the theme of cerebral inhibition. One of the speakers was Arturo Rosenblueth, a Mexican, Harvard-based physiologist who presented some of the research he had been doing with Wiener and the engineer Julian Bigelow. Wiener and Bigelow were engaged in war work, trying to develop an automated anti-aircraft gun. They had realised that the system they were studying involved feedback: the gun crew responded to the trajectory of the enemy plane by manoeuvring their weapon accordingly, firing it, then correcting their aim, and so on.[24]

By viewing machines and even nervous systems as loops transmitting positive and negative feedback, Rosenblueth, Bigelow and Wiener were able to describe how apparently purposive behaviour could emerge from the activity of a simple system. This was particularly the case with negative feedback, where a device stopped performing a particular function when it reached a preset goal. McCulloch was excited by Rosenblueth's talk and began to think about how feedback loops could explain a variety of phenomena, including mental illnesses such as neuroses.[25]

In 1943, the same year as McCulloch and Pitts published their article on the immanent logic of the nervous system, Rosenblueth, Bigelow and Wiener summarised their thinking in an article entitled 'Behavior, Purpose and Teleology'. They sought to explain teleology – purposive, goal-driven behaviour in non-human systems – by casting it in terms of positive and negative feedback. As they explained: 'the behaviour of some machines and some reactions of living organisms involve a continuous feed-back from the goal that modifies and guides the behaving object'.[26]

Using examples from machines and animals, and with not an equation in sight, Rosenblueth, Bigelow and Wiener explored a common framework for understanding all behaviour, using feedback as the key mechanism. They suggested that positive feedback could explain certain pathological symptoms, such as tremors in Parkinson's disease (McCulloch and Pitts had also theorised this, in the most complex of their theorems, entitled 'nets with circles', with circles representing the positive feedback loop). The great insight of Rosenblueth, Bigelow and Wiener was that they showed how negative feedback could produce apparent purposiveness in a machine or an animal – once a certain activity produced a given state, negative feedback would stop the activity, giving the illusion of purposive behaviour. This concept was absent from McCulloch and Pitts's vision. Although these ideas had been implicitly present in Seleno the robot dog, Uexküll's function circles and Lotka's wind-up beetle (no reference was made to any of these precedents), Rosenblueth, Bigelow and Wiener were the first to turn them into a general vision of the basis of behaviour.

At the same time, on the other side of the Atlantic, the Cambridge psychologist Kenneth Craik published a slim book entitled *The Nature of Explanation*. Most of the book was philosophical, but in the second half Craik focused on a hypothesis about what the brain does. Referring to Hull's models of conditioning, Craik explained that he preferred a more abstract approach, looking not at particular synaptic mechanisms, but rather at 'the fundamental feature of neural machinery – its power to parallel or model external events', something which he also argued existed in calculating machines.[27]

Rather than simply drawing a vague analogy between, say, the brain and a telephone exchange, Craik's approach was much more profound: he was interested in discovering what kind of calculations a mechanism would have to perform to 'play a part in thought'.[28] The essential sort of calculation that would be involved, Craik argued, was symbolic – representing aspects of external reality. Highlighting the greatest advances in technology of the time, such as calculating machines, telescopes and so on, which were effectively extensions of human sense-organs and bodies, Craik claimed that 'our brains them-selves utilise comparable mechanisms to achieve the same ends and that these mechanisms can parallel phenomena in the external world as a calculating machine can parallel the development of strains in a bridge'.[29] The key function of those symbolic representations in a nervous system, he said, was their role in exploring alternatives and in anticipation. A mental model, composed of the activity of neurons, could prepare the organism for future events.

When it came to imagining how these mechanisms were embodied in the brain, Craik was less confident, arguing that varia-tion in microanatomy meant that 'models of the brain – on the pattern of a telephone exchange – would be much more convincing if they did not postulate any particular connections'. Craik's version of the embodiment of the immanent logic of the nervous system was even more abstract than that of McCulloch and Pitts.[30] Two individuals with different neuronal structures could nevertheless show the same behaviour, if the same processes were represented in their brain, he argued. Craik imagined that a system of random connections 'would assume the required degree of orderliness as a result of experience', as long as there was sufficient plasticity.[31] The 'right' connections would eventually emerge, given enough time and experience. Tragically, Craik was unable to pursue his vision of the nervous system as 'a cal-culating machine capable of modelling or paralleling external events': in 1945 he was knocked off his bicycle in Cambridge and killed.

Craik's book did not cause much of a stir when it was first published, but in 1946 Adrian focused attention on it in a series of lectures that were published in book form the following year. Adrian summarised Craik's idea: 'the brain must contrive to model or par-allel external events by using something like the kind of symbolism which is employed in a calculating machine to represent a physical structure or process'. The implication of this is that 'the organism carries in its head not only a map of external events but a small-scale model of external reality and of its own possible actions'. Consciously or not, Adrian was contradicting one aspect of the argument at the heart of Leibniz's Mill – the suggestion that even if we could see the innermost workings of the brain, we would understand nothing. Adrian's view was that even if it did not reveal how thought arose, it would provide insight:

> Images and thoughts are then to be regarded as the finished products of an elaborate machine ... we could tell what someone was thinking if we could watch his brain at work, for we should see how one pattern after another acquired the necessary brilliance and definition.[32]

*

Looming behind all these approaches that came to fruition in 1943 were the ideas of Alan Turing. In 1936 the twenty-four-year-old Turing wrote a paper that used logic to describe an artificial device that could compute anything computable.[33] This hypothetical apparatus was generously dubbed a 'Turing machine' by the American logician Alonzo Church, who had been developing similar ideas and with whom Turing had just begun working at Princeton. The imaginary Turing machine consisted of a tape divided into squares, each of which carried a written symbol, a scanning head that could consider one square at a time, and a set of rules telling the machine what to do in response to each symbol. In principle, this machine could compute anything that is computable, which logically included imitating another machine.

The parallel between the elementary components of the Turing machine and Boolean neuronal circuits was made explicit by McCulloch and Pitts when they pointed out that if their neural nets were connected to suitable input, output and storage components such as a tape and scanners, the neurons would be able to compute the same things as a Turing machine. The two approaches were complementary, and the McCulloch and Pitts nerve nets provided 'a psychological justification of the Turing definition of computability and its equivalents'.[34] As McCulloch explained five years later: 'What we thought we were doing (and I think we succeeded fairly well) was treating the brain as a Turing machine.'[35]

Although Turing did not initially think in terms of artificial intelligence, or of links between organisms and his hypothetical devices, he soon began to do so. At the beginning of 1943 – that year again – Turing was at Bell Labs in New York, in their futuristic building on the Lower East Side of Manhattan, which had an overground subway line running right through it. Turing was there to work on the encryption protocols that would eventually allow the undersea hotline between London and Washington to operate securely during final phase of the war. One of the people he met at Bell Labs was the twenty-six-year-old mathematician Claude Shannon, who was working on theories of encryption and developing the mathematical conception of information. They regularly chatted over lunch and coffee, talking about their shared interest: building an electronic brain.

Working on his Master's thesis at MIT in 1937, Shannon had realised there was a link between Boolean logic, the Bell Company telephone circuits he had studied as a summer intern, and a mechanical analogue calculating machine built by Vannevar Bush at MIT. Shannon's insight was essentially the same as that realised by McCulloch and Pitts a few years later: it was possible to use logic to describe circuits in terms of symbols. In particular, those three basic operators AND, OR and NOT could be represented as electric circuits working on the basis of binary logic. This attracted the attention of von Neumann and helped to clarify his thinking about the future digital computer. It also provided obvious points of common interest with Turing, and in their chats the two men seemed to try to outdo each other in terms of imagining what the future might hold. Shannon recalled:

> We had dreams, Turing and I used to talk about the possibility of simulating entirely the human brain, could we really get a computer which would be the equivalent of the human brain or even a lot better? And it seemed easier then than it does now maybe. We both thought that this should be possible in not very long, in ten or 15 years. Such was not the case, it hasn't been done in thirty years.[36]

In turn, Turing was amazed by some of Shannon's ideas for

what they could do with an electronic brain. As Turing told the Bell Labs researcher Alex Fowler: 'Shannon wants to feed not just data to a Brain, but cultural things! He wants to play music to it!'[37]

*

A few months after the war was over, in March 1946, the Macy Foundation held the first of a series of meetings under the rather cumbersome title 'The Feedback Mechanisms and Circular Causal Systems in Biology and the Social Sciences Meeting'. In subsequent years that title was popularly abbreviated to the simpler 'Cybernetics Meeting', following the publication of Wiener's 1948 best-seller, *Cybernetics: Or Control and Communication in the Animal and the Machine*.

The ambition of these mini-conferences was stated in the title – they sought to unite biology and the social sciences (and the embryonic field of computing) through the study of common mechanisms, in particular feedback. At the first meeting, which involved barely more than a dozen people, von Neumann and the Spanish neurophysiologist Rafael Lorente de Nó explored the significance of electronic and neuronal digital systems. Lorente de Nó, who had worked with Cajal in the 1930s, described neurons as elements of an automaton in the flesh.[38]

But at the very moment when the field seemed to coalesce and to promise new insights into brain function, von Neumann began to have doubts. In November 1946 he wrote a letter to Wiener in which he suggested that the focus on parallels between computers and the brain was probably mistaken. He argued that 'after the great positive contributions of Turing-cum-Pitts-and-McCulloch is assimilated, the situation is rather worse than better than before'.[39] The problem, von Neumann realised, was that real nervous systems were far more complicated than those described by McCulloch and Pitts, and did not actually function in a digital manner beyond the elementary all-or-none aspect of a single action potential. In particular, as Adrian had shown, the neuronal code includes a key analogue element: firing rate increases with stimulus strength. When it comes to representing the outside world, neurons are not digital.

Von Neumann now considered that he and Wiener had made

an error in opting to study 'the most complicated object under the sun' – the human brain. Choosing a simpler nervous system, such as that of an ant, would not help, argued von Neumann: 'we lose nearly as much as we gain. As the digital (neural) part simplifies, the analogy (humoral) part gets less accessible ... the subject less articulate, and our possibilities of communicating with it poorer and poorer in content.' Von Neumann's solution was that they should abandon the study of nervous systems all together – the best possibility of successfully using logic to understand biology would come, he felt, from the study of viruses.*

Despite his pessimism, Von Neumann continued to participate in discussions about cybernetics and the brain, and in September 1948 he gave a talk in Pasadena as part of a conference on Cerebral Mechanisms in Behavior, in which he contrasted the structures of analogue and of digital computers, before comparing both with nervous systems.[40] Von Neumann recognised that neurons were not truly digital, not only because of the way they respond but also because the feedback loops they are involved in – for example those controlling blood pressure – contain both neuronal and physiological components. As he put it: 'living organisms are very complex – part digital and part analogy mechanisms'.[41] Von Neumann also explained that brains were far smaller than any computer and contained far more components (this was before the days of transistors, which had become a practical proposition just a year earlier, allowing for a first step in miniaturisation, but the point is still valid today). Above all, he put his finger on what remains one of the major questions in neuroscience, using a verb that is now commonplace, but which at the time was novel: 'how does [the neuron] encode a continuous number into a digital notation?'

After accepting that McCulloch and Pitts's approach proved that 'anything that can be exhaustively and unambiguously described ... is *ipso facto* realizable by a suitable finite neural network',[†] von

---

* That is another story, which you can read about in my previous book, *Life's Greatest Secret*.
† 'Neural network' first appeared in a review of a book about the Rashevsky group (the term does not appear in the book itself). Reiner, J. (1947), *Quarterly Review of Biology* 22:85–6.

Neumann went on to point out the practical problems in achieving this. Something as simple as a visual analogy – 'an object is like a triangle', for example – might conceivably involve 'altogether impractical numbers' of components. He concluded, pessimistically:

> It is, therefore, not at all unlikely that it is futile to look for a precise logical concept, that is, for a precise verbal description, of 'visual analogy'. It is possible that the connection pattern of the visual brain itself is the simplest logical expression or definition of this principle.

Von Neumann was arguing that even for some relatively simple mental processes, no model was possible, merely a part-by-part imitation of the actual nervous system devoted to that calculation. Any material representation of the human brain based on the approach outlined by McCulloch and Pitts, he worried, might prove 'too large to fit into the physical universe'.

At the same meeting, McCulloch gave a lecture provocatively entitled 'Why the Mind is in the Head'. (The bathetic answer, given in the final paragraph, was that that is where all the neurons are.) He, too, was now pessimistic about the possibility of modelling the brain – Pitts was engaged in mapping input/output relations for simple reflex arcs 'and has yet no very simple answer', McCulloch reported, before continuing: 'There is no chance that we can do even this for the entire cortex.'[42] This would come as no surprise to most neurophysiologists, then or now.

A few weeks after this meeting, Wiener published *Cybernetics or Control and Communication in the Animal and the Machine*, which changed everything. As well as coining the term cybernetics, which came to encompass a whole field (Wiener derived the word from the Greek for steersman), Wiener's book was an international bestseller, despite containing vast tracts of equations that were incomprehensible to most readers (and were full of errors), becoming a key work for scientists and for the general public. Wiener himself was a larger-than-life figure who attracted media attention. Portly, with thick glasses and a Van Dyke beard, Wiener had studied with Bertrand Russell as a teenager – the first volume of his autobiography was

accurately, if immodestly, entitled *Ex-Prodigy*. He also had a complicated home life: although his father was a Russian Jew, his wife was
an anti-Semite and Hitler supporter. Taffy McCulloch recalled that
Wiener would visit her parents' country house and join in with the
skinny-dipping in the nearby lake: 'He was a character. He looked
like a frog with bulging eyes. I remember him floating in the lake
with his tummy sticking up, just talking away, waving his cigar in
the air, and slowly sinking in the water.' Weiner's daughter also
remembered those naked swimming trips, and the potential domestic explosion they represented: 'Oh, if Mother had got wind of that I
can just see her going into orbit.'[43]

In *Cybernetics*, Wiener explained the new mathematical concept
of information that had been developed during the war both by
himself and by Claude Shannon and emphasised the role of negative
feedback in producing apparently purposeful behaviour in animals
and machines. He also explored the analogies between brains and
computers. Like von Neumann, he took McCulloch and Pitts's identification of the action potential with a digital signal as his starting
point, recognising the fundamental influence of Turing as he did so.
He used this framework to discuss a number of models of memory,
including what appears to be a correct intuition: 'it is quite plausible
that information is stored over long periods by changes in the thresh-
olds of neurons, or, what may be regarded as another way of saying
the same thing, by changes in the permeability of each synapse to
messages'.[44]

Wiener also compared brains and computers, focusing on one
particularly significant difference – the potential role of hormones as
messages within the body, affecting brain and behaviour. As he put
it, these physiological signals are not hard-wired, so must be labelled
in some way as 'for whom it may concern', for they circulate freely
within the body but only affect certain sets of neurons. That was very
different from how a computer worked.

At the 1950 Cybernetics conference, the Chicago physiologist
Ralph Gerard took the long view, warning his fellow attendees about
the danger of grandiose claims and 'overoptimism' with regard to
using this approach to understand the brain, given the lack of knowledge about how the nervous system really functions. He emphasised

that, irrespective of the digital nature of a single action potential, the way that neurons communicate information is essentially analogue, and that neuronal networks did not function like an electronic machine.[45] McCulloch, who probably had most to lose, given his investment in the question, stuck to his guns, insisting that as far as signal transmission is concerned 'a single click will do it'. The discussion gradually became increasingly pernickety and ended up with a long argument about definitions which got nowhere. The gulf between the theoreticians and the practical biologists was growing.

In von Neumann's final remarks on the subject, published posthumously in 1958 as *The Computer and the Brain*, he repeated many of the arguments that he had developed a decade earlier, before conceding that the problem was not simply that brains were far more complicated than machines, but that they appeared to function along very different lines from those he had initially expected: 'there exist here different logical structures from the ones we are ordinarily used to in logics and mathematics', he wrote.[46] His conclusion was that 'the outward forms of *our* mathematics are not absolutely relevant from the point of view of evaluating what the mathematical or logical language *truly* used by the central nervous system is'. Theory was powerful, but complex biological reality was even stronger.

*

The Cybernetics meetings continued from 1946 to 1953 but did not lead to any real progress in our understanding of the brain beyond the role of feedback, and the claim that some common processes might be involved in the behaviour of machines and organisms. The physicist and molecular geneticist Max Delbrück, never a man to mince his words, attended one of the meetings and scornfully recalled the discussions as having been 'vacuous in the extreme and positively inane'.[47] The problem was that the talks would cover anything from learning in the octopus to quantum theories of memory, but because most of the members were not experts in the field the discussion tended to drift off into platitudes, endless requests for clarification, or comments such as the plaintive remark by Ralph Gerard at the 1949 meeting: 'I just asked what we are supposed to be talking about.'[48] Having come to no very great conclusion, the group eventually petered out, its final phase marked by a mysterious row that profoundly alienated Wiener from McCulloch and Pitts, and which was apparently manufactured by Wiener's wife for her own malicious reasons.[49]

In the United Kingdom, a group of junior researchers formed a more informal gathering in 1949, which met in London under the title of the Ratio Club.* The key criterion for admission to this select circle was that, like a twenty-first-century hipster, you had to have been into cybernetics before Wiener published *Cybernetics* (the exact phrase was that membership was restricted to 'those who had Wiener's ideas before Wiener's book appeared'; professors were also forbidden).[50] Although McCulloch came to speak on a number of occasions, the Ratio Club lacked the stardust (and the financing) of the Cybernetics group, and it, too, eventually collapsed in 1958, after thirty-eight meetings.[51]

The problem for both the US and UK groups was that, in the end, there was less to all this cybernetics than met the eye. Turing, who was a member of the Ratio Club, was particularly sceptical of the grandiose claims made by some of the cyberneticians – he dismissed McCulloch as a 'charlatan'.[52] Eventually Turing's attention drifted away from the brain, as he focused his brilliant mind on how organisms develop and grow.

There was one exception to this rather downbeat end to what had been a pivotal moment in post-war science: the Cybernetics group, the Ratio Club and the global public were all entranced by attempts to embody the insights of the new science in semi-autonomous robots. For example, at the 1951 Cybernetics conference Claude Shannon presented a maze-learning robot – the device navigated a simple maze by trial and error and could remember the correct route, and even had an 'anti-neurotic circuit' built into it so that if it were thwarted from reaching its goal for too long, it would begin random movements to find a correct solution.[53] The initial version of this robot, built of seventy-five bulky electromagnetic telephone relays,

---

*Members' memories differ as to how this was pronounced – it was either Ratio as in the ratio of two numbers, or RAT-ee-oh, as in the Latin for compute.

was a large grid with a finger sensor that moved over the surface of the maze; this was later upgraded to a more crowd-pleasing 'mouse', which was moved by magnets. The mouse – called Theseus – was the subject of a brief film in which Shannon claimed that its maze solving 'involves a certain level of mental activity, something akin, perhaps, to a brain'.[54] The robot impressed everyone, from the participants at the Cybernetics meeting ('It is all too human,' said one, uncritically[55]), through the readers of *Time*, *Life* and *Popular Science* magazines, to Shannon's employers, who discussed making him a member of the board in recognition of his achievement.[56] Despite the excitement, Theseus was simply a more sophisticated version of the mechanical maze robot built by Ross and Smith in the 1930s, and it, too, provided no insights into learning at all.[57]

Norbert Wiener also built a robot – a three-wheeled 'moth' that was attracted to light. If the polarity of the neutron flow was reversed, the device became repelled by light and turned into a light-fleeing bedbug.[58] The moth version was displayed to the public in 1950, during a prologue to a Harvard University production of Capek's play *RUR*. Christened Palomilla for the occasion and covered in a biologically inaccurate carapace made of papier mâché, the moth was lured onstage by Wiener, using a torch. As *The Harvard Crimson* reported: 'Palomilla made mistakes; it ran back into the curtain once and stalled often. But it acted with at least as much decision and far more speed than an earthworm.'[59]

At about the same time, Grey Walter of the Ratio Club developed a similar device, in the shape of a pair of wheeled tortoises called Elmer and Elsie (these were vague acronyms for ELectroMechanical Robots, Light Sensitive).[60] Like Wiener's moth, the tortoises – which were eventually exhibited at the Festival of Britain and are now on display in the Science Museum in London – were attracted to light. In 1951 a breathless Pathé newsreel announcer informed British cinema-goers that Walter's tortoise (rebaptised 'Toby', apparently for the purposes of alliteration) contained 'an electronic brain that functions like the human mind'.[61] In fact all that Toby did was to navigate his way towards a light, and to move about randomly if he bumped into an object. When his batteries ran low, he would return to his recharging station, a kind of hutch. In reality, neither the moth/

bedbug nor the tortoises represented a conceptual advance over
Seleno the electric dog – these devices all used feedback.

More serious claims were made for a machine built by another
member of the Ratio Club, W. Ross Ashby, who used surplus Royal
Air Force electromagnetic bomb-aimers to create what he called the
Homeostat. This was a hybrid analogue/digital device that was
designed to respond to changes in its environment by searching for
a stable state through random choices. Its workings were obscure
– Pitts in particular had a hard time understanding it when it was
presented at the final Macy Conference in 1952 – but the Homeostat showed how random changes could incrementally contribute
to the emergence of adaptive behaviour.[62] Although this might have
been an interesting metaphor for how evolution by natural selection shapes our senses, it still remains unclear what insight – if any
– it might have provided with regards to brain function.[63] In reality,
whatever their impact on public interest in robotics, or even the possibility that behaviour might emerge from sets of instructions and
inorganic components, neither Theseus, nor Palomilla, nor Toby/
Elmer, nor the mysterious Homeostat had any consequence for scientific approaches to how the brain works.[64]

*

In one respect, however, the post-war world did see the appearance
of a significant consensus about brain function – the assumption
that in some way the activity of the human brain and the existence
of the mind are the same thing. Two key contributions to this new
certainty were produced almost simultaneously, in rather different forms. In 1949 the philosopher Gilbert Ryle wrote an accessible
account of the problem, *The Concept of Mind*, while in 1950 Alan
Turing wrote a dense academic article entitled 'Computing Machinery and Intelligence'.[65]

Turing's article was immensely influential, for it was here that
he proposed what has become known as the Turing Test as a way of
answering the question 'Can machines think?' Turing's idea – which
he called the imitation game (hence the name of the film about his
life) – was that if you had a device that could answer questions, and a

human could have a conversation with this device and not be able to detect that it was a machine, then to all intents and purposes it could think. Turing's article contained no mathematics and was essentially a work of philosophy that the likes of Locke or Leibniz would have grasped instantly. Turing was confident that technical developments would lead to machines passing the test:

> I believe that in about fifty years' time it will be possible to programme computers, with a storage capacity of about 10^9, to make them play the imitation game so well that an average interrogator will not have more than 70 per cent chance of making the right identification after five minutes of questioning.[66]

As to how was this to be done, 'the problem is mainly one of programming', he thought.[67] With the right approach, Turing's machine, composed of matter, would apparently be able to think. A similar conclusion was reached a year later by Donald MacKay of the Ratio Club: 'we have failed to find any distinction in principle between the observable behaviour of a human brain and the behaviour possible in a suitably designed artefact'.[68] For Turing the trick was finding appropriate programming, for MacKay it was a question of suitable design. For both, logic suggested that a machine could produce an output that was indistinguishable from that of a brain.

Ryle's *The Concept of Mind* did much to cement the conviction among many readers that the mind has a material basis. Ryle's readable argument makes no attempt to explain either how the brain works, nor how the activity of the brain results in the existence of the mind – the word 'brain' is barely mentioned. Ryle's main objective was to systematically dismantle Descartes's dualism, which he disparagingly characterised as 'the ghost in the machine'. His argument created a coherent philosophical basis for considering that mental life is identical with the physical activity of the brain, but it did not prove this to be the case.

In the UK, these ideas rapidly percolated into the public sphere, with radio playing an important role. In 1950 the BBC Third Programme broadcast a series of talks by speakers such as Sherrington,

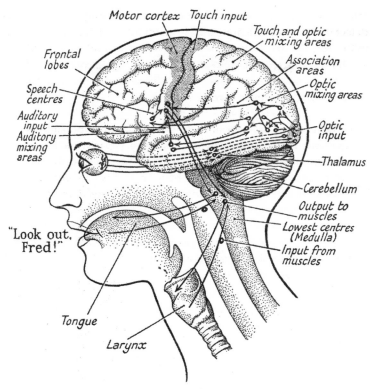

Motor cortex Touch input
Touch and optic mixing areas
Frontal lobes
Association areas
Speech centres
Optic mixing areas
Auditory input
Auditory mixing areas
Optic input
"Look out, Fred!"
Thalamus
Cerebellum
Output to muscles
Lowest centres (Medulla)
Input from muscles
Tongue
Larynx

*Schema accompanying J. Z. Young's 1950 Reith Lectures, showing some
of the pathways involved in shouting a warning at a child who is about
to cross a road. Compared to Charcot's schema this has more anatomical
detail, and, above all, the arrows represent information pathways.*

Adrian and Ryle, under the general title *The Physical Basis of Mind*,
while the Reith Lectures of that year were given by J. Z. Young and
were subtitled 'A Biologist's Reflection on the Brain'.[69] All these talks
were reprinted in the BBC's magazine *The Listener*, and were published as books. Both the zoologist Solly Zuckerman, in his talk 'The
Mechanism of Thought: The Mind and the Calculating Machine',
and above all Young in his seven lectures, popularised the new view
of brain and behaviour developed by Wiener, explaining his ideas

about negative feedback and highlighting the significance of information in the brain. As Young told his audience:

> Information reaches the brain in a kind of code ... of impulses
> passing up the nerve-fibres. Information already received is
> stored in the brain either by sending impulses round closed
> circuits, or in some form corresponding to a print. This is just
> what calculating machines do – they both store old informa-
> tion and receive new information and questions in coded form.
> The information received in the past forms the machine's rules
> of action, coded and stored away for reference ... The brain
> has an even greater number of cells than there are valves in a
> calculator and it is not at all impossible that it acts quite like
> an adding machine, in some ways ... However, we still do not
> know exactly how the brain stores its rules or how it compares
> the input with them. It may use principles different from those
> of these machines.[70]

The confidence shown by the likes of Ryle, Turing and Young
with regard to the physical basis of mind stood in stark contrast to
pre-war conceptions. An important strand within these older views
was represented by Sherrington, who in 1937, at the age of eighty,
gave a series of lectures in Edinburgh. These were published in 1940
under the title *Man on His Nature* and contained Sherrington's rather
rambling views on the links between mind and brain. In one of these
lectures, he used a quaint analogy to describe brain function, a metaphor that has become curiously famous among neuroscientists: the
'enchanted loom'. The phrase occurs when Sherrington describes
what happens when we awake:

> the brain is waking and with it the mind is returning. It is as
> if the Milky Way entered upon some cosmic dance. Swiftly
> the head-mass becomes an enchanted loom where millions of
> flashing shuttles weave a dissolving pattern, always a mean-
> ingful pattern, always a meaningful pattern though never an
> abiding one; a shifting harmony of subpatterns.[71]

The term 'enchanted' was not simply poetic – Sherrington's whole argument, sometimes expressed as a kind of lyrical meta- physics, was that although there was a correlation between mind and brain, that did not mean that the mind was in the brain, merely that the brain was a site of interaction between the two, just as Descartes argued. The actual nature of mind was unknown, and, to Sherrington, it seemed to have no physical basis at all. He repeatedly criticised the widely held materialist assumption that mind was some form of energy, pointing out that it was impossible to detect any difference in form or function of those neurons involved in areas of the brain that were supposedly the site of thought, and those found elsewhere in the body.[72] Mind, Sherrington claimed, was not based on material phenomena at all – the unstated implication was that the metaphor- ical loom was literally enchanted, because it worked by magic. As he put it in his typically poetic language: 'Mind, for anything perception can compass, goes therefore in our spatial world more ghostly than a ghost. Invisible, intangible, it is a thing not even of outline; it is not a "thing". It remains without sensual confirmation, and remains without it for ever.'[73]

McCulloch and Pitts had been well aware of the contradiction between the approach outlined in their 1943 paper and that of Sher- rington. At the end of their paper they made the following claim: 'both the formal and the final aspects of that activity which we are wont to call mental are rigorously deducible from present neuro- physiology ... in such systems "Mind" no longer "goes more ghostly than a ghost"'.[74]

The precise approach developed by McCulloch and Pitts turned out to be wrong, because nervous systems do not function in the way they presumed. But their focus on processes that could be conceived of as computations, and on the significance of the fundamental struc- tures of the nervous system, were both of enormous significance. Although few scientists would now argue that 'the brain is a com- puter', except as a way of expressing a very loose concept to the public, most would accept that the brain can be considered – among other things – as a computational organ which, as Craik argued, manipulates symbolic representations of the outside world so that it can explore alternative outcomes or solutions. But although during this period a consensus emerged about what the brain does, there was little agreement about how it might do it. Theories about this began to develop in the second half of the twentieth century, as the new way of looking at the brain was applied to all its functions, ushering in the scientific world we know today, bringing us into the present.

PRESENT

IN MAY 2018, I attended a conference on neural circuits at the Janelia Research Campus outside Washington DC. At lunch one day, sitting in the Virginia sun, a group of us were chatting with one of the resident researchers, Dr Adam Hantman. The previous evening I had given a talk based on some of the ideas in this book, and this led to one of those 'where are we going?' conversations. Adam's view was forthright: 'What conceptual innovations have we made in the last thirty years?' he asked, before answering his own question: 'None.'

Adam was wrong, but only because he did not go far enough. In reality, no major conceptual innovation has been made in our overall understanding of how the brain works for over half a century. This period has seen immense, Nobel Prize winning discoveries – astonishing new techniques have given researchers an amazing degree of precision and control of brain activity, massive computer simulations capture the activity of millions of neurons and we now appreciate the role of chemistry in controlling the activity of neural networks. All of this gives us a far richer understanding of what is happening where in the brain, compared with past generations, but we still think about brains in the way our scientific grandparents did.

According to that view, established by Craik, McCulloch and others, a brain contains symbolic representations of the outside world that it manipulates to predict what will happen and to produce behaviours; it does this using some kind of computational approach, but it is not like any machine we have yet constructed, because it bathes in a complex system of chemical communication and its activities are partly determined by its own internal states. The immense success of our understanding of what the brain does is built on this common approach that was established in the 1940s and early 1950s.

This period saw not only the creation of a powerful framework for understanding the brain but also an explosion of scientific interest that led to a new discipline and a new term – neuroscience. The word first began to appear in the 1960s, and by the 1970s it had gradually taken over what were once parts of psychology, physiology and neurology as well as creating its own unique field. The usual paraphernalia of academia – journals, learned societies, training schemes, prizes, university departments, research programmes and degrees

– soon coalesced around the new discipline. Above all, more and more scientists adopted this approach to studying the brain.

There are now tens of thousands of brain researchers around the world, beavering away in a bewildering range of new subdisciplines: cognitive neuroscience, neurobiology, theoretical neuroscience, computational neuroscience, clinical neuroscience, and many others – each with their own questions, methods and approaches.[1] Thousands of research articles relating to brain function appear each year. Massive government and private initiatives have been devoted to understanding the brain and its link to mental health problems, while neuroscience has played a major role in the development of computer technology and for a while had a modish influence on the humanities.

The following chapters all cover the same period – roughly from 1950 to today. Each looks at a different aspect of the brain (not necessarily the human or even the mammalian brain) – memory, neural circuitry, computer models of the brain, brain chemistry, brain imaging and finally the renewed interest in the nature of consciousness. This division of topics is artificial and is not entirely satisfactory – there are common ideas and methods that thread their way through these different chapters, and sometimes the same names recur as scientists move from one field to another. There are also recurrent themes, in particular the shifting arguments over whether particular functions are localised in the brain, and if so, to what extent and at what level. Rather than being a complete history, these chapters therefore constitute a kaleidoscopic description of our current understanding of how the brain works, exploring how our knowledge has developed over the last seventy years.

They also reveal changes in who has been doing the science. There were very few women in the previous chapters; that begins to change from this point, particularly in the passages dealing with the last three decades. Other aspects of diversity in science – in particular where it is done, by which socio-economic and ethnic groups – remain pretty much as in previous centuries. It is not clear if these structural biases affect our understanding of what the brain does and how it does it, but this uncertainty is primarily because no one has studied the question.

In dealing with this recent period, history bleeds into current trends and interests, and as a result there are views in these chapters that will provoke some readers. My colleague at the University of Manchester's Centre for the History of Science, Technology and Medicine, the late Jeff Hughes, pointed out that it was particularly hard to write the history of contemporary science – scientists and historians often find they have contradictory objectives.[2] In the present case these problems may be amplified, given that I am both an observer of and, to a very minor extent, a participant in some of what follows. Experts will surely be frustrated that a given area, experiment or researcher is either not mentioned or is passed over too quickly – for example, sleep research, non-visual perception, hormones, emotions, the development of the brain and the way that genes affect brains are all dealt with in little detail. My apologies to researchers in these fields and others, but it is impossible to do justice to the full range of work being done on the brain, and there is often little agreement among these different subdisciplines about where we are going.

Paradoxically, despite the immense progress, it is not clear if we have the theoretical tools necessary to face the challenges of understanding the brain in the twenty-first century. But to know where we are going, and what the future might hold, we first need to know where we are now, and how we got here.

- TEN -

MEMORY

1950 TO TODAY

From the 1930s onwards, the Montreal neurosurgeon Wilder Penfield carried out hundreds of brain operations with the aim of relieving chronic, debilitating temporal-lobe epilepsy.[1] To identify which part of the brain to remove, Penfield gently stimulated his conscious patients with electric currents carried by delicate electrodes. If stimulation of a particular brain location led to indications that a seizure was imminent, Penfield knew that site was a candidate for removal. This procedure revealed something rather eerie: sometimes the stimulation led the patient to relive very precise events. These experiences were vivid and detailed, like a waking dream. Often the patients heard sounds – a piano being played, someone singing a well-known song, or a telephone conversation between two family members. In one case, when the electrode was left in place and the current flowed, the music continued in the patient's head, and she sang along. In another, each time a particular region was stimulated, the patient heard an orchestra playing a popular song of the time, 'Marching Along Together'. In yet other examples a patient saw a man and a dog walking along a road near his home, another saw jumbles of lights and colours, while yet another relived the recent experience of his mother telling his young brother he had his coat on backwards.

These strange sensations would occur only when the relevant area was being stimulated; if the electrode was removed, or if the patient was told that stimulation was occurring when it was not, nothing happened. As Penfield put it 'recollections which are clearly derived from a patient's past memory can sometimes be forced upon him by the stimulating electrode'.[2] These oneiric* experiences were remarkably constant for a given individual – repeated stimulation at the same location evoked exactly the same sensation in the patient. For Penfield these data suggested that memory might have a very precise location in the brain. For the patients, it was something they often found rather disturbing.

*

Penfield's dramatic findings heralded the reopening of a long debate about the localisation of brain function that reached back into history and which continues to the present day. In the middle of the twenti-eth century, one of the dominant views in studies of the neural bases of memory was expressed by Karl Lashley, whose experiments on animals seemed to show that deficits in learning produced by sur-gical operations were proportional to the extent of the damage to the cortex. He explained this effect in two ways: first, all cells had equal capabilities; second, the whole of the brain contributed to the making and recollection of memories – what he called 'mass action'. For Lashley, like Flourens in the nineteenth century, the activity of the brain could be understood only as a whole.

In 1950 Lashley gave a talk in Cambridge summing up his life's work on memory, with the title 'In Search of the Engram'.[3†] At the peak of his influence (he fell ill in 1954 and died four years later, aged sixty-eight), Lashley argued that memory was distributed across the brain. Reviewing his lifelong search for the engram, he suggested that it had been in vain and concluded wryly:

---

*Look it up. It's exactly the right word.

† 'Engram', meaning the physical trace of a memory, was coined in German in 1904 by the zoologist Richard Semon, and first appeared in English in a 1921 translation of his book *The Mneme*.

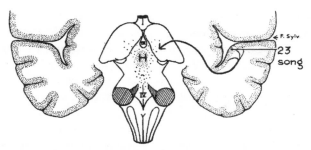

Figure by Penfield showing a vertical slice through the brain indicating where stimulation produced the perception of a song.

> This series of experiments gives a good bit of information about what and where memory is not. It has discovered nothing directly of the real nature of the engram. I sometimes feel, in reviewing the evidence on the localization of the memory trace, that the necessary conclusion is that learning just is not possible ... Nevertheless, in spite of such evidence against it, learning does sometimes occur.[4]

Lashley's distributed view of memory was soon apparently contradicted by Penfield's eerie findings, which he first reported at a meeting in 1951. Lashley was in the audience. Penfield explained the weird experiences of his patients in the following terms: as we pay conscious attention to events in our lives, we are 'simultaneously recording it in the temporal cortex'.[5] This record contained both visual and auditory stimuli, and was stored somewhere in a region below the cortex, in the middle of the brain, which is connected to the cortex by a complex set of nerve fibres. During stimulation, the impulses represented by these sensations 'pass in the reverse direction to those which created that pattern'. To put it another way, the experience is played back through the same networks that recorded it. It would appear that Penfield had activated an engram.

In the discussion of Penfield's presentation, Lashley had to admit he was stumped: 'I have no clear alternative to offer in explanation of Dr Penfield's data,' he said. Nevertheless, he did his best to undermine Penfield's observations by emphasising the complexity of memories before concluding, somewhat lamely, 'I do not see how it is possible for the small number of cells in the centrencephalic system to mediate, or even transmit, these complexities.'[6]

Whether Lashley liked it or not, and irrespective of the number of cells in the region stimulated by Penfield (it in fact contains millions of cells, at least), the data were the data. Somewhere deep in the brain, linked to very particular parts of the temporal lobe, memories could be evoked by electrical stimulation of a very precise area.*

As Penfield recognised, what he called 'evoked recollections' were very different from ordinary memories, containing far more detail. Our everyday memories do not involve replaying a second-by-second account of an event – they are generally quite vague and are constructed by the brain, containing false elements, or components that are guessed at from the context. It seemed likely that the experiences Penfield was evoking did not simply involve activating engrams, but also introduced other elements that related to different aspects of brain function, producing the eerie dream-like quality the patients reported. One thing was clear – there was nothing special about the particular memories that his electrodes could conjure up. As Penfield explained: 'The events recollected are often unimportant and uninteresting.'[7]

The result was soon replicated, and more recent studies have confirmed the accuracy of Penfield's experimental work.[8] The degree of localisation of function that Penfield had discovered was extraordinary, but the exact nature of that function was not so clear. In 1951 Penfield described the area that he was stimulating as 'memory cortex', suggesting that it was the location of memory, but by 1958 he accepted that memories were not actually stored in the site he was stimulating. Instead, the area seemed to be able to trigger the activity of a distant part or parts of the brain where the memory was actually located.[9] Localisation began to seem less localised.

---

* This finding entered popular culture – in Philip K. Dick's novel *Do Androids Dream of Electric Sheep?* (this was the basis of the film *Blade Runner*), people use a Penfield Mood Organ to induce emotions in themselves and other people.

This argument between localised versus distributed function did not only relate to memory. In 1937 Penfield had published some simpler results of stimulation of patients undergoing brain surgery, which at heart were like those of Fritsch and Hitzig and of Ferrier, but on conscious humans.[10] Sometimes the patients would report very specific sensations when particular parts of their brain were stimulated – tingling in the fingers, odd tastes on the tongue, feelings of warmth down one side of the body. On other occasions eyelids flickered, legs jerked, and some patients grunted. To summarise their findings, Penfield asked a medical illustrator, Hortense Cantlie, to make a drawing.[11] The result was a grotesque image of the human body, showing the various parts in proportion to their representation in the brains of their patients. This image, which Penfield called a homunculus, suggested how the brain sees the body. As might be expected on the basis of every day experience, the tongue, hands and face were particularly well represented. Other very sensitive parts of the body, such as the genitalia and the rectum, were not shown.

In 1950 Penfield presented a more sophisticated diagram, separating out the sensory regions (on the left) and the motor regions (on the right) of the brain, which was shown in cross-section.[12] This revealed that sensory and motor cortexes showed different representations of the body – to take a trivial example, teeth and gums are well represented in sensory cortex, but barely in motor cortex. More interestingly, the hand dominates the motor cortex, while the lower face is most significant in the sensory cortex. Ultimately, this uneven representation of the parts of the body across the brain is a consequence of our evolution and our ecology – other primates show different patterns.

Despite their long-lasting influence, these diagrams were deceptive, for they implied that there was a strict one-to-one correspondence between a particular brain area and a particular part of the body, which was constant across individuals.[13] In fact, the Penfield homunculus represented the average response of all the patients – any individual might show a somewhat different relation between brain region and body part to that seen on the figure. Nevertheless, Penfield's work was taken to show that the brain contained both a precise map of the body and an intensely detailed system for storing

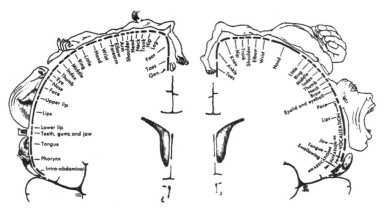

*Penfield's 'homunculus' diagram of the representation of the human
body in motor cortex (right) and sensory cortex (left).*

and retrieving extremely particular events. For most scientists at the
time, function appeared to be highly localised.

*

All these findings, which undermined Lashley's anti-locationist
views, chimed with the ideas of one of Lashley's students, the
Canadian psychologist Donald Hebb. In 1949 Hebb published
The Organisation of Behaviour, which set out key elements of what
became the modern biological framework for understanding how
the brain functions.[14] Hebb's starting point was straightforwardly
materialist – the mind was simply a product of the activity of the
brain. Although he accepted that this was only 'a working assumption', he distanced himself from those scientists who either accepted
dualism and viewed mind and brain as distinct and made of different stuff, or who claimed that the nature of mind was unknowable.
Hebb's response to such pessimism was robust: 'Our failure to solve
a problem does not make it insoluble. One cannot logically be a
determinist in physics and chemistry and biology, and a mystic in
psychology.'[15]

Hebb's book explored all the aspects of the study of the human brain that have dominated our investigations ever since, including learning, perception and mental illness. One of his insights was a conception of how learning occurs at a cellular level. Opposing his one-time teacher Lashley, Hebb insisted 'memory must be structural'.[16] According to Hebb, that structure involved two levels – a complex 'tridimensional lattice-like assembly of cells' (less poetically, a network), and the way those cells were connected. What Hebb described as his neurophysiological postulate of memory was as follows: 'When an axon of cell A is near enough to excite a cell B and repeatedly or persistently takes part in firing it, some growth process or metabolic change takes place in one or both cells such that A's efficiency, as one of the cells firing B, is increased.'

Hebb was saying that synapses can develop and grow stronger when neurons are activated together (this is often snappily summarised as 'cells that fire together wire together'). According to Hebb, the fine structure of the nervous system – the network of the connections between cells – is formed through experience. This concept was, as Hebb admitted, an old one – the fundamental idea went back at least to the eighteenth-century associationist thinkers, such as David Hartley – but Hebb had recast it in the light of modern neuroanatomy and neurophysiology, giving it a far more precise form.

Hebb argued that the complexity of many cell assemblies meant that 'at each synapse there must be a considerable dispersion in the time of arrival of impulses, and in each individual fibre a constant variation of responsiveness'. This meant that the same assembly can function in a different way in different circumstances, and that distinct patterns of activation corresponding to distinct stimuli or memories therefore occur not only in space, but also in time. The lattice-like assembly of cells that constituted Hebb's version of the engram was four-dimensional.

Hebb also emphasised that real neural assemblies show spontaneous activity in the absence of any stimulus, meaning that the brain has to continuously distinguish signal from background noise. To do this, he said, they are organised with complex, non-linear and conditional connections that enable the nervous system to carry out the necessary computations. This was a less abstract view of what neurons do than McCulloch and Pitts had suggested – it was more like engineering than logic.

Finally, despite his focus on learning, Hebb dismissed the widely held view that learned and instinctive behaviours were fundamentally different, asserting that 'Ultimately, our aim must be to find out how the same fundamental neural principles determine all behaviour.' For some researchers that is still the goal; others consider it to be a pipedream, because no such principles exist.

*

Within a decade of the publication of Hebb's book, dramatic evidence shed light on the way that fundamental brain processes linked with memory can be associated with particular structures. This came about through an entirely chance and tragic event that occurred to a man who, as far as the scientific community was concerned, was known simply by his initials, HM. In 2008 HM died, his identity was revealed, and his full story has now been told.[17] His name was Henry Molaison, and he was the most famous patient in the history of brain science.

In 1935, aged nine, Henry was knocked over by a bicycle. Shortly afterwards, perhaps as a consequence of this accident, he began to have debilitating epileptic seizures. By the time he was a young man, Henry's condition was so severe he had to give up his job in an engineering factory; drugs were unable to help him, and it seemed that surgery was the only option. In the hands of someone as painstaking as Penfield, psychosurgery could have some success in relieving epilepsy following the very precise removal of a limited area. But many other surgeons used grosser techniques and, in cases of severe mental illness such as schizophrenia, they often removed whole lobes of the brain (hence 'lobotomy').

The American surgeon William Scoville was an enthusiastic advocate of psychosurgery – by the early 1950s he had carried out around 300 lobotomies on patients with severe schizophrenia. Despite his lack of experience in treating epilepsy, on 1 September 1953 Scoville operated on the twenty-seven-year-old Henry Molaison. He used the same procedure he employed on his schizophrenic patients, even though there was no precedent for such a drastic intervention in epilepsy – it was, as he later admitted, a 'frankly experimental operation'.[18] Scoville operated on both of Henry's temporal lobes, creating holes in his skull above his eyes, each about 2.5 cm wide, and then removed about 8 cm depth of each hemisphere of the brain, including most of the hippocampus, the amygdala and the entorhinal cortex on each side. The operation was classed a success, and HM began his recovery.

Except Henry never really recovered. Indeed, as far as he was concerned, he never got beyond that day in 1953. Poor Henry now had profound memory deficits, although his other mental faculties remained intact; he could remember his childhood and many events in the period up to the operation, but for the rest of his life he could not develop new memories. Until his death in 2008 Henry lived in a permanent now, unable to recall something that happened an hour earlier. Even the terrible consequences of the operation had to be perpetually explained to him. Every moment, he said, was 'like waking from a dream', and every day was 'alone in itself, whatever enjoyment I've had, whatever sorrow I've had'.[19]

The significance of the catastrophe that had occurred to Henry was discovered after Wilder Penfield met Scoville at a conference in 1954 and learned of Scoville's operations on the temporal lobe. Penfield and his colleague, the young psychologist Brenda Milner,[20] had already noted links between damage to the hippocampus and problems with memory formation, so they decided to study Scoville's patients.

Milner tested a dozen or so patients and noticed that the three people who had suffered the most extensive damage showed similar complete losses of what is called episodic memory formation – the ability to create memories of events that occur to the individual. As well as Henry Molaison, patients DC and MB had both been operated upon in order to relieve severe psychotic problems; both showed no memories since their operations, and both failed all tests of memory formation.[21]

As to the point of all this surgery and cerebral destruction, Henry's fits became slightly less debilitating and his medication could be reduced, while DC and MB became less violent, but their underlying problems remained. As the neuroscientist Gordon Shepherd later remarked drily, 'Hardly a dramatic achievement.'[22]

To be brutal, what was a catastrophe for Henry Molaison became a godsend for science. Over the next half-century, Henry cheerfully participated in a unique long-term study of brain function – he of course had no memory of this and had to have everything explained to him anew each time he carried out the tests. A one-sided relationship grew up between Henry and the experimental team, because although they knew him well – after his death Milner said it felt like she'd lost a friend – he never knew that he had met them before.[23]

These endless tests and discussions – always carried out with gusto by Henry, for whom it was a perpetually novel event – revealed that Henry's inability to form memories was not absolute: he would sometimes refer to events or people who became famous after his 1953 operation (astronauts, the Beatles, President Kennedy), but these memories were fleeting and could not be reliably evoked.[24] Equally, on some tests he would show improvements if he was repeatedly tested over several days, even though he could not recall ever having done the test before. However, these were exceptions. Basically, Henry was stuck in the present tense.

Brenda Milner's first report on HM's behaviour, co-authored with Scoville, became a classic in brain science.[25] Over subsequent decades there were scores of studies of HM, ranging from psychological investigations (many of which were carried out by Milner's student Suzanne Corkin, who studied Henry throughout her career), right down to the post-mortem analysis and 3-D reconstruction of his brain.[26] They all showed that the destruction of Henry's hippocampi was responsible for his inability to form new memories. That does not mean that memories are stored in the hippocampus, but rather that this structure is required for the brain to create them. HM's tragedy did not reveal the location of the engram, but it showed localisation of function for a decisive aspect of memory formation.

According to Milner, Scoville did not feel any guilt about what happened to Henry, nor did she think he should have done. The operation was a last resort – 'HM was so desperate. He was having an absolutely miserable life,' as she put it. She did recall, however, that Scoville was deeply affected by the similar damage he caused to patient DC, who, unlike Henry the factory worker, was a fellow physician.[27]*

*

In March 1947 the psychologist Edward Tolman gave a light-hearted, far-ranging and self-deprecating lecture at the University of California, in which he described his work on animal learning. He focused on rat maze-learning experiments, and in trying to understand what was happening in the animal's brain, Tolman came up with a telling metaphor:

> we assert that the central office itself is far more like a map control room than it is like an old-fashioned telephone exchange. The stimuli, which are allowed in, are not connected by just simple one-to-one switches to the outgoing responses. Rather, the incoming impulses are usually worked over and elaborated in the central control room into a tentative, cognitive-like map of the environment. And it is this tentative map, indicating routes and paths and environmental relationships, which finally determines what responses, if any, the animal will finally release.[28]

For example, if a rat is allowed to explore an empty maze several times and is then presented with a reward at the end of the maze, it will be able to find its way through the maze far more quickly than a rat with limited experience of the maze. The rat has obviously been paying attention to its environment and memorising the maze, even though there was no reward. Similarly, if a rat is shocked at a

---

*In 2016 the journalist Luke Dittrich (Scoville's grandson) raised a series of ethical questions about the way that HM was treated by scientists, focusing on the role of Suzanne Corkin. These issues related primarily to conflicts of interest, consent issues and ownership of data, including the alleged destruction of data relating to HM by Corkin. Psychologists responded by strongly defending Corkin's conduct. Dittrich, L. (2016), *Patient H. M. – A Story of Memory, Madness, and Family Secrets* (London: Chatto & Windus).

particular location in its cage, it will thereafter avoid that place. Tolman's explanation was that the rat has a map in its brain. Somehow, it represents the outside world in its neurons.

The first proof that Tolman may have been right came in the late 1960s, when John O'Keefe at University College London was studying the activity of cells in the rat's thalamus as the animal moved about. One of these cells gave a very strong response when the animal moved its head – this was the first time that O'Keefe had seen anything like this, and he was intrigued. After the experiment was over, O'Keefe killed the rat and sliced up its brain to see exactly where the cell he had been recording from was located. To his great surprise he discovered that by mistake he had inserted the electrode into the rat's hippocampus. This error changed O'Keefe's life and the course of brain science.[29]

In 1971 O'Keefe, together with his student Jonathan Dostrovsky, reported data from eight cells in the hippocampus that were each activated when the rat was in specific locations in the cage. But it was not only location that was significant: the strongest response came from a cell that fired when the rat was in a particular place, was being held by an experimenter, and the lights were on. If any one of those factors was absent, the cell stopped firing, indicating that it required a very specific set of stimuli to be activated. As O'Keefe and Dostrovsky put it:

> These findings suggest that the hippocampus provides the rest
> of the brain with a spatial reference map. The activity of cells
> in such a map would specify the direction in which the rat was
> pointing relative to environmental land marks and the occur-
> rence of particular tactile, visual, etc., stimuli whilst facing in
> that orientation.

O'Keefe and Dostrovsky went further, suggesting that their hypothesis would enable the animal to predict what would happen if it moved:

> The internal wiring of the hippocampus, on this model, would
> be such that activation of those cells specifying a particular

orientation together with a signal indicating movement or intention to move in space ... would tend to activate cells specifying adjacent or subsequent spatial orientations. In this way, the map would 'anticipate' the sensory stimuli consequent to a particular movement.[30]

If the animal was deprived of the map, by having its hippocampus removed, 'it could not learn to go from where it happened to be in the environment to a particular place independently of any particular route'.

O'Keefe's research revealed that as well as the ability to encode episodic memories, the hippocampus contains a literal map of the environment – the representation in the brain is isomorphic to the environment, to use the jargon. This map, consisting of what are called place cells, also contains information about how to get from one location to another, enabling the animal to navigate the world and to predict what it will find in different places. It is a map, but as Tolman had brilliantly intuited, it is a cognitive map, involving multiple sensory modalities and based on associations and predictions, not a simple one-to-one representation of the outside world. In species with different ecologies these hippocampal maps have different forms – for example, while the maps are 2-D in rats they are 3-D in bats, describing the animal's position within a sphere – but they are always cognitive, not simply spatial.[31]

How spatial information gets into the hippocampal place cells became apparent partly through the work of the wife-and-husband team May-Britt and Edvard Moser, who discovered cells in the entorhinal cortex, adjacent to the hippocampus, which fired when the animal was in several locations. The activity of these cells produced a grid-like network that formed the raw data used by place cells in the hippocampus – you can predict the activity of a given place cell from the activity of these grid cells.[32] Other cells in this part of the cortex, which records the animal's head direction, its speed of movement and the presence of a border in the environment, also contribute to the cognitive map in the hippocampus.

For their work on these discoveries – which eventually involved hundreds of researchers around the world – O'Keefe and the Mosers were awarded the Nobel Prize in 2014.[33] Although all this work was done on small mammals, there is said to be a real link with human behaviour. We also appear to use our hippocampus in navigating around the world. One famous claim is that London taxi drivers, who are required to learn routes through the capital by heart, have larger hippocampi than normal, and that this effect increases with their time in the job.[34] The exact way that this apparent effect worked – increased number of neurons, or simply increased volume – is not clear, because the technique used to measure the size of the hippocampus was too imprecise. Although there have been a number of studies by the same research group on London cabbies, the finding has not yet been replicated.

Despite all these findings pointing in the direction that the hippocampus contains a map of the animal's surroundings, as O'Keefe and Dostrovsky's initial report hinted, the brain creates something far richer than a mere spatial map, and place cells and grid cells do not simply provide a kind of biological GPS. Cognitive information regarding rewards, smell, touch, vision and time is integrated into their activity.[35] Place cells in rats and bats are also involved in the processing of social information, specifically the location of others.[36]

After rats have investigated a new place, the hippocampal cells corresponding to their mental map of the location are reactivated during sleep, as the rat's brain consolidates that memory (perhaps the rat is dreaming of the place and the situation).[37] If an unexplored location – such as a blocked-off arm of a maze – is associated with a reward, then place cells in the rat's brain corresponding to that site are activated, as though the animal is anticipating visiting the location; some researchers now argue that this predictive function is the true role of place cells.[38] The human brain also replays events relating to non-spatial learning as the subject is at rest, with the hippocampus being the focus of this activity, apparently enabling us to abstract new knowledge from previous experiences.[39] All these intriguing results show that the hippocampus integrates information of various kinds with different objectives, including decision making and generalisation from one task to another. At the same time, other brain regions are involved in the creation and recall of a particular memory, suggesting a mixture of localisation and distributed function.

In 2016, this complex situation led the late Howard Eichenbaum, a leading researcher on the hippocampus, to claim that research had validated Lashley's ideas.[40] Not many researchers would put it so robustly, but Eichenbaum's argument highlights the fact that the memories that are processed by the hippocampus involve distant regions of the brain. The hippocampus is not the site of the engram, it is the encoder and the gateway. There is localisation of memory but we have yet to identify exactly where it is found; distributed information is also involved, and we do not fully understand how the encoding and recall of memory take place in the hippocampus and its associated areas.[41]

Recently, a German study of humans learning a spatial task, using a novel brain-imaging technique that reveals microstructural changes in tissues during learning, suggested that the hippocampus did not play such a vital role as was expected.[42] The key changes associated with spatial learning took place in the posterior parietal cortex, not the hippocampus. These changes appeared rapidly, lasted over twelve hours and were apparently linked to memory-related functional activity of the brain. All this reinforces the suggestion that the hippocampus does not contain 'the' engram, and that different regions of the brain are involved in creating our memories.

Such ideas have repeatedly surfaced as scientists have found evidence for both distributed and localised changes associated with learning. The two stances may not be so contradictory. In 1986 Timothy Teyler and Pascal DiScenna suggested that the anatomical links between the hippocampus and various parts of the brain indicated that the hippocampus generated episodic memories by indexing them with the various features associated with the episode.[43] Searching for one of more of these indices would activate the engram. This contrasted with a hard version of the cognitive map theory, which argued that place alone provided the key to unlocking memory. Despite the discovery of place cells, experimental evidence to distinguish between the two theories has been lacking – perhaps because, in many people's eyes, they are not strictly counterposed. We now know that hippocampal cells can maintain their engram function but adopt new links to different place cells, dependent on experience. Spatial coding can be separated from the engram, suggesting that the broader, index function of the hippocampal engram may be correct.[44]

The complexity of the hippocampus can be seen in the intriguing link between the cognitive maps it contains and its role in forming episodic memories. The ancient Greek method for memorising a large number of pieces of information involved imagining putting the thing to be remembered in a particular room in a 'memory palace' – it may be that this is what we are doing all the time, through the hippocampus, when we learn and remember things. Olfactory information can also be encoded via the hippocampus, through the entorhinal cortex; this may explain how smells can evoke not just the memory of an event, but also give you the powerful sense of the location where that event occurred, a phenomenon that may be similar to that experienced by Penfield's patients during electrical stimulation.[45] Because of the damage to his hippocampi, Henry Molaison could not accurately compare two odours, nor identify common foods based simply on their smell; he found it equally difficult to read a map.[46] In humans, olfactory perception and spatial memory are fundamentally intertwined, based on the hippocampus and the frontal cortex.[47]

Despite the apparent degree of localisation revealed by these studies, they have not told us much about how the brain might work on a cellular or circuit level, nor what kind of computations might be carried out by the cellular networks. Scientists have had to be satisfied with assigning general functions to different regions containing millions of cells, creating functional flow diagrams in which the identity of the components is based primarily on detailed regional anatomy, rather than on identified individual cellular function, or even the activity of large populations of cells.[48] How exactly memory works is not revealed by such studies.

*

In 1957 a young researcher at the National Institutes of Mental Health (NIMH) in Bethesda, just outside Washington DC, picked up Milner and Scoville's paper describing the effects of Scoville's operations on HM and the other unfortunate patients. As the young man later recalled, he was so impressed by the article that, in an instant, 'the question of how memory is stored in the brain had become the next meaningful scientific question for me'.[49] His name was Eric Kandel, and he was not quite twenty-eight years old. A widely read intellectual – he obtained a degree in History and Literature from Harvard before turning to medicine and maintained a lifelong interest in psychoanalysis – Kandel would help create our modern understanding of how neuronal activity changes during learning, work which culminated in the Nobel Prize in 2000.

After initially studying the electrophysiological activity of neurons in the cat's hippocampus, Kandel soon realised that getting to the heart of the cellular changes that interested him would require a much simpler system than a vertebrate brain. He found the solution and a hint as to the kind of phenomena he would need to study in the work of two researchers in Adrian's group in Cambridge, Alan Hodgkin and Andrew Huxley. In 1952 they had revealed the physiology of the action potential – how the neuron sends its message – finally providing decisive evidence for a theory that developed and extended Bernstein's ideas from the beginning of the century. Hodgkin and Huxley's work had been interrupted by the war, but once hostilities were over they were able to show that the action potential is transmitted through changes in the permeability of the neuron membrane, which alters the concentration of sodium and potassium ions, leading to a rolling wave of depolarisation that passes rapidly down the cell.[50] They also correctly hypothesised that tiny pores within the membrane – ion channels – allowed the neuron membrane to alter its permeability. In 1963 Hodgkin and Huxley won the Nobel Prize for their work, sharing the award with Eccles.

For Kandel, how Hodgkin and Huxley made their discovery was just as important as what they had found. Their work was not done in Adrian's basement laboratory in Cambridge, but instead in a marine biology facility in faraway Plymouth – Hodgkin and Huxley studied the responses of the giant axons of the squid and the cuttlefish. This system, which was pioneered by J. Z. Young in the 1930s, consisted of large identifiable neurons that could be reliably studied. The lesson – which had long been known by physiologists – was that to get at fundamental processes, you should choose a simple system that will give clear answers.

With this principle in mind, in 1959, after six months of hard thinking, Kandel decided he would investigate the cellular basis of learning and memory by studying a giant shell-less marine snail of the genus *Aplysia*, which is found off the Californian coast. This animal, which grows to over 30 cm long, has large neurons that can be seen under the microscope, a very simple brain – a mere 20,000 neurons grouped into nine clusters – and a simple set of behavioural reflexes. At the time, only a handful of people in the world were studying *Aplysia*, and when Kandel made his decisive choice he had never dissected a snail nor recorded from its neurons; he was not even certain that they could learn.[51] His ambition, which he set out in his first grant application and which he fully realised over subsequent decades, was 'to study the cellular mechanisms of electrophysiological conditioning and of synaptic usage in a simple nerve network'. This meant he was going to investigate how the activity of the *Aplysia* nervous system, in particular its synapses, changed as a result of learning.

Kandel focused on an easily measurable behaviour, the snail's gill withdrawal reflex – a light touch on parts of the animal's body caused it to retract its gill in a basic protective response. Kandel's group demonstrated that this reflex could show very simple forms of learning and short-term memory – habituation (a decline in the response with repeated stimulation) and sensitisation (an increase in the response if a light touch was associated with a brief electric shock) – and they eventually showed that the snail could learn in a classical conditioning set-up, much like Pavlov's dogs.

Over the years, Kandel and his colleagues identified the neural circuits involved in these behaviours and proved that Hebb's neurophysiological postulate was correct – learning involves a change in the strength of synapses in small circuits of neurons. In short-term memory, that change involves enhanced release of a neurotransmitter; in long-term memory, which is induced by the repeated association of stimuli, this increased release of neurotransmitter was accompanied by the growth of new synaptic connections between the two cells. The engram, as Hebb had predicted, was ultimately nothing more than a change in the activity of a synapse.

At the beginning of the 1980s Kandel's group joined the molecular revolution that was transforming biology, which made it possible to describe complex molecular cascades within cells and to identify the genes that produced the components of those systems. Eventually, Kandel's group, along with many others, was able to identify the molecules within a neuron that are involved in the creation of memory – cyclic AMP, various enzymes and a protein called CREB that effectively turns the cyclic AMP gene on and off, enabling the organism to decide if it wants to remember what it has learned. These molecules, generally called second messengers, because they relay the message initially carried by the neurotransmitter or by a hormone, interact in a submicroscopic ballet that takes place rapidly inside the cell as learning occurs, leading to the creation of new growth in the neuron and the creation of new synapses. Most satisfyingly, it soon became apparent that this model of learning applies to all animals – for example, the *dunce* memory mutation in the *Drosophila* fly, which set me on my career when I read of it in 1976, turned out to code for the enzyme that degrades cyclic AMP. The same biochemical system is being used in your head right now. Not everything about the biochemical basis of memory is resolved; other molecules apart from neurotransmitters are involved in synaptic activity and consolidation (over 5,500 different proteins can be found at a human synapse), but we now understand in broad terms how memory is created.[52]

*

There was a brief but significant diversion in this apparently endless progress in understanding memory. In the 1960s and 1970s a series of studies claimed that learned behaviour could be transferred from one animal to another by injecting brain extract, or RNA, or protein. The Swedish biochemist Holger Hydén suggested that during learning specific forms of RNA were produced, which could then be transferred; this was supported by a large number of studies showing that learning could be inhibited by molecules that blocked protein synthesis or that affected RNA molecules. This apparently worked in a range of animals from rats to goldfish, and included studies of the planarian flatworm, which can grow a new brain if its head is cut off. In 1959 James McConnell at the University of Michigan reported that planarians that had regrown their elementary brains showed a learned avoidance of light if they had received an electric shock in the presence of light before having their heads removed; later studies showed naive flatworms could even acquire this behaviour by eating bits of trained planarians.[53] The effect in rats was less grotesque, but congruent – injecting brain material from rats that had been trained to avoid light led to an apparent transfer of learning, suggesting that some kind of biochemical substance was involved.[54]

There was widespread media interest in the idea that an engram might be composed of a single molecule that could be transferred from individual to individual – perhaps it could lead to pills that would enable humans to learn by swallowing a tablet. It soon became clear, however, that both the behaviour and the biochemistry involved in the transfer of learning were less certain than first appeared. Many of the behavioural studies were based on very small sample sizes or used quite subjective methods for determining if an animal had learned. Then, in 1966, a brief article appeared in *Science*, signed by twenty-three researchers from eight different laboratories, stating that they had been unable to replicate the transfer of learning by RNA.[55] The nucleic acid explanation of the engram was dead.

A French pharmacologist at Baylor College of Medicine in Houston, Georges Ungar, came up with an interpretation that reconciled both the claims of transfer of learning and the criticisms of those who showed that RNA was not involved. It was possible, he argued, that the RNA extracts contained small proteins called peptides that were in fact responsible for the effect. In the late 1960s and early 1970s Ungar pursued the substance involved in the transfer of learning, eventually identifying it in extracts from the brains of over 4,000 trained rats. He gave it a name, scotophobin (*skotos* means dark in Greek), and researchers at the universities of Illinois and of Michigan claimed that synthetic scotophobin induced dark-avoidance in naive mice, strengthening confidence in the discovery.[56] Between 1968 and 1971, there were at least fifteen major press articles on Ungar's work, including in *Time* magazine, the *New York Times* and the *Washington Post*.

However, the reality of scotophobin soon began to evaporate. In July 1972 Ungar published an article in *Nature* in which he claimed that scotophobin induced dark-avoidance and speculated that there might be many such behaviourally active molecules in the nervous system.[57] The article had been submitted to the journal seventeen months earlier; the reason for this long delay between submission and publication was that one of the referees involved in peer review of the paper, Walter Stewart, felt strongly that the whole thing was nonsense. In a highly unusual step, *Nature* eventually decided to publish Ungar's article, but to accompany it by a long paper from Stewart detailing his criticisms of Ungar's biochemical claims, including with regard to the synthetic version of the molecule. Stewart argued that, despite having already published seventeen scientific articles on the topic, covering more than a hundred pages, Ungar's group had not provided the necessary experimental detail to allow the work to be replicated and its claims to be tested. His conclusion was that 'the authors' conclusions are more likely false than true'.[58]

Despite a brief riposte from Ungar in the same issue of *Nature*, the effect of Stewart's critique was devastating. Improved behavioural measures and better biochemistry soon revealed that there was no transfer of learning, and that if scotophobin did exist it was probably a polypeptide, perhaps produced as a consequence of stress when the animals were shocked, and had nothing to do with learning.[59] Funding for transfer-of-learning experiments dried up virtually immediately, as something that had excited the scientific community and beyond for years turned out to be an illusion.[60] Scotophobin became neuroscience's version of N-rays, a form of radiation that briefly obsessed physicists at the beginning of the twentieth century but which turned out not to exist.[61] The exact origin of the widespread behavioural effects that were reported remain unclear, but neither engrams nor the fear of light can be transferred from one animal to another in a syringe. Nevertheless, recent research confirms that memories can indeed reappear in the regenerated head of a planarian, suggesting that not everything done at this time was complete nonsense.[62]

*

While work on invertebrates such as *Aplysia* and *Drosophila* was helping to uncover the biochemical basis of learning, researchers on vertebrates developed a way of indirectly studying the manner in which synapses develop during memory formation. In 1973 two researchers in Oslo, Tim Bliss and Terje Lømo, reported that they could change the structure of the neuronal pathways into the hippocampus of the rabbit by stimulating the pathway with a very rapid series of electrical pulses.[63] What they called the potentiation effect produced by stimulating the pathway – effectively imitating a strong stimulus experienced in real life – created changes to the synapses in the pathway that could be detected for hours.

Lømo had first observed the effect in 1966 and had subsequently worked on the subject with Bliss in 1968–69, but problems of replication led them to continue fiddling around, trying to be confident of what they had discovered.[64] Eventually, even though they had not resolved these problems, they decided to publish anyway. Although Bliss and Lømo both left the field for a while (nearly a decade for Bliss, thirty years for Lømo), other researchers picked up on the effect, which became known as long-term potentiation, or more simply LTP, and the number of papers on the topic soon grew exponentially. By stimulating the animal's brain very precisely and then observing biochemical and structural changes, it was possible to reveal the complexity of different types of synaptic change, using slices of tissue – including from the human brain – rather than whole animals.[65] In a major review published in *Nature* to mark the twentieth anniversary of the first LTP paper, Tim Bliss and Graham Collingridge highlighted that the number one unresolved problem in the field was the true physiological significance of LTP, in particular, 'is it a central component of the synaptic machinery of memory?'[66] Researchers were uncertain as to the link between real memory and the effects seen in laboratory studies of LTP. This issue still remains unresolved – in 2006 Bliss would go no further than to say LTP is 'a compelling physiological model of these processes'.[67] Recent demonstrations that LTP and its 'negative' equivalent, long-term depression, can deactivate and reactivate memories in rats support the existence of a causal link, but that does not mean that LTP is itself memory. The continued lack of clarity over the precise biochemical basis of

LTP, and problems such as the fact that learning can take place after a single event whereas LTP requires repeated stimulation, continue to prompt some scientists to doubt that LTP fully represents how the brain encodes memory.[68]

When Penfield tried to explain the weird evocation of memories he produced by stimulating the brain, he suggested that the same pathways were activated during recall as during learning. This has now been demonstrated using the latest piece of neuroscientific kit – optogenetics. This technique, developed in the early years of the century by a number of researchers, including Gero Miesenböck, Karl Deisseroth and Ed Boyden, now dominates many areas of research on animal brains and neurons. It involves introducing a gene that encodes a light-detecting molecule into a cell of interest; this molecule can then be activated using light, thereby making the cell respond. Optogenetics provides a way of precisely identifying and stimulating neurons, and has been used to show that cells involved in learning exhibit some changes that are typical of LTP, and that the same cells are activated during memory recall.[69] These are now generally known as engram cells, although they are not the only component of the engram, which involves many, many neurons.[70]

In 1982 Francis Crick, the co-discoverer of the DNA double helix, suggested that structures known as dendritic spines – tiny outgrowths of the dendrite, the input part of the neuron – might play a key role in synaptic activity by changing their shape during learning.[71] Crick was right about their significance, but the exact mechanism turned out to be simpler than he imagined – as long-term memory is created, new synaptic connections are established through the creation of new spines, rather than existing spines changing their shape. New spines are observed following learning in a wide range of animals, and in 2015 a study used optogenetics to shrink spines that had been created following learning; memory of the specific task that had been learned was disrupted, indicating that dendritic spines are a key component of engram formation.[72] The situation is not quite so simple, however, as it is becoming apparent that neurons do not create new synapses on their own – other cells called astrocytes, which respond to neurotransmitters, seem to promote synaptic plasticity and enhance memory. If the activation of astrocytes in the hippocampus is blocked, then memory is impaired.[73]

There will be more discoveries and clarifications to come, but overall the results support Hebb's understanding of learning.[74] Despite the intensity of the debates on localisation versus distribution that have punctuated the history of brain science, it now seems that memories are not found in a single place, although individual cells may still play a key role in memory formation and recall. Memories are often multimodal, involving place, time, smell, light and so on, and they are distributed across the cortex through intricate neural networks.

Some research is taking our understanding of the physical nature of memory in a direction that you might consider to be disturbing. In 2009 researchers led by Sheena Josselyn at MIT specifically deleted cells in the mouse amygdala that had expressed high levels of the protein CREB during a learning task.[75] The result was that the mouse forgot what it had learned. The engram had been erased. The development of optogenetics has enabled researchers to manipulate memory in mice even more profoundly. In Nobel Prize winner Susumu Tonegawa's lab at MIT, false memories have been created in the rodent hippocampus, leading an animal to freeze in a particular part of the cage as though it had previously been shocked there, although it had never had any such experience.[76] The group has also turned an aversive memory into a positive one, leading the animal to be attracted to a location where it had previously been shocked. The meaning of the engram had been transformed.[77] They have even activated positive engrams in the mouse, leading to a decline in behaviours that look like human depression.[78] Other researchers have created a memory entirely from scratch by optogenetically activating both the olfactory bulb and centres involved in reward and aversion; as a result the mouse remembered something about a smell it had never encountered.[79] All this precision might seduce us into thinking that only those particular units that were manipulated in each of these experiments are involved in memory formation – in reality, underlying these single cells there are large numbers of cells that contribute to the activity of the network and produce the behaviour.

1. Imaginary portrait of the second century physician, philosopher and poet Galen, from a sixteenth century book. We have no idea what he actually looked like, but almost certainly not like this.

2. Galen's gruesome second century experiment on a pig, which showed that the brain, and not the heart, controls movement. Taken from the title page of a sixteenth century collection of his works.

NICOLAVS STENONIVS

3. The Danish physician Nicolaus Steno (1638–86) who laid out much of our modern approach to brain function. He also founded geology, showed how muscles work and was the first to realise that women have eggs. He featured heavily in my first book, *The Egg and Sperm Race*.

4. Portrait of Julien Offray de La Mettrie (1709–51) on the frontispiece of one of his books. He looks like he would be good fun down the pub.

5. Self-portrait of the pioneer Spanish neuroanatomist, Santiago Ramón y Cajal (1852–1934), in his laboratory. Cajal won the Nobel Prize in 1906.

6. The 1923 London performance of *RUR* – the play by Czech author Karel Čapek that gave the world the word 'robot'.

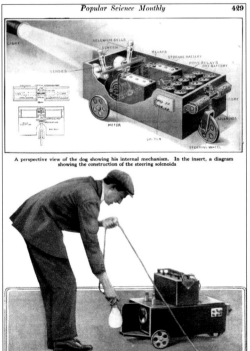

A perspective view of the dog showing his internal mechanism. In the insert, a diagram showing the construction of the steering solenoids

7. A 1918 description of Seleno the Electric Dog, which homed in on its target. Seleno's co-creator, Benjamin Miessner, saw in it the future of mechanised war: 'The electric dog which is now but an uncanny scientific curiosity may within the very near future become in truth a real "dog of war", without fear, without heart, without the human element so often susceptible to trickery, but with one purpose; to overtake and slay whatever comes within range of its senses at the will of its master.'

8. Walter Pitts (1923–69), the brilliant young mathematician who helped shape the idea that nervous systems carry out computations. Pitts was so odd that one friend recalled that people who had not met him could imagine he was the product of some kind of collective delusion.

9. David Hubel (left) and Torsten Wiesel, who made decisive breakthroughs in understanding how the brain processes visual stimuli. The photo was taken in 1981, after the announcement that they had been awarded the Nobel Prize.

10. Eve Marder of Brandeis University. Marder has spent her glittering scientific career trying to work out how the few dozen neurons of the lobster's stomach do what they do. Even this apparently simple system is currently beyond our grasp.

11. Doris Tsao of the California Institute of Technology, whose Twitter bio reads 'cortical geometer'. Tsao has studied how single neurons in the visual cortex process images, in particular how they combine to detect faces; she hopes to be able to explain the famous face-vase illusion.

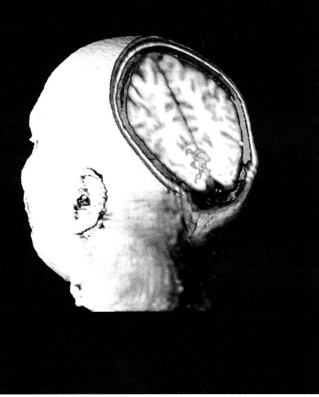

12. Cover of *Science* magazine from 1991, showing the first use of fMRI to reveal brain activity. This breakthrough, and the striking image, marked a major shift in brain research.

13. Patient S3 (Cathy Huthchinson), who became tetraplegic following a stroke, eating an apple by using her brain to control a robotic arm in John Donoghue's laboratory at Brown University in the USA. This was the first time she had fed herself in over a decade. 'Controlling the robotic arm felt natural,' she later said.

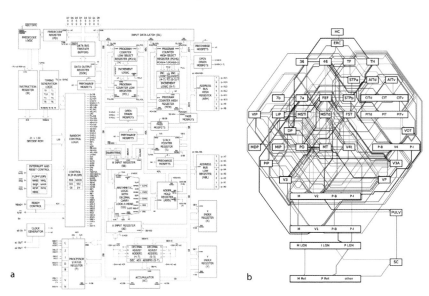

14. Figure from a 2017 article in which Eric Jonas and Konrad Paul Kording tried to 'reverse engineer'a computer processor. *a)* The clearly hierarchical organisation of the processor; *b)* a 1991 representation of the complexity of the visual system in the macaque (it was this work that prompted Francis Crick to call for intensive research on the anatomical organisation of the human brain).

To bring the story of learning full circle, researchers recently employed optogenetics to activate neurons that use the neurotransmitter dopamine – long associated with reward in vertebrates, and also implicated in addiction – when an otherwise neutral light was turned on in the animal's cage.[80] Even as little as four one-second activations of these dopaminergic neurons, eighty seconds apart, were enough to create a Pavlovian response in the animal – it would subsequently move towards the light as soon as it was turned on. The reflexes shown by Pavlov's dogs were presumably functioning in the same way.

All this work showing that memories can be precisely created, altered and erased has been done on genetically manipulated mice. These techniques cannot be used for changing memory in humans, although psychologists are interested in seeing how insights into the process of memory formation might be transferred into clinical practice.[81] Nevertheless, the broader ethical implications of this research have not been lost on those involved: in 2014 one article was entitled 'Inception of False Memory', in a nod to the mind-boggling 2010 Christopher Nolan science fiction film *Inception*, in which the viewer is never quite sure what is real.[82] An even more appropriate reference would have been Philip K. Dick's 1966 short story 'We Can Remember It for You Wholesale', later made (and remade) as a feature film under the title *Total Recall*. In Dick's typically paranoid story, set in the 'not too distant future', a bored clerk called Douglas Quail ('Quaid' in the film) has fake memories implanted about being a secret agent who has visited Mars. These memories – and others that surge into his brain, involving an alien plot to invade Earth – turn out to be true. Or do they?

The research on the cellular basis of memory highlights what many psychological studies had previously demonstrated: memory is malleable. It is not simply a recording of ongoing events, it is constructed, and it can be false. Above all, however, it has a material basis.[83] We have found elements of the engram, and it is not like memory on a hard disk in a computer. Biological memory is rich, unreliable and highly interconnected, with access to it taking place via multiple routes, not through a single address.

The link between our ordinary sense of how we remember

things and the complex and precise memories that could be evoked by Penfield is not clear. We do not appear to be perpetually recording our whole lives and yet those experiments reveal that highly specific and apparently inconsequential moments can be recalled, either by some external event, or by the pulse of an electrode. The engram has given up a few of its most basic secrets, but we are still far from understanding what is happening when we remember. Our brains might be like a computer in terms of how they sometimes process information, but the way we store and recall our memories is completely different. We are not machines – or rather, we are not like any machine we have yet built or can currently envisage.

These advances in identifying the physical basis of memory raise the question of how sensory information – the stuff of memory – is processed by the brain in the first place. Memories are held in particular sets of neurons, but that does not explain how the brain is able to work out what is in the world, and what exactly is being remembered. As with so many of the events described in these pages, a key moment in uncovering the secrets of perception, and in grasping the degree to which it is localised or distributed, came about by accident.

ELEVEN

CIRCUITS

1950 TO TODAY

In early 1958, two researchers at Johns Hopkins University in the USA – a Swede and a Canadian, both in their early thirties – were investigating how cells in the cortex of a cat respond to visual stimuli. The anaesthetised animal lay on the operating table while the scientists recorded from a single brain cell with an electrode. They projected various shapes of light onto the animal's retina, using microscope slides onto which they had glued metal discs, producing a dark spot against a light background. To no avail. The cell's electrical responses were weak, transformed into feeble cracks and pops over the loudspeaker in the lab. Then, as the pair recalled:

> Suddenly, just as we inserted one of our glass slides into the ophthalmoscope, the cell seemed to come to life and began to fire impulses like a machine gun. It took a while to discover that the firing had nothing to do with the small opaque spot – the cell was responding to the fine moving shadow cast by the edge of the glass slide as we inserted it into the slot. It took still more time and groping around to discover that the cell gave responses only when this faint line was swept slowly forward in a certain range of orientations. Even changing the stimulus orientation by a few degrees made the responses much weaker,

and an orientation at right angles to the optimum produced
no responses at all. The cell completely ignored our black and
white spots.[1]

The cell was activated by a very particular stimulus – a vertical
moving line. It was not interested in stationary lines or in horizontal lines at all. Completely by accident, the two researchers, David
Hubel and Torsten Wiesel, had made a discovery that would help
change our view of how sensory stimuli are processed in the brain,
revealing the existence of sometimes surprisingly complex representations of the environment in single cells.

Over the following years, Hubel and Wiesel showed that some
brain cells responded to the orientation of visual stimuli, while
others required a specific kind of movement. As they moved the
electrode down through the cat's brain, they found that the visual
cortex was organised in columns and layers, with each column corresponding to a particular object (line, dot and so on) and each layer
to a particular orientation of that object. These basic elements would
send their information to a cell at the next level of the brain, where
they apparently began to form a more complex representation of the
visual world.

Hubel and Wiesel's discovery fitted in with a number of previous findings. In 1953 the Cambridge physiologist Horace Barlow
– a member of the Ratio Club who was also Charles Darwin's great-
grandson – had shown that the cells in the frog retina were organised
in groups, each of which covered a small part of the frog's visual
field.[2] Each of these circuits enabled the frog to identify small dots
about the size of a fly. In principle, linking such circuits together in
the frog's retina enabled the system to detect a moving insect – a
group of cells would fire as the image of the fly passed over it, then
stop as it moved away. Although in some respects Barlow's discovery prefigured that of Hubel and Wiesel, there was an essential
difference – Barlow's work was in the peripheral nervous system,
not in the brain.

There were also more direct precedents. In 1938 Lorente de Nó
had suggested that the neuroanatomy of the visual cortex seemed
to be organised in columns, with groups of interconnected cells

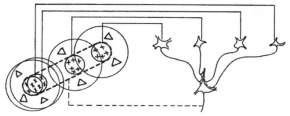

Text-fig. 19. Possible scheme for explaining the organization of simple receptive fields. A large number of lateral geniculate cells, of which four are illustrated in the upper right in the figure, have receptive fields with 'on' centres arranged along a straight line on the retina. All of these project upon a single cortical cell, and the synapses are supposed to be excitatory. The receptive field of the cortical cell will then have an elongated 'on' centre indicated by the interrupted lines in the receptive-field diagram to the left of the figure.

Diagram from Hubel and Wiesel showing how they envisaged a detector for a line being produced by simple neuronal connections.

extending into the brain from the surface, although nobody had imagined what this columnar organisation might imply about brain function.[3] And a year before Hubel and Wiesel's experiment, Vernon Mountcastle found that, in the cat's cortex, cells responding to the same kind of stimulus (for example, touch) from different parts of the body were organised vertically, while cells in the same layer of the cortex responded to different kinds of sensory stimuli from the same part of the body.[4] Shortly afterwards, McCulloch and Pitts, along with Pitts's friend Jerry Lettvin and the Chilean cyberneticist and neurophysiologist Humberto Maturana, reported the presence of analogous cells in the brain of the frog.[5]* Similar results were soon reported from many other vertebrate species, and it became clear that this was a general principle of how the vertebrate visual system was organised.

The implication of this high degree of localisation was that the brain first identifies precise elements of the environment – lines, moving objects and others – and then somehow assembles them into

---

*This was one of the last scientific contributions from Walter Pitts. Apparently deeply perturbed by the split with Wiener years earlier, Pitts increasingly turned to drink and died in 1969, aged only forty-six.

a recognisable overall image. Despite there being clearly defined areas of the brain devoted to each sensory modality (smell, hearing and so on), it emerged that there is an important degree of integration: auditory signals are integrated in the visual regions of the cat's brain, while visual signals are also analysed in the auditory cortex of the mouse.[6] This interaction between sensory modalities is presumably related to precise identification of important stimuli, such as the association of a rustle with the movement of a predator or prey.

In the late 1960s and early 1970s, studies of the cat's brain showed that many of these structures required experience to be established. Experiments by Cambridge physiologist Colin Blakemore and others revealed that if you rear a cat in an environment composed solely of vertical stripes, it cannot detect horizontal stripes – the cells in the brain that would normally respond to such features do not fire.[7] This effect had real behavioural consequences – Blakemore reported that kittens reared in a horizontal-stripe-only environment ignored a stick that was waggled vertically. They literally could not see that part of the world: 'despite their active, and as time went on increasingly frenzied, visual exploration of the room, they often bumped into table legs as they scurried around'.

Furthermore, if these kittens continued their lives in an environment full of horizontal stripes, they would never be able to fully respond to vertical lines in later life. Their brains had not received the necessary stimulation during what is known as a critical period. The situation is somewhat similar in humans. Long-standing evidence from adults who have been cured of congenital blindness suggests they have to learn to see faces, and even to recognise simple figures such as a triangle.[8] In many cases, they never attain normal levels of recognition because they have missed the relevant critical period.

The discovery of the organisation of the brain's visual processing system strengthened the view that the brain carries out computations, but the developmental effects revealed that the brain is not fully hard-wired; within certain limits it makes its structures as a consequence of experience, and of the animal's exploration of its environment. This is a computer, Jim, but not as we know it.

*

Hubel and Wiesel's work suggested that visual processing in the brain is organised along some kind of hierarchical structure within which objects are identified with increasing levels of precision by increasingly localised structures as the hierarchy is ascended. This led to an argument about how detailed the information encoded in those higher single cells could possibly be, focusing attention once again on the contrasting claims of whether function was localised or was distributed across the brain. In 1969 Jerry Lettvin decided to highlight the problems with the ultra-locationist view by satirically claiming that his non-existent second cousin, a Russian neurosurgeon called Akakhi Akakhievitch, had treated a patient who suffered psychologically from an overbearing mother by removing the cells responsible for recognising her (there were allegedly 18,000 of these 'mother cells', responding to her in various orientations and guises). The punchline was that Akakhievitch, having successfully treated his patient, had now turned to his next challenge – grandmother cells.[9]

This story was in circulation when I was a student and – shorn of its joke scaffolding – the phrase 'grandmother cell' was used as a shorthand way of underlining the inherent silliness of suggesting that every object we recognise, whatever its orientation or context, is represented by the activity of a particular cell or group of cells. Taken to its absurd conclusion, there would have to be a cell for your grandmother sitting, your grandmother standing on her head, your grandmother playing the ukulele, and all possible combinations of the infinite variations in which you could recognise your grandmother. Add to your grandmother a list of everything else you could possibly see, and there would have to be an infinite number of cells in our brains to explain our perceptual abilities. This is clearly wrong.

But truth is often stranger than fiction. Two years before Lettvin's jape, the Polish neuropsychologist Jerzy Konorski took Hubel and Wiesel's discovery of precise feature detectors in the brain to their logical conclusion. In his 1967 book *Integrative Activity of the Brain*, Konorski argued that the brain contained what he called 'gnostic neurons' that identified very precise stimuli, such as cats, goats, or the same word written in different styles.[10] There were no grandmothers there, but that was just because Konorski had not included them in his list.

Soon afterwards, Charles Gross and his colleagues at Princeton described something very like a grandmother cell – neurons in the monkey's brain that responded preferentially not to the animal's grandmother, but to the shape of a monkey's hand. Like Hubel and Wiesel, they apparently made their discovery by accident. One of the researchers, frustrated that the cell they were recording from did not respond to any of the visual stimuli they were using, waved their hand in front of the stimulus screen. The cell fired strongly.[11] The finding was first reported in 1969, almost shamefacedly, at the end of an article in *Science* – 'one unit that responded to dark rectangles responded much more strongly to a cut-out of a monkey hand, and the more the stimulus looked like a hand, the more strongly the unit responded to it'.[12] Many scientists who accepted the 'grandmother cell' critique found this hard to swallow, but the results were clear.

At the end of the 1970s matters became even odder, when researchers in Oxford showed that some cells in the monkey's brain responded only to faces in all sorts of orientations. Scientists at Cambridge soon extended this finding to sheep, where cells responded either to pictures of other sheep of the same breed, to the size of horns or to pictures of potentially threatening stimuli, such as humans or dogs.[13] The Cambridge group noted drily that sheep 'do not respond to upside-down faces, which seems reasonable since, unlike monkeys, sheep do not usually view other sheep upside down'.[14]

Over the subsequent decades, the precision with which some cells respond to visual stimuli has been extended to a degree that is unsettling for anyone who, like my younger self, thinks that the grandmother cell is a killer argument against localisation of visual perception. In 2005 researchers led by Itzhak Fried at the UCLA School of Medicine and Christof Koch at Caltech described a study of eight patients who had electrodes inserted into their brains as part of the initial stage of surgery for intractable epilepsy. Visual images were presented to the patients and the responses of individual cells in the hippocampus were recorded. These cells sometimes responded with bizarre degrees of exactitude:

In one case, a unit responded only to three completely different images of the ex-president Bill Clinton. Another unit (from

a different patient) responded only to images of The Beatles, another one to cartoons from *The Simpsons* television series and another one to pictures of the basketball player Michael Jordan.[15]

Further investigation showed that one patient possessed 'a single unit in the left posterior hippocampus activated exclusively by different views of the actress Jennifer Aniston'. The cell did not respond if Aniston was pictured with her then partner, Brad Pitt. In another patient, a cell responded consistently to pictures of the actress Halle Berry, including when she was dressed up as Catwoman (a role she had recently played on screen), while intriguingly yet another cell responded preferentially to images of the Sydney Opera House and to the written words 'Sydney Opera'. To show that our brains are not just full of rubbish, a cell in one patient responded to the Pythagorean theorem $a^2 + b^2 = c^2$. This patient was an engineer who was interested in maths.[16]

This might seem to show that our brains contain precisely focused grandmother cells, firing when we see someone or something we recognise, but the authors were more cautious – although the cells responded consistently to Aniston or Berry or Clinton, that did not mean that these were the only stimuli that could potentially excite these cells – the patients had been shown only a very limited range of pictures. In subsequent papers, the group argued that the cells they had detected may represent a concept – hence the activation by both a picture of the Sydney Opera House and the words – and play a key role in memory.[17]

Most significantly, the researchers recognised that just because a single cell responded to an image, that did not mean it was the only cell involved in recognising the image, merely that it was the only cell they had recorded from that belonged to the relevant network. They estimated that perhaps a million neurons would be activated by each stimulus – many of those neurons are responding to aspects of the image or concept that would also be activated by another stimulus, but with a slightly different overall network.[18] This explains how the researchers got so lucky, finding the one cell that responds to Jennifer Aniston: there is not just one cell, there are millions. In reality, while we are amazed by the precision of the responses of individual cells that scientists happen to record from, as Rafael Yuste has pointed out, we should really be focusing on the complexity of the underlying circuits, and the shifting patterns of multicellular activity that occur when we see an image we recognise.[19]

The existence of higher-level dedicated circuits in vision was highlighted in 1992, when David Milner and Melvyn Goodale suggested that there are two separate visual processing streams in the mammalian brain, with different output functions.[20] After initial processing in the visual cortex at the back of the brain, visual information is split into two pathways, one of which goes to the top of the brain – this is called the 'where' pathway or dorsal stream, and is thought to encode information about the spatial location of the object that has been detected, projecting into regions involved in motor control. The other pathway, which goes deeper into the base of the cortex, is called the ventral stream, and is sometimes called the 'what' pathway. It is involved with identifying the objects that have been seen; this pathway projects to brain regions associated with memory and social behaviour. There are connections between the two streams – at some point, looking at a cat and wanting to stroke it, you need to put those two things together.

The distinctions between the two streams – where and what, dorsal and ventral, identity and action – underline the complexity of functional localisation in the brain.[21] The function that is localised is not just a physical aspect of the stimulus, but also some aspects of a stimulus that require the organism to be ready to respond in a particular way – reaching out for it, or remembering it.[22] This is a much less rigid way of thinking about functional localisation than imagining that all aspects of our grandmother are located in the same area.[23] But as the number of interconnections grows, and the involvement of different sensory modalities in similar neural tracts is discovered, the idea that function is fully localised is gradually becoming weaker. Our understanding of what exactly is localised is becoming more confused – or richer, if you prefer.

*

The ultimate kind of localisation, as suggested by both Hubel and Wiesel and – to a limited extent – by the existence of Jennifer Aniston cells, can be described in terms of the activity of single cells. In 1972 the Cambridge physiologist Horace Barlow outlined what he called 'five dogmas' about the relation between the activity of single neurons and sensation.[24] These 'dogmas' were in fact propositions or hypotheses that could focus thinking about how the nervous system worked, and lead to future experiments. Barlow explicitly took his approach, and the term 'dogma', from Francis Crick's hypotheses about the genetic basis of protein synthesis, given in a lecture in 1957, which had helped create the incredibly successful framework of molecular biology.[25] Barlow's paper was enormously influential – the cognitive scientist Margaret Boden has called it revolutionary.[26]

Barlow's starting point – his first dogma – was that a full description of the working of the nervous system required a description not only of a cell's activity but also of its role as a node in a network. The underlying principle of how such networks functioned, argued Barlow, was that 'at progressively higher levels in sensory pathways information about the physical stimulus is carried by progressively fewer active neurons'. To explain this, Barlow referred to an idea put forward in 1890 by William James, who argued that the brain must contain a 'pontifical cell' which, like the Pope, presided over all other brain cells.[27] Although there is no such anatomical structure, the term 'pontifical cell' stuck as a way of describing a highly hierarchical theory for the organisation of the brain. Barlow suggested that although there was no 'pontifical' cell, there might be what he jokingly called 'cardinal cells'. As in the Catholic Church, these would be lower in the hierarchy and present in large numbers, although only a few of them would be active at any particular time.[28]

Barlow emphasised that neurons respond to features in the environment in a way that has been selected during evolution, but which involves both genetic and environmental factors, and that the frequency with which a neuron fires can be taken as a measure of 'subjective certainty' – the higher the firing rate, the more likely that the cause of the neuron's activity is actually present in the real world. As to what was happening during all this processing, Barlow argued, objects were represented in neuronal activity as a symbolic abstraction. This idea, which Barlow took from Craik's work three decades earlier, suggests that certain elements of a stimulus are encoded in neuronal activity, enabling the brain to process only these key abstractions.

In a challenge to what it feels like to be conscious, Barlow insisted that 'There is nothing else "looking at" or controlling this activity'. To understand how nervous systems control behaviour there is no need to introduce some kind of homunculus sitting in our heads observing the output of neural circuits. As Barlow put it in his fourth dogma, 'the active high-level neurons directly and simply cause the elements of our perception'.[29] The activity of the neurons in a network determines behaviour and perception, including in humans. That is all there is in our heads, whether we are fly or human. In 2009 Barlow wondered whether this claim was too strong – not because he felt it was wrong, but because despite all the intervening research, there is still no proof of this – 'I now have difficulty even imagining anything scientific that would explain the personal and subjective aspects of perception', he said.[30] No matter how difficult it might be to imagine how it works, the facts are stubborn: there is no evidence for anything immaterial in our heads, or in the heads of any other animal.

Overall, Barlow's dogmas have fared well in the intervening years. In particular, his idea of cardinal cells has been recast in terms of what is called 'sparse coding', whereby the higher the level of representation, the fewer the number of cells involved and the sparser their activity, but the more significant it is in terms of both the overall activity of the system and the representation of the stimulus.

Barlow's dogmas reflected a new reductionist approach to the brain, in which scientists tried to understand apparently simple nervous systems in order to shed light on more complex forms. One of the earliest justifications for this had been provided by his own 1953 study of the fly detectors he discovered in the wiring of the frog's retina, and the snapping behaviour they would release from the frog if the cells were activated. Complex and evolutionarily significant behaviours can emerge from very simple neural networks, which may not involve the brain to any real extent. To explore this, a series of different approaches were adopted, all using the same reductionist logic. Around the same time as Eric Kandel pursued the engram by studying *Aplysia,* in the early 1960s some of the giants of the golden age of molecular biology turned to the study of the structure and function of the nervous system.

The most significant of these changes of direction came from the good friends Sydney Brenner and Seymour Benzer, who both created what are now major areas of neuroscience. Brenner focused on the tiny worm *Caenorhabditis elegans,* and ambitiously decided to understand the complete organisation and development of all its 900-odd cells, including 302 neurons.[31] Although the worm barely has anything resembling a brain, it can navigate along chemical gradients, detect pheromones and learn. Using electron microscopy and primitive computers, the work of Brenner's group and then of a global community of worm researchers eventually led to insights into how animals develop and saw the field – and Brenner – rewarded by the Nobel Prize in 2000.[32]

Seymour Benzer decided to study genetic factors in behaviour, creating behavioural mutants in the fruit fly, *Drosophila melanogaster.* Although *Drosophila* had been used to lay the basis of genetics at the beginning of the twentieth century, interest in this tiny insect had waned in the post-war years following the rise of molecular genetics, which focused on bacteria and viruses. Benzer's approach played a key role in reviving work on the fly – within a decade of launching his research programme, Benzer and his young research group had identified genes involved in circadian rhythms (this work eventually led to the Nobel Prize in 2017) and learning.

From the 1980s onwards, with the development of molecular techniques for studying and manipulating the genes of these species and others, there was a transformation in our ability to study the brain. New tools made it possible to visualise neurons and their organisation in ways that could not have been dreamed of in the past. New maps of brains and nervous systems were drawn up, most recently enabling the identification of previously unsuspected types of brain neurons on the basis of the genes that they express, rather than their shape. Whole new organisms became the focus of intense study, such as the tiny zebrafish, chosen as a model of vertebrate development. New ways of manipulating neurons became available, from 'knock-out' mice in which a particular gene was deleted, to a system in *Drosophila* for expressing any gene from any organism in virtually any tissue. The most recent developments have been the use of optogenetics to literally turn neurons on and off and the arrival of the CRISPR gene-editing technique which makes it possible, in principle, to manipulate any known gene in virtually any animal. But a fundamental problem remains. We do not understand the detail of how the brain is put together, except in the simplest of organisms such as *C. elegans*.

*

In 1993, seven years after the publication of the full wiring diagram of the worm *C. elegans*, Francis Crick and Edward Jones published an article in *Nature* which, from its title onwards, bemoaned the 'Backwardness of Human Neuroanatomy'.[33] They were struck by a study of the major pathways in the macaque cortex, published two years earlier by Daniel Felleman and David Van Essen.[34] This article included a dramatically complex and much-reproduced summary diagram showing 187 high-level links between thirty-two identified visual areas. In contrast to the detailed understanding of the macaque brain, Crick and Jones described the paucity of contemporary knowledge about the human brain as 'shameful':

> We can provisionally make the assumption that the connectional map for the visual areas of the human cortex will be similar to that for the macaque, but this assumption will have to be checked. For other cortical regions, such as the language areas, we cannot use the macaque brain even as a rough guide as it probably lacks comparable regions.
>
> It is intolerable that we do not have this information for the human brain. Without it there is little hope of understanding how our brains work except in the crudest way.

At the time, there was no word for what Crick and Jones were describing, but in 2005 two researchers separately came up with a term to denote 'a comprehensive structural description of the network of elements and connections forming the human brain'.[35] The word was connectome, one of a series of -ome and -omics words that scientists have coined in the wake of the arrival of the genome and genomics into everyday language.[36] Simply put, -omes are things that group together every example of a particular biological phenomenon, while -omics is the study of a particular -ome.

Partly as a result of Crick's initiative, a whole series of projects have been launched to provide scientists with a framework for describing the neuroanatomy of the brain in a range of organisms from flies to leeches to mice to humans. With larger animals, including humans, the word connectome is often used quite loosely, referring to a map of large-scale connectivity between brain areas, like that in the macaque which provoked Crick and Jones, rather than a true connectome, which would be based on individual cells and their synapses. There are four different levels of connections in such maps – macroconnections between brain regions, mesoconnections between neuron types, microconnections between individual neurons, and nanoconnections at synapses.[37] Each one tells us different things about what might be going on, but it is not always clear what kind of connectome scientists are referring to when they describe their research.

For example, in 2009 Thomas Insel, the head of the NIMH, claimed that the US Human Connectome Project would 'map the wiring diagram of the entire, living human brain'.[38] In reality the project does not study neuroanatomy, but instead uses brain scans (a very imprecise method) to look at nerves (so bundles of large numbers of neurons) that connect regions of the brain. This connectome is composed of macroconnections. The first findings of this project revealed that the brains of people with what the research team called more 'positive' variables, such as education, endurance and good memory performance, were more 'strongly connected' than those of people with more 'negative' variables such as aggression, smoking or alcohol problems.[39] Whether these differences – if they exist – are the causes or consequences of the supposed behavioural differences is impossible to determine from these data. Claims that this study revealed differences in the brains of men and women have been hotly contested.[40]

In terms of really explaining how the brain works, these broad measures provide little insight – the resolution of the imaging technique used in the Human Connectome Project is of the order of millions of neurons. As two connectomics researchers put it, somewhat acidly, 'none of its many goals relate to describing the synaptic connectivity of the brain'.[41]

We are still a very, very long way from having a complete connectome, based on micro- and nanoconnections, of a mammalian brain. Mouse retinal connectomes have been described, as have cell-level connectomes from tiny parts of the mouse brain, but these reinforce the mistaken impression that our brains are composed of anatomically distinct modules. In reality, there are neurons that link whole areas, and sometimes the whole brain – a recent imaging study of five neurons in the mouse showed that they threaded their way round the brain with such complexity that their total length was over 30 cm.[42] Analysing the function of even such a small number of neurons, which do not simply act as relays but interact with many different areas of the brain, will be both technically and intellectually challenging. For example, researchers have shown how long-range connectivity is effected by what are called projection neurons, in the shape of a connectome of 1,000 neurons in the mouse brain, which form a network over seventy-five metres long (there are millions of such cells in your brain).[43]

There are no plans to create a synapse-level full connectome of any mammal. The technical challenges are too vast. Even in the mouse brain, which has been intensively studied, we have measures of the number of cells (not their interconnections) for a mere 4 per cent of the estimated 737 brain regions and figures for these locations can differ wildly from study to study – up to thirteenfold. Recent attempts to use algorithms to provide brain-wide estimates of the number of cells in each region of the mouse brain give some insight, but nothing will replace cell-level knowledge of how even a single mouse's brain is organised.[44] And for the moment that is not even on the horizon. As to the human brain, with its 90 billion neurons, 100 trillion synapses and its billions of glia (these figures are all guesstimates), the idea of mapping it to the synapse level will not become a reality until the far distant future.

Nonetheless, the prospect of having connectomes of some kind, in some species, is enticing. In 2013 the leading US neuroscientist Cori Bargmann, who works on *C. elegans*, wrote an essay in which she concluded:

> Defining the connectome is like sequencing the genome: once the genome was available, it was impossible to imagine life without it. Yet both for the genome and for the connectome, structure does not solve function. What the structure provides is a better overview, a glimpse of the limits of the problem, a set of plausible hypotheses, and a framework to test those hypotheses with greater precision and power.[45]

*

As Bargmann explained, the widely held assumption – or dogma or simply hope – that underpins work on connectomes is that describing the wiring diagram of a particular organism or of one part of its brain will enable new insights into how behaviour and sensation emerge from the activity of neural circuits. This implicit hypothesis has been there from the outset – in Brenner's laboratory, the 302 reconstructions of each of the worm's neurons were placed in a series of notebooks that were jokingly labelled 'the mind of a worm'.[46] Except it was not just a joke.

Some researchers believe that connectomes will provide our understanding of the brain with the grand explanatory theory that it has been lacking. In 2016 Larry Swanson and Jeff Lichtman claimed that the creation of 'a biologically based dynamic or functional wiring diagram model of the nervous system' would in and of itself represent 'a powerful conceptual framework analogous to the periodic table of the elements for chemistry, the double-helix model of DNA for molecular biology, or Harvey's model of the circulatory system for physiology'.[47] It would be an extremely significant step, but work by researchers on simple nervous systems suggests that connectomics, while providing essential anatomical background to our understanding of the brain, will fail to explain what is going on unless accompanied by both experimental and modelling approaches.[48]

Even in a simple circuit, each neuron is connected to many other neurons both by chemical synapses and by what are called gap junctions, which directly connect two cells and allow electrical signals to pass (they are also called electrical synapses for this reason; they were first anatomically identified in the 1950s, and their function demonstrated in the 1960s).[49] Furthermore, a neuron can secrete several different types of neurotransmitter into the synapse. Simply looking at the surfaces of separation between two neurons will not tell you what happens in that synapse – whether it is excitatory or inhibitory, nor how many neurotransmitters are involved.

Because of these kind of factors, even very simple systems can involve astonishing degrees of complexity. For example, in the body wall of a maggot there are cells that respond as the maggot stretches when it moves, forming part of a circuit that controls movement. Each of these cells has eighteen input synapses and fifty-three output synapses; most if not all of these synapses can involve more than one neurotransmitter.[50] All that just to tell a maggot muscle movement circuit – not even its brain – that its skin has stretched. Researchers have recently described a single inhibitory neuron in a region called the visual thalamus of the mouse – it has 862 input synapses and 626 output synapses.[51] What exactly the cell does is not clear, beyond the fact that it is involved in many different functions. The complexity of the nervous system – any nervous system – is simply astonishing.

The same problems can be found even in the simplest parts of our brain. In 2018 Sophie Scott, a neuroscientist from University College London who is interested in how sounds are processed in the brain, in particular how sounds with a similar frequency are represented in neighbouring structures (this is called tonotopy), despairingly tweeted a picture of the high-level wiring diagram of the first stages of the human auditory system.

The starting point of studying connectomes is an updated version of Cajal's neuron doctrine – neurons are individual units that show unidirectional transmission – but scientists are increasingly recognising that things are more complicated. The nervous system contains star-shaped glial cells called astrocytes, or astroglia, which wrap around synapses; these cells help to keep neurons alive, but over the last two decades it has become clear that these cells can

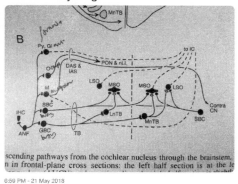

Prof Sophie Scott
@sophiescott

Spent today reading abt subcortical auditory
processing. Like this bastard. ONE NEURON
AWAY from the cochlea. All hell breaks loose
- 8 different cell types, 5 different parallel
processing streams, all preserving tonotopy.
And that's JUST THE START. How do we
ever hear anything!

6:59 PM - 21 May 2018

Tweet from Professor Sophie Scott of University College London.

alter the activity of neurons, releasing calcium or neurotransmitters
and altering the activity of the brain in a mouse.[52] Quite how significant astrocyte activity is in natural situations is still the subject of
intense argument, but recent results suggest that, in the zebrafish at
least, they have a computational role that modulates the activity of
neurons, accumulating sensory input that reveals that a given action
is ineffective and, by altering neuronal activity, rendering that action
less likely in the future.[53] Nervous system function is undoubtedly
more complex than Cajal imagined.[54]

Furthermore, we have known for over two decades that neuronal
activity can sometimes have effects upstream as well as downstream.
In 1997 researchers in California studied a simple network of isolated hippocampal cells and found that depression of the activity
at the output synapse of a cell propagated back up the neuron and
affected first the input synapses and eventually the whole network.[55]

Although action potentials are generally transmitted in only one direction down the axon, this is not always the case. Cells are not electronic components in a computer, and a wiring diagram will not reveal how they function in a network.

These discoveries have led some researchers to suggest that the neuron doctrine is not adequate for understanding the complexity of brains, and that unknown collective properties that emerge from the activities of groups of neurons – integrative emergents, in the jargon – may turn out to be significant. As a group of leading neuroscientists put it in 2005:

> the complexity of the human brain and likely other regions of the nervous system derive from some organizational features that make use of the permutations of scores of integrative variables and thousands or millions of connectivity variables and perhaps integrative emergents yet to be discovered. The answers extend well beyond explanation by the neuron acting as a single functional unit.[56]

Ten years later, Rafael Yuste of Columbia University similarly argued that the neuron doctrine is now being if not surpassed, at least supplemented. Many parts of the brain appear to be organised in networks, such as the sets of inhibitory neurons that, as Yuste put it, 'are often linked to each other by gap junctions, as though they are designed to work as a unit', and the ability of some inhibitory neurons to broadcast their neurotransmitters into tissues, rather than simply releasing them into a synapse, suggests they 'appear to be designed to extend a "blanket of inhibition" onto excitatory cells'.[57] Yuste went on to point to some concrete examples of function emerging from the activity of a network of neurons that could not be identified at the level of single cells, such as a 2012 report of the activity of cells in the brain of a mouse as it navigated a virtual maze.[58] The pattern of activation in the network could explain the mouse's behaviour, but the activity of individual cells could not.

Although Yuste argued for the need to develop a theory of how neural circuits operated, like his predecessors he was unclear as to what the next steps forward might be, beyond insisting that it would not be enough to simply record from lots of networks and expect an explanation to emerge. Instead, it might be necessary to take account of all levels of the system – from the molecular, through the activity of single cells, to the network behaviour and the behavioural output – in order to fully understand.[59] Although this is probably correct, there is little detail here. This is not Yuste's fault – this inability to go beyond general principles is typical of current thinking about how the brain works: we do not have an appropriate theoretical framework, nor the experimental evidence that could point to an answer. We cannot yet see how to take the next step.

One potential exception is the work of György Buzsáki, who has been attempting to apply Hebb's ideas about cell assemblies to modern data sets, in particular in terms of fluctuating interactions between networks of cells during brain activity.[60] This has led him to argue for what he has called an 'inside-out' view of brains, which he sees as systems for taking action, rather than for simply receiving and processing information. The activity of cell assemblies needs to be seen in terms of their output and its implications for the organism, rather than simply representing the outside world. However, while this view forms part of the welcome recognition that brains are not passive structures, it has yet to be widely accepted.

An increasingly popular framework uses complex mathematical explorations in order to reduce the multiple dimensions present in the rich data sets that are produced by brain studies. By describing data in terms of what are known in the jargon as low-dimensional attractor-like manifolds, researchers claim that they can identify different states in the activity of a given network and see how the system shifts from one state to another as the animal is stimulated.[61] This population-level approach to the activity of neural networks is both ambitious and welcome – many animal brains are composed of unimaginably large numbers of neurons and this provides a way of describing the activity of these networks.[62] However, these analyses take us far from the component neuronal activities, and the approach is generally descriptive. A number of studies have recently indicated that such methods may lead to more functional insights. The basis of learning is the plasticity of individual synapses – when considered at the level of a complex brain-driven behaviour, the function of this plasticity may be to enable new patterns of activity to be generated by the network of neurons. Although the activity of individual neurons may be quite variable over time, the activity of a synchronously active network can be very stable. Furthermore, it appears that there are clear parallels between such analyses and how the brain actually produces and controls movement.[63]

*

Part of the problem in analysing how brains work is the complexity that can be found in the behaviour of even the simplest neural circuits. This is the lesson of the work of Eve Marder of Brandeis University, who has devoted her glittering scientific career to studying the crustacean stomach.[64] This structure grinds up food using two rhythms that are produced by around thirty neurons (the exact number differs from species to species) organised in three circuits. Each circuit contains an example of what is known as a central pattern generator – a set of components that will spontaneously produce a repetitive output with no sensory input, and above all without that rhythm being specified either externally or within any individual unit.[65] The rhythm emerges from the activity of the network.

Despite having a clearly established connectome of the thirty-odd neurons involved in what is called the crustacean stomatogastric ganglion, Marder's group cannot yet fully explain how even some small portions of this system function. The problems associated with understanding such apparently simple central pattern generators have long been realised: in 1980 the neuroscientist Allen Selverston published a much-discussed think piece entitled 'Are Central Pattern Generators Understandable?', in which he argued that a key problem was identifying the nature and function of the components in such circuits.[66] Despite the increase in computing power for modelling, and the greater level of precision in our ability to identify and record from neurons, the situation has merely become more complex in the last four decades.

Marder's work has revealed that neuronal activity can be altered by neuromodulators – neuropeptides and other compounds that are secreted alongside neurotransmitters and which function as relatively slow-acting mini-hormones, locally altering the activity of neighbouring neurons.[67] Furthermore, the activity of each neuron is affected not only by its identity (that is, by the genes that determine its position and function), but also by the previous activity of the neuron.[68] In *C. elegans* worms, neuromodulation can also explain long-term individual differences in behaviour between animals with identical wiring diagrams – personality, if you like.[69] The same neuron in different animals can also show very different patterns of activity – the characteristics of each neuron can be highly plastic, as the cell changes its composition and function over time. As Marder has put it, a neuron is like an aeroplane that is flying at altitude while simultaneously replacing all its pre-manufactured components with elements it has created onboard.[70] There are not many computers that can do that.

The tight link between circuit structure and a particular output, which was long assumed by many researchers in this field, turns out not to exist. Using computer simulations of real electrophysiological data, Marder's group showed that there were many different sets of activity in individual neurons that could produce the same overall pattern when they were connected together.[71] You cannot simply presume that the same behaviour involves the same structure or pattern of neuronal activity. Furthermore, the function of a given circuit can switch from one mode to another, as the multiple connections between the same pair of neurons are altered by the activity of the cells in the circuit – the same circuit can produce radically different behaviours, while the same behaviour can be produced by very different circuits.[72] Decades of work on the connectome of the few dozen neurons that form the central pattern generator in the lobster stomatogastric system, using electrophysiology, cell biology and extensive computer modelling, have still not fully revealed how its limited functions emerge.[73] That brutal, frustrating fact is the benchmark for all claims about understanding the brain.

Even the function of circuits like Barlow's bug-detecting retinal cells – a simple, well-understood set of neurons with an apparently intuitive function – is not fully understood at a computational level. There are two competing models that explain what the cells are doing and how they are interconnected (one is based on a weevil, the other on a rabbit); their supporters have been thrashing it out for over half a century, and the issue is still unresolved.[74] In 2017 the connectome of a neural substrate for detecting motion in *Drosophila* was reported, including information about which synapses were excitatory and which were inhibitory.[75] Even this did not resolve the issue of which of those two models is correct.

A connectome is not enough to explain how the system works. The description of the 302 neurons in the *C. elegans* nervous system has led to the identification of neurons involved in a variety of behaviours including foraging, feeding and egg-laying. And yet because the worm wiring diagram was simply that – an anatomical description – it was not immediately possible to describe how those cells interact. Understanding the chemical and electrical connections between the cells is necessary for the many alternative functional outputs of the neuronal circuits to be hypothesised and tested.

In the future, such studies may be able to produce functional maps of limited areas of the mammalian brain. A recent reconstruction of a tiny part of the mouse brain by researchers at the Max Planck Institute in Germany used both artificial intelligence and one hundred student annotators to identify inhibitory and excitatory neuron subtypes from the connectomic data. The complexity they discovered was mind-boggling. The little block of brain they were studying was slightly less than one tenth of a millimetre on each side. There were just 89 neurons that had their cell bodies in this space, making up less than 3% of the total 'wiring' (about 70 mm long) they observed. But crammed in alongside these cells there were 2.7 m of neuronal 'wires' from other cells that had their cell bodies outside the studied volume – in total this tiny part of the mouse brain had 6979 pre- synaptic and 3719 postsynaptic sites, each with at least 10 synapses, making a total of 153,171 synapses. Remember that there are perhaps 70 million neurons in the whole of the mouse brain.[76]

The challenge of understanding how even simple nervous systems work is massive. Marder's group has shown that the pattern generators in different crabs of the same species, with exactly the same wiring diagram, can respond differently to changes in acidity, while individual *C. elegans* worms, with the same connectome and at the same developmental state, produce different changes in the activity of their electrical synapses in response to starvation, leading to behavioural plasticity and different responses.[77] Worms, despite their identical, seemingly robot-like structure, do not all behave in exactly the same way, unlike a set of machines with exactly the same wiring diagram.

In 2015 a multinational group led by Manuel Zimmer in Vienna directly measured the activity of around 130 sensory and motor cells in the worm's head region.[78] The researchers did not reveal the worm's mind, but they did show that waves of activity sweep round the nervous system, activating different groups of neurons – the circuits that are involved in determining speed of movement, for example, even though the animal was immobilised. As the journal expressed it, 'internal representation of behaviour persists when decoupled from its execution'. In other words, the worm was thinking of moving. Intriguingly, despite the findings from more complex animals, the group did not find any single-cell representations of sensory stimuli (touch, smell) beyond the immediate receptor cell. For the moment, there are no grandmother cells in the worm.

*

At the time of writing, the only full synapse-level brain connectome (with the exception of *C. elegans*) is that of the larva of a sea-squirt, an animal that is a chordate, so despite appearances is more closely related to you and me than it is to an invertebrate.[79] The tiny brain has only 177 neurons and 6,618 synapses, and yet, even in such a small structure, it shows left–right asymmetries although the numbers of cells on either side of the brain are identical. The next step up in connectome complexity will probably be the completion of the cell-level study of the brain in an animal I have spent much of my career studying, the *Drosophila* larva. For years, a team of researchers from twenty-nine laboratories around the world, led by Albert Cardona of the Janelia Research Campus and the University of Cambridge, have been slowly describing the synapse-level wiring diagram of the brain of a maggot.

The 'a' at the end of that last sentence is deliberate – analysing electron microscope images of slices is a painstaking process, even with the aid of modern computers, and for the moment the group have information on just a single maggot. They already know that even maggots show variation between individuals, so this connectome, just like a genome, will not truly represent every member of the species. And as with genomes, inter-individual variation is not an annoyance, it is a fascinating source and explanation of differences in behaviour and may also reveal the evolutionary history of the species. We need to understand how and why the different connections develop in different individuals, and what the consequences are in terms of brain function.

One of the techniques that may provide some insight into this issue is the advent of single-cell transcriptomics (another -ome word, referring to the activity of the genes in a cell). Researchers recently described the identity of 30,000 cells from the mouse brain by identifying a partial transcriptome of those cells based on the activity of 1,000 genes.[80] This technical tour de force was nonetheless merely a proof of principle – the mouse genome contains over 20,000 protein-encoding genes (slightly more than a human*), while the mouse brain contains around 70 million neurons, so the study looked at 4 per cent of mouse genes and perhaps 0.04 per cent of the neurons in a mouse brain. Nevertheless, this research shows that new ways of classifying all the neurons in a brain, not simply on the basis of their anatomy and their location but also the genes that are expressed in them, will eventually be available.

Already, one group of researchers has been able to identify the gene activity profile of individual cells in a region of the mouse brain known as the preoptic hypothalamus and relate these differences to behaviours, while another has shown that two areas of the mouse cortex house 133 different types of cell, identified by the genes that are active within them rather than their morphology.[81] In some way, these 133 types reflect different functions, because they indicate that the cells of a particular type were responding to their environment in a similar way by activating and deactivating their genes. This revealed, for example, that some neurons using the glutamate neurotransmitter had a particular profile of gene activity that was linked

---

* Yes, this is true.

to their long-range connections within the brain, while transcriptome profiling enabled the researchers to identify two new types of neuron involved in the control of movement.[82]

It is not clear if this kind of minutely detailed study will generate insights, or simply more information. There are many scientists who feel we are drowning in a tide of data about the structure of brains, while what we really need are some clearer theories and ideas about how it all fits together. As the pioneer neuroscientist Vernon Mountcastle put it in 1998: 'in and of itself knowledge of structure provides no direct understanding of dynamic function. *Where is not how.*'[83]

While this is true, modern techniques mean that sometimes our maps are not simply diagrams – they can also be tools for investigating function. So, for example, in the *Drosophila* maggot brain, 'where' and 'how' are being explored at the same time. At the time of writing, the connectome of the 10,000 neurons that make up the maggot brain is 70 per cent complete. The current draft describes an astonishing two metres of neurons and 1.36 million synapses; the final draft will probably contain information on around 2 million synapses – all that is packed into a structure the size of the dot on this i. Because of the way the cells have been identified genetically, this preliminary map of 'where' has been used by Marta Zlatic's group at Janelia Research Campus to study 'how' – the neural basis of key behaviours controlled by the maggot brain (for example, rolling away from a sharp prod, or eating) – providing elegant descriptions of the function of each component in the system.[84] Insight can also be gained from studying exactly how the activity of each cell alters its activity – single-cell transcriptomic data are being collected that show how the networks of the maggot's brain change as it is stimulated in different ways.

Despite this astonishing progress, and the promise of what is to come, my view is that it will probably take fifty years before we understand the maggot brain – that is, we can fully model its working and accurately predict how a change in the activity of one neuron will affect the whole system, in a wide variety of conditions. That also tells us quite how far away we might be from understanding the human brain. Not everyone is so pessimistic. In 2008 Jerry Rubin, who created the Janelia Research Campus where much of the work on the *Drosophila* connectome is being coordinated, said that understanding the adult fly brain – substantially larger and more complex than the maggot's – would take around twenty years. And then what? 'After we solve this, I'd say we're one-fifth of the way to understanding the human mind,' he said.[85]

The first steps are being taken to produce atlas toolkits that enable researchers to both identify and manipulate cells in vertebrate brains, both in the mouse and in the zebrafish.[86] These techniques have not yet produced a full-brain connectome, but they do point the road to the future, when brain maps will include ways of manipulating individual cells as well as tracing their connections.

Nevertheless, there are plenty of reasons to imagine that the insights provided by a single connectome may be limited – to fully gain the benefit of such research, we need more than one connectome. Studies of *Drosophila* reveal that random developmental effects produce tiny differences in the wiring of each fly's visual system; these differences predict how individual flies will respond to objects. It is not only inter-individual differences that will be significant; the evolutionary and comparative approach will be essential to show what is general and what is particular about any given map. In 2016 the French neuroscientist Gilles Laurent argued for a comparative, inter-specific approach to be introduced into connectomics, in order to reveal common mechanisms and algorithms in widely differing animals.[87] True to his word, Laurent's group, which had previously focused on insects, has published a comparative cell-level transcriptomic study of the cortex in turtles, lizards, mice and humans.[88] Meanwhile, other researchers have been using high-level connectomes to compare humans and macaques, much as Crick and Jones recommended a quarter of a century ago.[89]

In justifying his call for a broader approach to understanding the brain, Laurent quoted the view of the twentieth-century Russian condensed-matter physicist Yakov Frenkel:

> A good theoretical model of a complex system should be like a good caricature: it should emphasize those features which are most important and should downplay the inessential details. Now the only snag with this advice is that one does not really know which are the inessential details until one has

understood the phenomena under study. Consequently, one should investigate a wide range of models and not stake one's life (or one's theoretical insight) on one particular model only.[90]

Not all brains belong to humans, mice, flies or worms.

*

The researchers who work on the various connectome projects are well aware of these problems. In 2013 Joshua Morgan and Jeff Lichtman explored the 'top ten arguments against connectomics', many of which have been outlined above.[91] In most cases, their response to each of these arguments was fundamentally the same and quite valid. They argued that even if a theory of brain function will not simply pop out of a wiring diagram, detailed neuroanatomy, accompanied by and providing a framework for sophisticated electrophysiological measures, will obviously advance our understanding of brain function. As they put it, echoing Crick and Jones two decades earlier, 'neuroscientists cannot claim to understand brains as long as the network level of brain organisation is uncharted'.

The most surprising aspect of Morgan and Lichtman's article was the argument against connectomes that they put at number one. They highlighted a quote from a radio interview with Francis Collins, the head of the National Institutes of Health (NIH). Collins complained about the static nature of connectome representations: 'It'd be like, you know, taking your laptop and prying the top off and staring at the parts inside, you'd be able to say, yeah, this is connected to that, but you wouldn't know how it worked.'[92]

Neither Collins, nor Morgan and Lichtman, recognised it, but this could be seen as Leibniz's Mill argument updated for the computer age. The problem with this criticism, as for Leibniz's original version, is that although simply observing components and their relations will not explain how a system works, describing the nature of the relations between those components, and how they affect each other, does indeed provide the basis for explaining how the system functions. Whether it will explain consciousness, the original target of Leibniz's argument, is another matter.

This highlights the fact that in order to explain how the brain works, we need something more than a map, be it ever so functional. We need a theoretical explanation for how at least some parts of the system work, in order to interpret how the components interact. This was an approach favoured in the 1970s by the British mathematician and theoretical neuroscientist David Marr. After initially being 'fully caught up' in the 'vigour and excitement' of Barlow's five-dogmas paper from 1972, Marr realised that 'somewhere underneath, something was going wrong'. What was missing from Barlow's approach, Marr felt, was an understanding of the overall significance of what cells were actually doing at a circuit level. As he wrote:

> Suppose, for example, that one actually found the apocryphal grandmother cell. Would that really tell us anything much at all? It would tell us that it existed – Gross's hand-detectors tell us almost that – but not *why* or even *how* such a thing may be constructed from the outputs of previously discovered cells …
> The key observation is that neurophysiology and psychophysics have as their business to describe the behaviour of cells or of subjects but not to explain such behaviour.[93]

This 'so what?' faced with the potential discovery of a grandmother cell was not simply another version of Leibniz's Mill; it was somewhat more sophisticated. Marr was shifting the question away from simply describing the components of brain activity and towards trying to fit them into an overall model. To do that, he argued, we needed to copy key abilities of the brain. 'The best way of finding out the difficulties of doing something is to try to do it,' he wrote. And that is what he attempted, by investigating what would be required to build a machine that could see. In taking this path, Marr was following not only the researchers who had attempted to physically model the brain in the 1930s, but also the early computer pioneers who had become interested in the brain in the 1950s. These researchers had spawned a whole new field that changed our approach to the brain and is now altering the whole of society.

TWELVE

COMPUTERS

1950 TO TODAY

At the very beginning of the computer age, scientists were struck by the parallels between these new machines and brains and were inspired to use them in different ways. Some ignored biology and focused simply on making computers as smart as possible, a field that became known as artificial intelligence (AI – the term was coined by John McCarthy in 1956). But in terms of understanding how the brain works, the most fruitful approach came from those who did not attempt to create a super-intelligent machine, but instead tried to model the functions of the brain by exploring the rules governing the interconnections in the model – a neuronal algebra, if you like.[1]

An early attempt to simulate the nervous system came in 1956, when researchers at IBM tested Hebb's hypothesis about neuronal assemblies being a basic functional unit of the brain. They used IBM's first commercial computer, the 701, a valve-based machine composed of eleven large units that literally filled a room (only nineteen were sold). The team simulated a net of 512 neurons; although these components were not initially connected, they soon formed assemblies that spontaneously synchronised their activity in waves, just as Hebb suggested.[2] Despite the limits of what was a very crude model, this suggested that some aspects of nervous system circuits simply emerge from very basic rules.

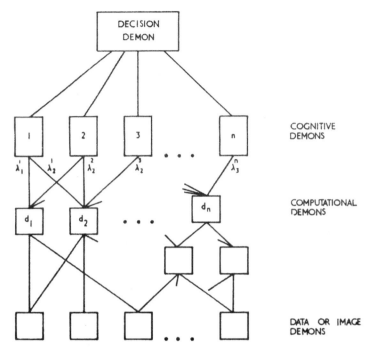

Schema outlining Pandemonium.

One of the first people to use a computer model to shed light on how the brain might function was Oliver Selfridge, a mathematician who was one of Wiener's students and was close to Pitts, McCulloch and Lettvin. In 1958 Selfridge unveiled a hierarchical processing system called Pandemonium, which had developed out of his work on machine-based pattern recognition. The starting point was the creation of simple units ('data demons') that would recognise elements in the environment by comparing a feature, such as a line, to some predetermined internal template. These data demons would then tell the next layer up ('computational demons'), what they had detected. Selfridge explained what happened then:

At the next level the computational demons or sub-demons

perform certain more or less complicated computations on the data and pass the results of these up to the next level, the cognitive demons who weigh the evidence, as it were. Each cognitive demon computes a shriek, and from all the shrieks the highest level demon of all, the decision demon, merely selects the loudest.[3]

The end result would be that a complex feature – say a letter – would be recognised by the decision demon.

At first sight, this appears to be merely an electronic version of previous hierarchical views of sensory processing, going back to Smee. But Pandemonium was different – it could learn as it went along. The program continually noticed how accurate it was in classifying objects (in the initial stages, this information was provided by human observers); by using what Selfridge called 'natural selection' of the demons – they were retained if their classification was correct – and by running the program repeatedly, the system would become increasingly accurate over time. It could even recognise things for which it had not been designed.[4] According to the cognitive scientist Margaret Boden, the influence of Pandemonium was incalculable – it showed that computer programs could model quite sophisticated sensory processes, and that, if it was provided with appropriate feedback about its success, the program's functions could change over time.[5]

At the same time, another US scientist, Frank Rosenblatt, presented a slightly different model, the Perceptron. This was also focused on pattern recognition, using the same idea of flexible hierarchical connections – an approach that became known as connectionism.[6] Rosenblatt argued that a brain and a computer shared two functions – decision making and control – both of which were based on logical rules in both machine and brain. Brains, however, carry out two further, intertwined functions: interpretation and prediction of the environment. All these functions were modelled in the Perceptron – 'the first machine which is capable of having an original idea', claimed Rosenblatt.[7]

In reality, the Perceptron, like Pandemonium before it, simply learned to recognise letters – in the case of the Perceptron these were

about half a metre high.[8] The crucial difference was that the Perceptron could do this without a pre-existing template, by using parallel
processing – simultaneously carrying out different calculations
– much like a brain. This was no accident: Rosenblatt was just as
interested in coming up with a theoretical explanation of brain function as he was in developing what, at the time, was a jaw-dropping
piece of technology.

The media loved it. When Rosenblatt's funder, the US Navy,
announced his work in 1958, the *New York Times* crowed: 'The Navy
revealed the embryo of an electronic computer today that it expects
will be able to walk, talk, see, write, reproduce itself and be conscious
of its existence.'[9] These claims did not come from an overexcited
journalist, they were quotes from Rosenblatt himself. One scientist
recalled of Rosenblatt: 'He was a press agent's dream, a real medicine man. To hear him tell it, the Perceptron was capable of fantastic
things. And maybe it was. But you couldn't prove it by the work
Frank did.'[10]

Despite his calculated hype to the press, Rosenblatt was relatively level-headed about the true significance of the Perceptron. As
he wrote in a 1961 report, *Principles of Neurodynamics*:

> Perceptrons are not intended to serve as detailed copies of any
> actual nervous system. They are simplified networks, designed
> to permit the study of lawful relationships between the organ-
> isation of a nerve net, the organisation of its environment,
> and the 'psychological' performances of which the network
> is capable. Perceptrons might actually correspond to parts of
> more extended networks in biological systems ... More likely,
> they represent extreme simplifications of the central nervous
> system, in which some properties are exaggerated, others
> suppressed.[11]

By the mid-1960s experts accepted that even the Perceptron was
not all it was cracked up to be.[12] In 1969 the pioneer of artificial intelligence, Marvin Minsky, published a highly negative book on the
Perceptron model, together with his colleague Seymour Papert. They
presented a mathematical analysis of the power of the Perceptron

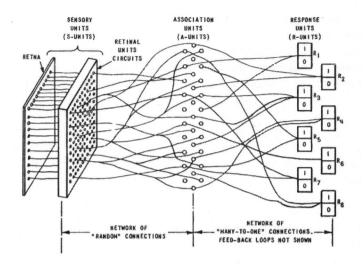

A schema describing the Perceptron

which suggested that the approach was a dead end, both for AI and for understanding the brain – because of the way Perceptrons were constructed, they could not internally represent the things they were learning.[13] Following this critique, and a slowdown in progress shown by these models, US funding for connectionist approaches dried up and the field withered away.[14] Rosenblatt – who began studying the transfer-of-learning phenomenon that would culminate in the scotophobin debacle – died in a boating accident in 1971, on his forty-third birthday.

Despite the failures of both Pandemonium and the Perceptron to produce insights that could be applied to biological pattern recognition systems, both these programs changed the way researchers thought about the brain – they showed that any effective description of perception, in humans or machines, had to include a substantial element of plasticity. They were therefore utterly different from the old models that were based on mechanical or pressure metaphors. Furthermore, there was a seductive parallel between the structure

of these connectionist programs and Hubel and Wiesel's discovery
of the hierarchical organisation of simple feature detectors, which
clearly influenced Barlow's 1972 idea about 'cardinal cells'. For some,
this implied that these new models were not simply using metaphors
to explain how the brain worked. They were actually revealing real
mechanisms.

*

As the excitement over Pandemonium and the Perceptron died
down, a different approach to computational models of brain
function was developed by David Marr. Marr had already made
a name for himself at Cambridge, publishing a series of articles in
which he proclaimed he had discovered how the brain worked. He
soon disowned these mathematical models, dismissing them as 'a
simple combinatorial trick', as he realised that a completely different approach was needed.[15] In 1973 Marr moved to MIT in Boston,
to work with Minsky. His aim was to create a machine that could
see and thereby to understand how human vision works. Four years
later, he developed leukaemia, and hurriedly began work on a book
entitled *Vision* that would summarise his insights. As he put it in the
Preface, 'certain events occurred that forced me to write this book a
few years earlier than I had planned'.[16] Marr died in 1980, just thirty-
five years old; *Vision* was published in 1982.[17]

Perhaps the recognition of his mortality gave Marr's prose
a grander perspective than the details of a model of vision might
otherwise have warranted. He put his view of how the brain worked
into a far broader, ethical context, telling us something about how we
have evolved and the origin of our most deeply-held views in the
effects of natural selection:

> To say the brain is a computer is correct but misleading. It's
> really a highly specialised information-processing device
> – or rather, a whole lot of them. Viewing our brains as
> information-processing devices is not demeaning and does
> not negate human values. If anything, it tends to support
> them and may in the end help us to understand what from

an information-processing view human values actually are, why they have selective value and how they are knitted into the capacity for social mores and organisation with which our genes have endowed us.[18]

It has been said that the density of Marr's mathematics means that his work has more often been cited than understood.* This quip reveals that Marr's significance lies not in the precise detail of his computational models of vision – even his most ardent supporters accept that much of his book is now largely of historical interest – but rather in his overall approach.[19]

Unlike Barlow, Marr did not think that the activity of single neurons could explain how circuits function and how perception works. As Marr put it in a somewhat barbed justification of his new method:

Trying to understand perception by studying only neurons is like trying to understand bird flight by studying only feathers: it just cannot be done. In order to study bird flight, we have to understand aerodynamics: only then do the structure of the feathers and the different shapes of bird wings makes sense.[20]†

Marr's approach to understanding how a particular function is executed in the brain (or a computer) involved dividing the problem into three parts. First, the problem to be solved has to be stated logically; this theoretical approach frames how the problem is explored experimentally or is modelled. Second, the way the input and the output of the system are represented has to be determined, along with a description of an algorithm that could get the system from

---

*Guilty.

† Marr's point about the bird's wing, which has become famous among neuroscientists, was in fact a sideswipe at Barlow. In 1961 Barlow argued in favour of a more neuron-centred view of function by pointing out that without a full understanding of a bird's musculature and the strength and lightness of its feathers, it would be impossible to know that it could fly. Barlow, H. (1961), in W. Rosenblith (ed.), *Sensory Communication* (Cambridge, MA: MIT Press), pp. 217–34.

one state to another. Finally, it has to be explained how the second level could be implemented physically – in the case of brain activity, in the nervous system. Marr's assumption was that the constraints on producing a network that can see, be it in a machine or in a brain, would be basically similar in all cases and would therefore lead to similar algorithms being used, even if there would be major differences in the way those algorithms were implemented in flesh or in silicon. By resolving the problem of vision in a machine, he argued, we would get a better grip on vision in our heads.

Marr used the findings of Hubel and Wiesel as the basis for his ideas about how something as simple as an edge could be identified, but his approach involved a far richer computational scheme than a mere hierarchy that stuck bits of line together to fit a template, as in Pandemonium or the Perceptron. As Marr put it at a meeting in Cold Spring Harbor in 1976, 'This contour is not being detected, it is being constructed.'[21] This view – which can be traced back to Helmholtz – emphasises that the brain is not simply a passive observer receiving sensory information. Perception involves assembling and interpreting those stimuli. This approach is essential for any model of vision, for nothing will happen if the machine (or the retina) merely identifies degrees of light and shade at each point of an image. That is what a camera does, and cameras cannot see.

Despite these insights, Marr's machine approach did not transform our understanding of machine vision, nor of how the brain sees. To the extent that we understand what is happening in our visual cortex, the same algorithms have not been found in brains and in computers.[22] Just as problematically, the specific approach Marr used to understand seeing could not be generalised to other aspects of brain function.

Although there have been massive advances in computer facial recognition and other approaches to artificial scene analysis, machine vision remains far behind what is occurring in your head right now. Similarly, our understanding of what happens when we see is still very poor. Everyone accepts that there must be some kind of symbolic representation of the scene in our brain, but no one is very sure how that takes place. On the thirtieth anniversary of the publication of *Vision*, Kent Stevens, who was one of Marr's students, surveyed

his supervisor's contribution and concluded that while the significance of symbolic representation in vision was undoubted, 'we don't
fully understand the place of symbol systems in biological vision'.[23]

Insights into this problem may have been provided by studies of
the face-detecting cells in the monkey's brain. In 2017 two researchers at Caltech, Le Chang and Doris Tsao, presented macaques with
a range of faces and studied the responses of individual cells in the
monkeys' brains.[24] The cells detected fifty dimensions of faceness
(each dimension was composed of multiple physical features), with
each face cell interested in only one of those dimensions. To show how
this information might be combined to produce an accurate overall
representation, Chang and Tsao recorded the responses of 200 of these
cells to a series of photos; they then used a computer to accurately
reconstruct the original images, on the basis of the electrical activity
of those neurons. Interestingly, they found no evidence for a monkey
Jennifer Aniston cell, or, as they put it, 'there are no detectors for identities of specific individuals'. Instead, as shown by another study from
a different group, there seems to be an area in the temporal lobe that is
involved in recognising the faces of familiar monkeys.[25]

Tsao, whose brief Twitter bio reads 'cortical geometer', suspects
that the kind of feature-extraction she has been able to reveal with
regard to face detection may be a general process occurring across
the visual cortex – 'We think the entire inferior temporal cortex may
be using the same organisation into networks of connected patches,
and the same code for all types of object recognition.'[26] Her current
challenge is to try and understand the neuronal basis of visual illusions, such as the well-known vase/face illusion; as she points out,
a decade ago no one would have known where to start. Now we do.

As to how humans recognise faces, including those of our grandmothers, it seems probable that, like the macaque, there is some kind
of distributed face recognition network in the brain.[27] This is very
different from the face recognition algorithms that allow your phone
to identify you, or enable the security services to screen photos for
suspects, which is entirely tailored to comparing varying biomet-
ric landmarks in a fixed presentation of a face, such as the distance
between the eyes, shape of the face and so on. Face recognition in biological vision is much more complex and abstract and is ultimately

based on the kinds of elements that were identified by Hubel and Wiesel – lines, blobs and so on – rather than bits of facial anatomy and their relationships. These are somehow organised into a complex hierarchical system of the general kind imagined by Marr, and which equally applies to other features in the environment, not just faces.

Further insight into exactly what such hierarchical cells might be interested in was recently revealed in an unsettling tour de force of computing and electrophysiology carried out on monkeys by Margaret Livingstone's group at Harvard, when the researchers projected pictures onto a screen and recorded the activity of single cells in the inferotemporal cortex of awake animals.[28] So far, so banal. But the images were not static: they were synthetic, ever-changing and fluid, 'evolved' by an algorithm called XDREAM that continually changed what was presented to the monkey, so as to get the maximum response from the cell. The results of this approach, which was pioneered a decade earlier by Charles Connor and his colleagues, are eerie. Over more than a hundred iterations, the evolved images went from a greyish blank slate to dream-like, surreally mangled versions of parts of a monkey's face, with something recognisable as eyes here, formless blurred bits of body there, with the different components at different orientations.

This kind of strange image was what the cells were really interested in, not a straightforward portrait. If something similar was happening in the brains of those people with the Jennifer Aniston cells, that would imply the cells were not tuned to anything like a photographic representation – the photo was simply close enough to produce a response. Similar, though less uncanny, results were reported at the same time by researchers from MIT, who used the same kind of procedure on cells in a part of the monkey visual cortex that is not involved in detecting faces.[29] These cells seemed to be particularly excited by weird, semi-organic geometrical images that resemble something out of a bad migraine.

Although it is tempting to imagine that these odd conglomerations are actually what the monkey is seeing when it looks at another individual, remember that there are millions of cells contributing to the perception of a face, and above all, as Barlow said, there is no mini-monkey peering at the output of these individual

*Synthetic images produced by the XDREAM algorithm, each
of which optimally stimulated a different cell in the monkey
visual cortex. They are supposed to look like this.*

cells. Somehow the whole system produces perception, not one cell
or even a small group of cells.

Work on mice has recently provided a powerful route to understanding the neuronal basis of visual perception. Within weeks of each
other in the summer of 2019, Rafael Yuste's group at Columbia and
Karl Deisseroth's group at Stanford both showed that it was possible
to recreate patterns of activity in the mouse brain that occurred during
visual perception, using a complex kind of optogenetics.[30] When these
patterns were activated, the mouse would show an appropriate behaviour, even though there was no visual stimulation present. In both cases,
the mice had been trained to lick when they saw a pattern of stripes.
The two groups used slightly different techniques – Deisseroth's team
precisely reproduced activity in a dozen or so neurons, while Yuste's
group focused on as few as two well-connected neurons that were able
to activate an ensemble of neurons in the visual system of the brain.
These impressive studies do not yet enable us to say whether those patterns of activity actually were visual perception in the mouse, or if they
represent some kind of necessary precondition for that perception to
occur, through the activity of other sets of neurons. Despite decades
of effort by both computational scientists and neurobiologists, we still
only dimly understand what is going on when we see.

*

In the mid-1980s, neuroscientists and psychologists became very interested in new computational approaches that made it possible to overcome the limitations of Pandemonium and the Perceptron. This new method, called parallel distributed processing (PDP), was announced in a two-volume book that described innovative computer models of behaviour and their potential psychological and neurobiological equivalents.[31] Amazingly for an academic book, it sold over 50,000 copies, and became extremely influential.[32] The development of this approach was the work of a number of researchers, including David Rumelhart, James McClelland and Geoffrey Hinton (now a leading researcher at Google), and also involved Francis Crick. It led directly to neural networks and deep learning, which have transformed computational neurobiology and artificial intelligence and regularly produce headline-grabbing results.

PDP networks all share the same basic three-layer structure they inherited from the Perceptron: two of these are an input layer that responds when some feature triggers a given unit and an output layer that informs the outside world when the previous layers have done their work. The magic lies in the intermediate layer (generally called the hidden layer), which uses various systems of interconnection and algorithms that generally follow Hebb's law: connections that are simultaneously activated are subsequently favoured.

The ability of these programs to mimic aspects of behaviour caused what Francis Crick described as a heady sense of euphoria in the scientific community.[33] Crick had been involved in the PDP Group that produced the breakthrough book, although he later described his role as 'that of a fringee, or perhaps a gadfly'.[34] That closeness to the work did not stop him from initially sharing the euphoria. He was particularly impressed by a program called NETtalk, written by Terry Sejnowski and Charlie Rosenberg, which learned to correctly pronounce written English – Crick thought the result was 'remarkable'. Although the program was able to correctly pronounce a novel text with up to 80 per cent accuracy, it was not explicitly learning the rules of English pronunciation (to the extent that they exist).[35]

The ability of PDP networks to perform tasks so effectively is largely based on the use of what is called back propagation (generally abbreviated to backprop), which involves information going both ways between layers in a form of feedback loop. This enables the program to refine its behaviour, leading rapidly to a more accurate output. Military and academic funders soon became excited by the possibilities, and research in the intervening decades, coupled with the growth in computing power, has led to the current massive interest in the subject from private sector companies such as Google.

From the outset, when these programs were set running, they took on a life of their own, producing unexpected results. This feature, which flows from the way the algorithms in the hidden layer are set up, can obviously cause disappointment if the software falls over and crashes or if it simply fails dismally (we do not hear much about these examples, which must be legion). It can also produce pleasant surprises. One of the earliest PDP programs was created by Rumelhart and McClelland to model learning of the past tenses of English verbs. It not only successfully fulfilled the task but also mistakenly took rules it had developed for regular verbs and generalised them to irregular verbs, at exactly the same point in the learning process as children do. For example, the program eventually said 'go/goed', despite having initially learned the correct form for this irregular verb ('go/went').[36]

Something even more extraordinary happened in 2012, when Google created a program containing 1 billion connections, which ran for three days on 1,000 machines and trawled through 10,000,000 images from different YouTube videos. It had no preset templates nor any intended output.[37] And yet, over the hours that the processors churned, the program developed a unit that responded to the faces of cats. A virtual grandmother cell for virtual cats. This was not the intended outcome of the project – the program was not looking at a photo of a cat and then getting excited because it had been told to look for cats. The images were presented as a one-dimensional data stream, and the program was simply learning to recognise data sequences that it encountered regularly in its YouTube training data set. Hence cat. These data sequences would have corresponded to components of a cat's face – eyes, triangular ears, etc. – which recurred in all those videos. Some perspective is needed on this extraordinary result: to naive eyes like mine, the essence of catness detected by the program is underwhelming (see Figure 6 of the paper), and when tested on a new set of pictures, the program correctly identified cats only 16 per cent of the time (a substantial improvement on previous values, but still).

This program used the latest thing in the field – a deep learning network. These are the systems behind so many of the extraordinary breakthroughs in computer technology, tasks which when I was a student were dismissed as being impossible for a machine – face recognition, scene analysis, driverless cars, natural language recognition, translation, games such as chess or Go and so on. Deep learning systems are adept at identifying the content of massive data sets, in particular those of natural objects, such as cats. Recently, these networks have been reinforced by an explicit nod to the way that the brain is organised – the introduction of a module that can remember things. This idea, first touted in 1997, is called long short-term memory (LSTM), and it massively increases the speed and efficiency of deep learning, enabling machines to extract information in a way that is truly remarkable.[38]

In 2018 researchers at University College London and at Google used deep learning and LSTM to track a virtual rat's position in a virtual space. To their amazement, as the program ran, they observed the spontaneous emergence of hexagonal patterns of activity much like those seen in the mammalian grid cells that underlie place cells in the hippocampus. Even more impressively, the output of these simulated cells was used by the simulated rat to navigate a virtual maze, including taking shortcuts that the authors said were 'reminiscent of those performed by mammals'.[39]

Despite the surprising nature of these unexpected outputs, just because a program generates something that is similar to a brain-produced behaviour, that does not mean that the two systems share either structure or function. As Eve Marder's work has shown, the same output can be produced by very many different structures. And as to Marr's assumption that the same algorithms might be involved in artificial and natural processes, that did not turn out to be the case in the past-tense learning program, which shed no light on how children learn language.

A recent attempt to compare how animals and deep learning networks identify visual objects confirmed what many biologists would have assumed. Although machine, monkey and human were all able to identify pictures of dogs, bears and so on, the computer program made very different errors from those made by the animals, suggesting it was not processing the images in the same way. Furthermore, tweaking the program did not improve matters, suggesting some fundamental differences in the processes taking place in the machine and in the animals.[40]

In 2015 Gary Marcus, who has spent his whole career studying such things, expressed a nuanced view: 'the utility of neural networks as models of mind and brain remains marginal, useful, perhaps, in aspects of low-level perception but of limited utility in explaining more complex, higher-level cognition'.[41] It is true that although most researchers in AI draw inspiration – or a challenge – from biology, there are only a few cases of these kinds of models shedding light on biological processes.[42] One example comes from learning. Many of the most effective programs use what is called temporal difference – differences in the accuracy of successive predictions – to make their remarkable achievements (this is essentially what lies behind the program that recently beat a human at Go).[43] In a study from 2003, the activity of dopamine-producing neurons during learning in humans was found to mirror exactly what was predicted by temporal difference models, providing strong evidence that natural learning involves this kind of process.[44] Given that temporal difference models were first derived from studies of animal learning, this is perhaps not entirely surprising.

A better example, although yet to be fully exploited, came in 2013, when Sophie Caron and Vanessa Ruta in Richard Axel's laboratory at Columbia University showed that the structure of the *Drosophila* olfactory processing network follows essentially the three-level structure of a neural network, with the 'hidden layer' corresponding to the mushroom body, the brain structure insects use to learn about odours.[45] The organisation of the mushroom bodies differs from fly to fly, with what appears to be a random structure. Working with their theoretical neuroscientist colleague Larry Abbott, Axel's group suggested that this random structure could provide the basis for the fly's ability to learn, something that Abbott and Axel went on to explore in a collaboration with researchers from Janelia Research Campus.[46] Each fly's mushroom body is wired differently, and the random organisation of this structure, together with feedback circuits (essentially the same as backprop, but involving a series of cells), apparently enables the fly to learn the significance of odours and to tune its behavioural output appropriately. This insight, in probably the best-understood animal brain, would not have been possible without the work of theoreticians using neural networks. Whether this is indeed how the fly brain works remains to be seen.[47]

Potential insight into how to improve artificial intelligence comes from the observation that animals can learn remarkably quickly, sometimes on the basis of a single example, whereas programs generally require extended learning and large training sets.[48] Animals can perform such feats because their nervous systems are evolved to respond to certain stimuli – their Kantian sensory a prioris mean that the brain is pre-prepared to make some links rather than others. For example, a rat that is made to feel sick after eating a novel food will learn from that single experience to avoid that food thereafter.* The same thing does not happen if you shock the rat, or if you associate feeling sick with a novel sound – the a priori apparently involves only the link between taste and sickness, for fairly obvious evolutionary reasons. By building such pre-existing structures into artificial network models, it may be possible to improve their performance even further.

Despite these examples, in general these astonishing computer programs do not produce clear biological hypotheses, and as a result they have shed little light on how real brains work. Part of the problem with trying to use neural network programs as pointers for biological investigations is that it is not clear exactly how the programs produce their results. Not only is this a mystery to people like me, it also bemuses the researchers themselves. It has always been this way: in 1987, the authors of NETtalk admitted that although they

---

*You may have had similar experiences – this happened to me when I was eight years old and was sick after eating cauliflower. I was in my thirties before I could eat the stuff again.

were able to understand 'the function of some hidden units, it was not possible to identify units in different networks that had the same function'; today's programs are far more complex and even more difficult to deconstruct.[49]

In December 2017, Google AI researcher Ali Rahimi argued that 'machine learning has become alchemy', because it is not clear what the algorithms are actually doing.[50] Another researcher went so far as to claim that the field was 'ridden with cargo-cult practices' and relied on 'folklore and magic spells'. In 2019, when interviewed by *Wired* magazine about neural networks, Geoffrey Hinton cheerfully admitted that 'we really don't know how they work'.[51] This should be a warning to any neuroscientist who looks to neural networks to provide a theoretical explanation of how the brain works. Many computer scientists realise that they, too, lack a theory to explain their complex systems.

*

Despite the breakthrough represented by the PDP approach, some critics soon wondered quite how useful it would be in understanding biological problems. In 1989 Crick wrote a four-page article in *Nature* entitled, with his usual patrician tone, 'The Recent Excitement about Neural Networks'.[52] In 1977 Crick had moved to the Salk Institute in California to work on neuroscience, where he was part of the PDP Research Group. But he soon chafed at what he saw as fundamental biological inadequacies in these programs, in particular their lack of anatomical and physiological accuracy. One thing that especially annoyed him was their reliance on backprop – 'it seems highly unlikely that this actually happens in the brain', he wrote.[53]*

Crick's criticism invoked more than mere biological inaccuracy. After a side-swipe at the motivations of some researchers (he suspected that 'within most modellers a frustrated mathematician is

---

* Although neuromodulators can produce backprop effects in biological systems in some circumstances, the overwhelming majority of neuronal activity is formed by classic, dendrite-to-axon action potentials; this is unlike PDP models, which generally use symmetrical backwards and forwards effects. Jansen, R., et al. (1996), *Journal of Neurophysiology* 76:4206–9.

trying to unfold his wings', and that their objective was to give 'an air of intellectual respectability to an otherwise low-brow enterprise'), Crick emphasised the gap between computer science and biology:

> Constructing a machine that works (such as a highly parallel computer) is an engineering problem. Engineering is often based on science, but its aim is different. A successful piece of engineering is a machine which does something useful. Understanding the brain, on the other hand, is a scientific problem. The brain is given to us, the product of a long evolution. We do not want to know how it might work but how it actually does work.[54]

For Crick, the implication of the brain's evolutionary past was that it had been constructed through a series of steps, each of which was not perfect, but merely adequate – 'anything will do as long as it works', as he put it. The brain was not designed, and as a result, we cannot be sure that it will embody 'deep general principles'. 'It may prefer a series of slick tricks to achieve its aim,' he said. Rather than a search for potentially non-existent logical principles, what was needed was 'a close inspection of the gadgetry'. He explained that scientists should 'look inside the brain, both to get new ideas and to test existing ones'. This was the road that, four years later, led Crick to argue for the development of a connectional map of the brain.

Needless to say, Crick's advice was ignored by most of those interested in computational approaches to behaviour. Some researchers have indeed made their models more realistic – for example by introducing the diffusing effect of signalling molecules such as nitric oxide (these programs are called GasNets[55]), or by showing that strictly symmetrical feedforward and feedback effects are not necessary for programs to perform effectively.[56] But most simply got on with their work, producing increasingly impressive pieces of software and quite understandably showing no interest in linking their work to the anatomy or physiology of the brain.[57]

Within a few years of Crick's complaint, some researchers took a different computational approach. Instead of modelling either a small subset of neurons, or of simply trying to reproduce brain-produced behaviour without regard for structure, they began to simulate nervous systems in computers, just as those IBM researchers had in 1956, but this time with a high degree of anatomical accuracy.

In 1994 Jim Bower and David Beeman produced a book that was part manifesto and part tutorial for programming a neural network simulator, cutely called GENESIS (GEneral NEural SImulation System).[58] The book – inevitably entitled *The Book of GENESIS*, and with chapter titles in gothic script – included floppy disks for running the system on a home computer. The program enabled researchers to simulate compartmentalised neurons with synapses on each compartment, and different densities of ion channels that functioned along the lines shown by Hodgkin and Huxley (a discovery that itself involved modelling), together with realistic synaptic potentials. These virtual neurons could then be linked together in realistic networks, depending on the neuroanatomy that interested the researcher.[59]

This relatively modest simulation environment was the forerunner of one of the most expensive scientific schemes ever undertaken, the Human Brain Project. This ten-year pharaonic programme, funded by the European Commission to the tune of over €1 billion, began in 2013. It covers 150 research groups in eighty institutions and twenty-two countries and will train 5,000 PhD students. The project initially claimed, ludicrously, that by 2020 it would be possible to produce 'cellular-level simulations of the complete human brain' if suitably powerful computers were to become available – implying the sole limiting factor was technological.[60] Presumably for this reason, a major part of the project was devoted to developing new computing approaches and database management systems.[61] This mixture of an overambitious claim and uncertainty over the biological relevance of many of the planned outputs led a number of European neuroscientists to decline to be involved, despite the unprecedented funding. Other researchers objected to the problem from a philosophical point of view, disputing that real insight can be gained from large-scale simulations ('epistemic opacity' is one term that has been bandied about).[62]

Further problems arose when, shortly after its launch, the project downgraded the significance of the cognitive and neurobiological work packages, which would seem to be essential to an explanation of the brain, in favour of the computational side of the project.[63] This led to an open letter to the Commission signed by over 750 scientists, and an opinion piece in *Nature* entitled 'Where is the Brain in the Human Brain Project?'[64] Since then, the three-man leading body – which included the creator of the project, Henry Markram – has been dissolved, and various governance issues have been resolved, but many neuroscientists still suspect that whatever its output in terms of computer science, the vast sums spent on this project will not provide any great insight into how the brain works.

In 2015 the first major output from the Blue Brain Project – another simulation approach also led by Markram – appeared, in the shape of three long papers.[65] These publications were based on data from a tiny cylinder of brain tissue, 2 mm long and 0.5 mm in diameter, taken from a part of the rat motor cortex that controls hind-limb movement – a minute fraction of the animal's brain. The 3-D structure of around 1,000 neurons was determined, and these were used to populate a model of this bit of brain with around 31,000 virtual neurons divided into 207 types, connected by around 37 million hypothesised synapses (this is a vast underestimate of what actually exists in a chunk of rat brain this size). The activity of the virtual neurons included in the model was taken from real data from over 3,000 cells. As the authors admitted, the model omitted 'many important details of microcircuit structure and function, such as gap junctions, receptors, glia, vasculature, neuromodulation, plasticity, and homeostasis'.[66] This long list of missing elements, coupled with the tiny proportion of rat brain that was modelled, helps explain why many neuroscientists view this approach as a waste of money and bristle at exaggerated press accounts of such projects.

Despite deliberately lacking so many key features, the system behaved roughly as a real set of neurons would do, showing synchronous activity, and apparently being able to shift between different states. The team had not simulated every neuron and every synapse in this tiny space, nor any of the missing cells and functions, and yet the model did not fall over and crash, but behaved in a way that was basically the same as that observed in a real section of brain. There was nothing earth-shattering in these studies, but it can be argued that the very fact that they existed, and that the models and data were now widely available, constituted a step forward.

Markram continues to argue robustly not only that the project is valid, but also that what he calls simulation neuroscience occupies a decisive place in the history of our understanding of the brain.[67] However, in 2019 science journalist Ed Yong surveyed a decade of work inspired by Markram and came to a rather limp conclusion: 'Maybe it's telling, though, that the people I contacted struggled to name a major contribution that the Human Brain Project has made in the past decade.'[68]

The Human Brain Project has a resolutely bottom-up approach. It has no overall theory about how the brain works. The idea is that by simulating part of the brain, function can be investigated by removing components in the simulation, altering their behaviour and so on, and seeing how that affects the overall system. A theory of how the brain works, if there can be such a thing, will emerge later on. A contrasting, top-down approach has been taken by a group led by Chris Eliasmith at the University of Waterloo in Canada. In 2012 they unveiled Spaun (Semantic Pointer Architecture Unified Network), a model containing 2.5 million neurons which was attached to a robotic arm. This was not a general simulation but was instead designed to carry out a very specific task: Spaun was presented with a series of images and was asked to draw one of them. This challenge therefore combined character recognition, memory and the tricky problem of controlling the arm to copy the desired character. The results were striking, with highly accurate recognition, including of handwriting, and accurate, child-level copying.[69]

Not everyone was impressed. Markram dismissed Spaun with a sweep of the hand: 'It is not a brain model,' he said.[70] Perhaps not, but maybe we do not need to model every neuron to be able to understand what is going on in the brain. That is the view of many of those who are not involved in major modelling projects, such as the neurobiologist Alexander Borst: 'I still don't see the need for simulating one million neurons simultaneously in order to understand what the brain is doing. I'm sure we can reduce that to a handful of neurons and get some ideas.'

*

Over the last two decades, many neuroscientists, in particular those working in cognitive and theoretical neuroscience, have become increasingly convinced that the brain works along the lines of Bayesian logic.[71] Thomas Bayes was an eighteenth-century British clergyman and statistician who looked at probability in terms of expectations based on existing knowledge or hypotheses. In 1980 the British psychologist Richard Gregory was an early advocate of this approach, using examples of visual illusions to back up his argument.[72] This view, which links with Helmholtz's ideas about the brain developing hypotheses about the environment, has an intuitive relevance to psychological processes – for example, in weighing alternatives, we often pay attention to strong evidence and downplay weak evidence, an essentially Bayesian process.[73]

In the early years of this century, the British neuroscientist Karl Friston used a complex mathematical model to develop Helmholtz's ideas using a Bayesian approach, known as the free-energy principle. Based on an aspect of Shannon's information theory relating to the prediction error in a signal, Friston boldly claimed that this principle transforms our understanding of how the brain works: 'If one looks at the brain as implementing this scheme … nearly every aspect of its anatomy and physiology starts to make sense.'[74] In particular, he emphasised that the hierarchical structure of the brain, with its relative weight of feedforward, feedback and lateral connections, would enable it to carry out the kind of iterative calculations associated with Bayesian probabilities.[75] All brains, Friston argued, seek to minimise error: 'biological agents must engage in some form of Bayesian perception to avoid surprising exchanges with the world'.[76]

The implication of Friston's view is that the computations underlying both perception and the predictions that are implicit in the feedback loops that are involved in control flow from simple physical principles that can be found in all living systems.[77] This idea goes back to Craik's suggestion in 1943 that the brain is 'a calculating machine capable of modelling or paralleling external events', which has proved enormously influential and fruitful.[78] The Edinburgh philosopher Andy Clark describes brains as 'prediction machines' and has used insights from Friston and others to develop a theory to understand both brains and artificial intelligence, while the Sussex psychologist Anil Seth has framed his understanding of human self-hood in terms of processes arising from the Bayesian functioning of what, following Descartes, he calls 'beast machines'.[79]

There is experimental evidence that our perceptions can be subject to the kind of top-down alteration of peripheral processing that is required by Friston's model and by Bayesian approaches in general. There are neural tracts that project down from higher brain areas back to the early processing V1 area; when these nerves were rendered inactive and unresponsive by a pulse of transcranial magnetic stimulation, human subjects were unable to perceive the illusory lights (phosphenes) that are normally induced by magnetic stimulation of another area of visual cortex, V5.[80] Changes to the activity of V1 neurons can therefore alter perception based in another area of the brain (it does not matter that the perception was an illusion). The brain can function in a 'top-down' fashion, rather than 'bottom-up' – it does not merely assemble simple descriptions of the outside world (lines, edges and so on) and allow perception to emerge.

However, despite the attraction of Friston's approach for the mathematically minded – I happily admit it is beyond my grasp – a fundamental problem remains. In 2004 David Knill and Alexandre Pouget described the activity of what they called the 'Bayesian Brain' as follows: 'the brain represents sensory information probabilistically, in the form of probability distributions'. They noted soberly that neurophysiological data in support of the hypothesis 'is almost non-existent'. Although prior beliefs can alter the activity of single neurons (this is effectively what learning is), we do not fully understand the computational logic underlying the way that populations of neurons perform Bayesian integration.[81]

Scientists have recently shown in monkeys that the activity of neurons in the frontal cortex is changed by their prior beliefs (in this case, about the expected interval between stimuli).[82] However, this study did not show exactly what groups of cells were doing and how they were making the inference, but instead demonstrated that prior beliefs altered a particular statistical property of groups of neurons ('low-dimensional curved manifolds') which contain an implicit representation of the optimal response. By using a model of the system, they were able to make predictions about how this property might change under various conditions, although these have yet to be tested in animals.

The gap between theory and neurobiological evidence of the precise activity of single cells can be seen in the study of an apparently simple predictive brain system, which does not require Bayesian calculations – the ability of some insects to intercept a mate or a prey while flying. This must involve detection of the position and movement of both perceiver and target, and at least two kinds of calculation – a measure of the initial relative locations of the two individuals and a prediction of the future relative position of the target – so that an intercept movement can be put into effect.

You can see this behaviour for yourself in the summer – hover-flies will gather in sunlit glades, flying around looking for mates. If you get an orange pip and squeeze it between your fingers, you can send it zipping past one of the flies, which will rapidly fly to meet it, deceived by the size and movement of the pip into thinking it is a potential mate, or a rival. In 1978 Tom Collett and Mike Land of the University of Sussex reported an experiment in which they fired peas from pea-shooters towards hoverflies and filmed the insects' movements in response (pea-shooters are more accurate than orange pips and fingers).[83] By mathematically analysing the behaviour of the flies, Collett and Land were able to describe the key parameters that the insects' tiny brains were computing, and also to show that although the intercept prediction was not continually updated through a tracking function, there was a feedback element that allowed the animals to break off in mid-flight.

I remember being entranced by this paper when it appeared; it is surprising, however, that over four decades later, despite the immense advances in the study of insect flight behaviour and our astonishing ability to precisely measure the activity of single cells in a fly's brain, the biological substrate of such banal predictions remains a mystery. Researchers are currently pursuing the even more complex calculations involved in predator–prey interactions (the prey can take evasive action), as seen in the gorgeous lions of the fly world, robber flies, or in dragonflies.[84] All this strongly suggests that these tiny insect brains contain a predictive model that represents the relative movements of predator and prey (and also external factors such as wind speed, which affects how both individuals may react), but for the moment we do not know how that model is actually embodied in the activity of the nervous system.

Our inability to identify exactly what kind of simple predictions are taking place in an insect brain reveals a problem with the apparent power of Bayesian theories to explain complex functions in the human brain (the failure of the rigid McCulloch and Pitts model of neuronal logic to translate into how real nervous systems function should also be a warning). The existence of something like Bayesian predictions taking place within the nervous system to explain perception seems certain. For the moment, the theoretical generalisation of this assumption to explain the whole of the brain remains speculative. Experimental evidence will always be the key determinant of the validity of any theory, no matter how elegant and seductive it might be.

*

The widespread view that 'the brain is a computer' is reinforced by the fact that for over a century we have been able to control its activity with electricity, just as we do with an electronic machine. In the 1920s researchers began using electrical stimulation of the brain to explore the anatomical and physiological bases of emotions. The US physiologist Walter Cannon showed that emotions had their origins in brain activity, not in the responses of the viscera and the autonomic nervous system. If humans were injected with adrenaline, this produced the usual visceral physiological responses associated with emotion, such as increased heart rate, but did not lead to the experience of that emotion.[85] For Cannon, emotional responses were coordinated by the hypothalamus, but were controlled by the action of the cortex; if the cortex was removed from a cat it would show continual aggressive responses – spitting and attack behaviour – even though there was no cause (Cannon called this 'sham rage').[86]

The Swiss researcher Walter Hess took this approach a step further and showed that electrical stimulation of the cat hypothalamus could cause spitting, raised hairs and dilated pupils, sometimes leading to a strike with the paw even though there was no threat. This suggested that emotions might be released by electrical stimulation of certain parts of the brain, and that the autonomic centres, involved in the basic physiological responses, could then activate the motor cortex.[87] Hess's work, which provided insights into how the various parts of the nervous system might interact, was rewarded by the Nobel Prize in 1949.

In one notorious experiment from 1965, Professor José Delgado of Yale University stepped into a bullring in Andalucía and waved a matador's cape at a young black bull called Lucero. The animal charged at him before suddenly stopping and turning its head in confusion. Delgado had earlier inserted an electrode into Lucero's caudate nucleus, an area associated with movement, and had a radio receiver in his hand that activated the electrode when he pressed a button (as Delgado later admitted 'one time there was a failure in the transmission circuit, and the bull managed to reach me, fortunately without more consequence than a good scare'[88]). This dramatic experiment, captured on film, also made the front page of the *New York Times*, where it was described as 'the most spectacular demonstration ever performed of the deliberate modification of animal behaviour through external control of the brain'. It was never written up as a scientific publication.[89] In fact it told us little more than previous extensive studies of the motor cortex – electrical stimulation of the brain can produce or stop movement. Away from the glare of publicity, Delgado admitted that his method of brain stimulation – he called the device a 'stimioceiver' – was 'a rather crude procedure'.[90]

Other researchers made even stronger claims and went far beyond the boundary of what was ethical, even for the time. From the 1940s onwards, Robert Heath, a psychiatrist at Tulane University in New Orleans, used electrical stimulation of the brain to treat psychiatric patients.[91] These included a male homosexual, known as B-19, whom Heath claimed to have 'cured' by stimulating the man's brain while he looked at pornographic images of women, a cure that was demonstrated by paying a female prostitute to have sex with him (the proceedings were recorded).[92] Heath also provided catatonic schizophrenic patients with a permanently implanted electrode and a portable battery that they could use to stimulate themselves, providing waves of pleasure and a relief from their symptoms.[93] He even used these electrodes to provide aversive stimulation, with horrendous results as the patients writhed in agony and threatened to kill the experimenter. These experiments were also filmed.

Heath's work formed part of an increased interest in using brain stimulation, following a report in 1954 by James Olds and Peter Milner, who worked with Hebb at McGill University. They inserted electrodes into the septal area of a rat's brain and discovered that the animal would do anything to stimulate the region.[94] A few years later, Olds reported that in order to obtain stimulation a rat would continuously press a bar until it was completely exhausted – after twenty-six hours of frenetic activity in some cases.[95] Brain stimulation reward, as it became known, revealed that there were areas of the brain associated with positive, rewarding feelings that could be stimulated by electrical activity. The technique is rarely used in a therapeutic context nowadays, not only because it is imprecise and highly invasive, but also because it poses obvious ethical issues when, as in one of Heath's studies, a patient provided with a way of self-medicating will happily indulge in brain stimulation for as long as possible.

Despite the ethical quagmire that brain stimulation has been mired in for much of its past, there is one area in which it has proven clinical benefits. The symptoms of Parkinson's disease, a degenerative disorder of the central nervous system that causes uncontrollable tremors and can also lead to depression, dementia and death, can be relieved – but not cured – by pharmacologically increasing the levels of the neurotransmitter dopamine. However, sometimes this treatment does not work, and since the early 1990s researchers have been using deep brain stimulation via implanted electrodes to reduce the symptoms. The effects are dramatic and the improvement in quality of life is remarkable.

A potentially less benign use of brain stimulation is suggested by the recent interest of the US government's Defense Advanced Research Projects Agency (DARPA). In 2017 DARPA announced a substantial research programme into 'targeted neuroplasticity training', with the ultimate aim of enhancing learning in soldiers, using non-invasive methods.[96] More alarmingly, another DARPA-funded project at the University of California, focused on post-traumatic stress, created an algorithm that will enable a computer to compare the current state of a subject's brain with a desired goal, and then automatically tweak their feelings by stimulating the relevant region.[97] With the prospect that nanoparticle optogenetic constructs might make it possible to produce such effects non-invasively, by a simple injection, you do not need to be Philip K. Dick to imagine how it could all go horribly wrong.[98]

Researchers are also engaged in some astonishing and extremely positive work that shows how the brain can control machines.[99] In 2012 John Donoghue's group at Brown University used electrodes implanted in the motor cortex of two tetraplegic patients – a fifty-eight-year-old woman and a sixty-six-year-old man, both of whom had suffered strokes many years earlier – to enable them to move a robot arm with their thoughts.[100] The female patient – Cathy Hutchinson – was able to use the arm to grasp a bottle, then slowly bring it to her mouth, drink coffee through a straw and then replace the bottle on the table. Hutchinson's joy at this extraordinary achievement – the first time in fourteen years she had been able to drink solely of her own volition – is evident in the video and images that accompanied the article.[101]

Donoghue and his colleagues have since implanted electrodes into the brain and arm of a patient who became tetraplegic following spinal cord injury. Signals from the patient's brain were translated into electrical stimulation of his muscles and, with the aid of a mobile arm support, he was able to feed himself.[102] This astonishing development has real potential to change lives.

These procedures did not involve feedback from the robot or human arm – this phenomenon is called proprioception and is an essential component of our control of bodily movement, telling us, for example, how tightly we are gripping something. This will soon come: researchers have recently developed a bionic hand for an amputee, controlled by electrodes implanted in his arm, and which provided him with to 119 sensory sources from the device that stimulated nerves in his skin, grouped into a number of categories – vibration, pain, movement and so on. Using this artificial proprioception, the patient was able to carry out quite subtle tasks such as moving an egg or picking grapes, as well as more personally significant gestures, such as touching his wife's hand.[103] Paraplegic patients would have to learn to interpret stimulation of part of their body that retains sensation in terms of proprioception from the device, but this would happen quite rapidly. Lives will be transformed by this amazing technology.

Eventually it may not be necessary to use invasive procedures. In 2018 Christian Penaloza and Shuichi Nishio from Kyoto reported that healthy patients could wear a skull-cap and then learn to use signals from their scalp muscles to control a third, robotic arm while they were doing something else.[104] So, for example, the subject would tilt a board with both hands, such that a ball on the board rolled around to different positions, and simultaneously command a robot arm to bring a drink to their mouth. Whether as a way of augmenting human work capacity, or transforming the lives of disabled people, this technology has an extraordinary future.

The first apparently successful attempts to connect an artificial eye to the brain were reported in 2000.[105] Electrodes connected to a video camera were inserted into the visual cortex of blind patients; however, this did not mean that the patients could see images directly. Instead, the electrodes activated the sensation of light (much as when you press your eyeballs); after a great deal of training, this electrical activity could be interpreted by the patient, enabling them to detect objects or even large letters. But nearly two decades later neither retinal implants nor brain implants have provided anything like real vision.

In the case of hearing, great progress has been made. Since the first cochlear implant in 1961, this approach has become routine, with hundreds of thousands of people around the world benefiting from this technology, and there are many heart-warming videos of the emotional responses of deaf people as they hear for the first time. However, although the results are transformative and far more effective than for artificial vision, the implants do not yet allow a full range of hearing.

Researchers from a number of different groups have recently begun work on a truly challenging frontier – producing synthetic speech directly from the activity of the brain.[106] Despite press excitement, these techniques do not involve 'mind-reading' – in reality the computer is learning to associate neuronal activity patterns linked to muscular control of speech with the actual noise that is being made. For the moment, the goal of translating neuronal activity associated with imagined speech into an artificial voice is far off.

Important as all these developments are, they do not imply that brains are actually computers, or that we know how they work. In reality, they highlight the plasticity of our brains – Donoghue's group has not cracked the neural code in the brain for volition and planning; instead their computer programs are able to translate patterns of neuronal firing in the brain into the movement of the robot arm, and the patients are able to rapidly tune the activity of their brains so as to manipulate the arm in the desired way.

Unexpected changes may occur to people who live with brain–computer interfaces. Bioethicist Frederic Gilbert of the University of Tasmania has described six patients in Australia who used electrodes implanted in their brains to warn them of an impending epileptic fit, thereby enabling them to take medication appropriately. Although this was a benign intervention, one patient ('patient 6') had a particularly extreme response, initially saying the interface 'was like an alien'. However, her attitude slowly changed: 'You grow gradually into it and get used to it, so it then becomes a part of everyday, it's there every day, it's there every night ... it follows you through the shower everywhere and it becomes part of you ... it was me, it became me ... with this device I found myself.' She went on to report that the device transformed her personality, making her more confident: 'With the device I felt like I could do anything ... nothing could stop me.'[107]

If you think that is disturbing, consider what happened next. The company that implanted the device into patient 6's brain went bust and the interface had to be removed. This had a profound effect on the poor woman – 'I lost myself,' she reported. Something that had been given to her had been taken away, because of the economic system. Gilbert bleakly summarised what had happened to patient 6 through her interaction with the implant, the new worlds she had briefly glimpsed, and the brutal reality of who was in charge: 'It was more than a device. The company owned the existence of this new person.'[108] In a future world where private companies are funding interfaces with our brains, we may lose control over our identity.

The lesson is that scientific research does not take place in a vacuum and that exciting discoveries and therapeutic opportunities may have profound, unforeseen consequences. This is evident throughout the past and present of brain science. Science and culture are deeply intertwined, in particular when it comes to scientific discoveries that affect our perception and our mood, some of which have had the most extraordinary cultural impact.

THIRTEEN

CHEMISTRY

1950 TO TODAY

On 19 April 1943 Albert Hofmann, a Swiss chemist working for the Sandoz pharmaceutical company in Basel, rode his bicycle home from work. Something was not quite right. As he later recalled: 'everything in my field of vision wavered and was distorted as if seen in a curved mirror'. When he got back to his house he experienced intense feelings of anxiety, which eventually gave way to a very strange sensation: 'kaleidoscopic, fantastic images surged in on me, alternating, variegated, opening and then closing themselves in circles and spirals, exploding in coloured fountains, rearranging and hybridising themselves in constant flux'.[1] Before getting on his bike, Hofmann had taken a massive dose of a seemingly innocuous molecule he had synthesised five years earlier: LSD.

Hofmann's momentous but accidental discovery – commemorated each year by acid heads as Bicycle Day – was typical of the transformation in our understanding of brain chemistry that was to occur over the next two decades. Hofmann had not intended to create a powerful psychoactive drug when he first synthesised LSD – he had been attempting to discover a compound that would help with breathing. Similar accidental breakthroughs soon changed how we think about the brain and how we understand and try to treat mental health problems.[2]

In the late 1940s the French drug company Rhône-Poulenc was developing antihistamine drugs in conjunction with a military surgeon, Henri Laborit. One compound, chlorpromazine, had very weak effects as an antihistamine but induced a strong calming effect. In 1952 psychiatrists at the Hôpital Sainte-Anne in Paris gave chlorpromazine to a number of manic or psychotic patients. The results were extraordinary. For example, a patient called Philippe Burg, who had been in a hopelessly psychotic state for several years, responded rapidly to treatment. One of the very French signs of Burg's recovery was that after a few weeks he was able to leave the hospital and dine with his doctors in a nearby restaurant. A series of similarly dramatic cases led to immediate global interest in the drug, which was soon marketed in Europe as Largactil and in the US as Thorazine, changing the lives of thousands of people. At around the same time, a drug with similar psychoactive effects, reserpine, was identified through another chance discovery. It had been developed to reduce blood pressure, using products found in traditional medicine, but it turned out to have psychological effects that were described as neuroleptic (the Greek roots of the word mean 'seizing the neuron'). In 1953 an employee of the CIBA drug company coined a simpler description of reserpine – it was a tranquilliser.[3]

At a time when psychiatry was dominated by psychoanalytic concepts, the discovery of LSD, which seemed to mimic some symptoms of certain illnesses, together with the appearance of new tranquillisers, represented a massive shift. Mood-altering drugs had been known for millennia, but these new substances were different: they had dramatic and very specific properties. Their discovery marked the beginning of a profound shift in approaches to mental illness, away from the psychoanalytic approach and towards today's medicalised, chemical view. As repeated waves of drugs have been developed over the decades each has been accompanied by great promises and enthusiasm, which have then turned into disappointment as important side effects have been discovered.[4] Nevertheless, these drugs have given scientists new ways of understanding the chemistry of brain function, in sickness and in health.

The initial steps on this new path involved some eye-popping experiments. In 1952 Humphry Osmond and John Smythies of the

National Hospital in London reported that mescaline, the active component of peyote, mimicked some of the symptoms of schizophrenia, and noted that it was structurally similar to a substance produced by the adrenal glands, noradrenaline (in US English, this is called norepinephrine).[5*] Two years later they suggested that adrenochrome, a naturally occurring oxidised version of adrenaline, might be responsible for the symptoms of schizophrenia. By this stage, Osmond and Smythies had moved to Saskatchewan, where they were pioneering the use of psychedelic drugs to treat the mentally ill.[6] In the best medical tradition, Osmond self-administered adrenochrome to see what happened. He reported the consequences in an article in the *Journal of Mental Science*:

> I closed my eyes and a brightly coloured pattern of dots appeared. The colours were not as brilliant as those which I have seen under mescal, but were of the same type. The patterns of dots gradually resolved themselves into fish-like shapes. I felt that I was at the bottom of the sea or in an aquarium among a shoal of brilliant fishes. At one moment I concluded that I was a sea anemone in this pool.[7]

It was not all fun, though. On another occasion, Osmond had a very bad trip, leading the researchers to warn against using adrenochrome except under highly controlled circumstances. (Perhaps for this reason, two decades later the drug-crazed gonzo journalist Hunter S. Thompson became obsessed with trying to obtain adrenochrome from the glands of a living human, according to his account in *Fear and Loathing in Las Vegas*.) Noting the similarity of this experience with those of people who had taken mescaline and LSD, the researchers suggested that studying adrenochrome and its metabolism might provide insights into the biochemical origins of schizophrenia.

Other researchers focused on newly identified chemical components of the nervous system. In 1955 Bernard 'Steve' Brodie and his colleagues showed that both reserpine and LSD affected the levels of serotonin, a substance of unknown function that was found in

---

*Osmond and Smythies coined the term 'hallucinogen'.

smooth muscles such as the gut and the uterus and which two years earlier had been identified in the brain by Betty Twarog.[8] Brodie's research showed that reserpine increased serotonin levels while LSD reduced them. Brodie's group soon suggested that serotonin had an important role in brain function and found that reserpine also altered two other substances present in the brain – noradrenaline and dopamine – which they thought might also affect neuronal activity.[9] The links between the psychological effects of the new drugs and their impact on brain biochemistry seemed to offer a clue to how the brain worked, and the possibility of developing new treatments for mental health disorders.

Using the structure of chlorpromazine as a starting point, researchers in Switzerland sought to develop a new treatment that might help schizophrenics. This apparently rational approach did not pan out as expected – one drug called imipramine was indeed highly psychoactive, but far from calming patients down, it was a powerful excitant. While this was useless for manic patients, it could help people with depression. In time a whole suite of drugs emerged, known as tricyclics because their molecular structure contained three rings, which for decades were the best pharmacological treatment of depression. Another drug, iproniazid, was created as a treatment against tuberculosis, but turned out to also have antidepressant qualities – it was a 'psychic energiser' claimed the researchers who explored this unexpected effect.[10] Because of its effectiveness, iproniazid was widely prescribed for depression until it was found to damage the liver and was withdrawn.

Finally, at the beginning of the 1960s, the anxiety-reducing benzodiazepines (Librium, Valium and many others) came on stream. Again, the psychoactive power of first of these – Librium – was discovered by accident. A researcher at Hoffman-Laroche stored an apparently useless compound after stabilising it with a chemical additive.[11] Two years later, the drug was taken off the shelf and, in its now-modified form, was found to be psychoactive. Benzodiazepines proved remarkably popular, and they are still widely prescribed for short-term anxiety reduction.

One exception to this opportunistic commercial programme of drug development was the discovery that common salts of lithium could help in the treatment of manic states. Lithium bromide had been used in the nineteenth and early twentieth century as a treatment for epilepsy, but the effective dose was also toxic, limiting its usefulness. In 1948 an Australian physician named John Cade discovered that dosing guinea pigs (the furry sort) with lithium caused them to become lethargic, so he tried the compound on ten patients with severe mania. The results were remarkable:

> W.B., a male, aged fifty-one years, who had been in a state of chronic manic excitement for five years, restless, dirty, destructive, mischievous and interfering, had long been regarded as the most troublesome patient in the ward. His response was highly gratifying ... He was soon back working happily at his old job.[12]

Lithium was not a chemical cosh – the patients were not sedated – but neither was it a cure.[13] If the patient stopped taking the drug (as WB did), the symptoms reappeared. However, lithium could not be patented and interest from the pharmaceutical industry was limited – it was only finally approved for use in the US as a mood-altering drug in the 1970s, following the appearance of a 'Lithium underground' of rebel psychiatrists who prescribed it anyway.[14] Remarkably, it is still not known how lithium exerts its very real effects.

All of these new drugs had two aspects: they were clinically significant, helping to alleviate profound suffering, and they held the promise of a radically new insight into how the brain – and even the mind – might work. As the historian Jean-Claude Dupont has put it: these results reinforced the fact that the brain 'was not only an electrical machine, but also a gland'.[15]

Despite this initial period of promise, it was hard to establish a link between the effects these drugs had on the mind and their physiological action. For example, it was initially thought that the psychedelic effects of LSD and the delusional symptoms seen in some schizophrenic patients were both mediated by serotonin. This hypothesis was abandoned when it was discovered that chlorpromazine, like reserpine, helped relieve delusional symptoms but had no effect on serotonin, while other drugs produced unpleasant psychotic effects but did not alter serotonin levels.

One much-touted chemical basis of schizophrenia simply evaporated, for bizarre reasons. In the 1950s Robert Heath, the psychiatrist who had used deep brain stimulation to 'cure' a gay man, suggested that a substance known as taraxein could be detected in the blood of schizophrenics and could induce schizophrenic symptoms if injected into healthy volunteers. Dramatic film evidence of the effect of the drug was shown at conferences, impressing the scientists who saw it. However, Heath's results could not be replicated, and eventually the hypothesis was abandoned.[16] According to the author Lone Frank, one of Heath's colleagues (the biochemist Matt Cohen) had deliberately withheld key parts of the relevant protocol from their scientific publications, rendering their work impossible to replicate. Cohen was in fact a fraud with no scientific training; he was a gangster on the run and had kept part of the taraxein technique secret as an insurance policy in case of discovery. He left the Heath laboratory abruptly in 1959 and was apparently killed in a mob shootout in Florida some years later.[17]

Also in 1959, Seymour Kety of the NIMH published a major two-part article in *Science* on biochemical theories of schizophrenia.[18] After advising his readers that the label schizophrenia might conceal a wide variety of underlying problems (this remains a major difficulty), Kety weighed up the evidence for various potential causes, including taraxein, but focused on the potential role of serotonin. The key stumbling point, he reported, was that 'the role which serotonin plays in central nervous function is blurred'.[19] Scientists could not explain what was happening when brain chemistry went wrong if they did not understand what it was supposed to be doing in the first place. New concepts were needed to explain the chemical complexity of the brain that was gradually being unveiled.

*

This astonishing period of pharmacological luck and creativity coincided with the definitive end of the war of the soups and the sparks in 1952, which was also the year that Hodgkin and Huxley showed how the action potential was propagated in the neuron. Scientists were converging on a common chemical view of nervous system function, but two major difficulties remained: how exactly the transmitter substance did its work, and what was happening in the brain. All those furious arguments over soups and sparks had taken place at the level of peripheral neurons, generally in the autonomic nervous system; no one could be sure if the same principles held for the central nervous system. What looks obvious to us – the brain works along the same lines as the peripheral nervous system, using neurotransmitters – was not at all clear in the 1950s and 1960s. Even the word neurotransmitter, which was a way of grouping a wide variety of compounds under a common functional heading, was not coined until 1961.[20]

According to Arvid Carlsson, who won the 2000 Nobel Prize for his work on dopamine, at the beginning of the 1960s the suggestion that there might be neurotransmitters in the brain met with considerable scepticism.[21] Even a few years later, when the idea seemed less outlandish, there was still no decisive proof – in 1964, Arnold Burgen, Professor of Pharmacology at Cambridge, complained in *Nature* about the lack of understanding of what was happening at the synapse:

> An even greater disappointment to all interested in synaptic physiology has been the failure to elucidate the nature of any chemical transmitter in the mammalian nervous system other than acetylcholine … despite considerable effort we are totally in the dark as to the chemical transmitter at the primary sensory afferent fibres, and both presynaptic and postsynaptic spinal cord inhibitory systems, not to mention other areas.[22]

Burgen's disappointment was soon assuaged as over the next decade the precise mode of function of neurotransmitters in the brain was revealed. A bewildering array of substances was discovered, grouped into three major types – amino acids (such as GABA), peptides (for example, oxytocin or vasopressin) and monoamines (noradrenaline, dopamine and serotonin) – along with the first neurotransmitter to be identified, acetylcholine. Among the more surprising findings was that nitric oxide, a gas, is produced by some neurons and diffuses through tissues and alters neuronal activity.[23] We still have not reached the end: according to the veteran neurotransmitter expert Solomon Snyder, there could be up to 200 different peptides acting as neurotransmitters in the brain.[24]

One of the key factors that convinced scientists of the existence of these new neurotransmitters was the use of fluorescence or radio-activity to create images that revealed their presence – Jean-Claude Dupont has even claimed that 'it was histochemistry, rather than pharmacology or electrophysiology, that finally led to the acceptance of neurotransmission by brain amines'.[25] After the first electron micro-scope images of the synapse were obtained in the 1950s, Bernard Katz showed that tiny vesicles in the presynaptic neuron released the neurotransmitter into the synapse, following the influx of calcium which underlies the action potential. Some neurotransmitters, such as GABA, were found to be inhibitory, resolving the problem of the nature of inhibition that had perplexed scientists for a century. It also became apparent that some neurons did not use neurotransmitters at all, but instead functioned with electrical synapses, or gap junc-tions. In 1970 three of the major contributors to this revolution in our understanding of brain chemistry – Ulf von Euler, Julius Axelrod and Bernard Katz – won the Nobel Prize for their work.

Many of the receptors involved in the postsynaptic response to a neurotransmitter were soon identified. There turned out to be two classes – some led to the immediate propagation of an action potential, whereas others caused a much slower response, through a cascade of second messenger molecules in the postsynaptic neuron. Paul Greengard's research on the slow synaptic response, which built on studies by Earl Sutherland and Ed Krebs in the 1960s, was rewarded by the Nobel Prize in 2000, which he shared with Carls-son and Kandel. This work is still not finished – the structure of the $GABA_A$ receptor, the target of Valium, has only just been described.[26]

*

The rich chemical world of the brain became even more complex as it was realised that cerebral activity involves not only the pulsing action of neurotransmitters, but also the effects of slower-acting neuro-hormones. These substances, often formed of peptides – short chains of amino acids – are released into the bloodstream or intracellular spaces and act as signalling molecules in the body, in particular in the brain. Much of this work focused on the role of the hypothalamus, which Edinburgh neurophysiologist Gareth Leng has called 'the heart of the brain'.[27] In the 1960s and 1970s, the hypothalamus and the hor-mones it produces were shown to be involved in coordinating complex physiological and behavioural responses, including those involved with stress responses and reproduction, and in 1977 the Nobel Prize was half awarded to Roger Guillemin and Andrew Schally for their discoveries relating to peptide production in the brain (the other half of the prize was awarded to Rosalyn Yalow, who developed radioimmunoassays for tracing peptide hormones). In the 1990s, the neuropeptides leptin and ghrelin were discovered; these are impli-cated in feeding behaviour and feelings of satiety. Neurohormones are therefore involved in the long-term control of essential physiological processes, many of which have a behavioural component.

These substances affect brain circuits involved in behaviour, either temporarily – for example, changing the responses of a female rat to pups, making her pick them up and create a nest for them – or permanently, for example in shaping a rat's brain to produce a more male set of behaviours. The way in which these peptides are secreted is very different from the action of neurotransmitters. Vesicles con-taining the neurohormones can appear anywhere on the body of a neuron, not just at a synapse; they are particularly prevalent on den-drites and can contribute to the functional reorganisation of parts of the nervous system during repeated stimulation.

This aspect of brain function is extremely complex – there are thought to be more than a hundred different neuropeptides that diffuse through the intracellular space of the brain, which makes up around 20 per cent of its total volume.[28] These molecules are released in large numbers – far higher than the number of neurotransmit-ter molecules, for example – in pulses that can go on for days. Each of these systems, which are influenced by the internal and external conditions affecting the animal's body, has its own feedback loops controlling how it changes the activity of the brain. Comparative studies show that these networks reach deep back in evolutionary time, appearing shortly after the Cambrian explosion, around 530 million years ago.

Although the general target areas that are the focus of the activity of these neurohormones can be identified, exactly how they alter the working of the brain to produce the observed changes in behaviour remains unclear. For example, neurons in the rat brain sensitive to oxytocin are involved in feeding, a variety of aspects of reproduction, social behaviour and the animal's sodium balance. Somehow, the same neurohormone coordinates these complex and very different behaviours. This complexity – recognised by von Neumann when he began to think seriously about the parallel between the brain and a computer – shows that the brain is a complex parallel processing organ. It can do more than one thing at the same time, using both near-digital and analogue neurotransmission, and continuous, analogue transmission through neurohormones.

One of the most intriguing discoveries relating to neuropeptides was the 1973 description of opiate receptors by Candace Pert, a postgraduate student in Snyder's group.[29] The existence of these receptors helped explain why mammals are so interested in opiate drugs – the research was funded by a US programme designed to respond to the growing use of heroin in inner cities and by army conscripts fighting in Vietnam. It also prompted the question of why there are such receptors in the first place – there must be some naturally occurring opiate-like substance in the brain that can bind with them. In 1975 John Hughes and Hans Kosterlitz of the University of Aberdeen discovered the presence in pig brains of two peptides with potent opiate activity, called endorphins.[30] A few months later, the same two endorphins were described in the rat by Snyder's group, who went on to locate these substances in areas of the brain that were involved in emotional responses, hence explaining the psychoactive effect of opiate drugs.[31] These endorphins are now known to be produced following injury and also after vigorous exercise, contributing to the 'runner's high'.

In 1978 Snyder, Hughes and Kosterlitz won the prestigious Lasker Award for their work on endorphins. Pert understandably felt that she had been slighted – the discovery was as much hers as it was

Snyder's – and publicly protested. She had also been overlooked for another major award in the previous year – the chair of that award panel subsequently recognised that this was 'a significant omission', but nothing was done.[32] Pert's role was never officially recognised.

*

All these discoveries in brain chemistry, together with growing public awareness of diseases such as Alzheimer's and Parkinson's – one of the few lasting consequences of President George H. W. Bush declaring the 1990s to be The Decade of the Brain – led to new approaches to mental health.[33] One influential aspect of these discoveries was the suggestion that the addictive powers of certain drugs might be based on their ability to release dopamine from neurons. In the 1990s a series of studies by Wolfram Schultz at the University of Cambridge showed that networks of dopaminergic neurons are associated with reward in animals. It is now realised that things are more complicated, and that these neurons help to measure the difference between predicted and actual conditions; they can also modulate the coding of aversive stimuli.[34] If an expected stimulus – including an aversive stimulus – does not occur, then dopamine neurons are involved in signalling that to the animal.[35] They also detect the temporal links between stimulus and reward or punishment that underpin learning, recognising the order of events and appropriately potentiating or depressing the activity at their synapses.[36]

In 1997 Alan Leshner of the NIH wrote an article in *Science* boldly entitled 'Addiction is a Brain Disease', in which he claimed 'virtually all drugs of abuse have common effects, either directly or indirectly, on a single pathway deep within the brain', referring to the dopamine system.[37] By reframing addiction in this way, Leshner sought to highlight the importance of neuroscience in understanding mental health, and to help create more effective policy – if addiction was due to brain disease, he argued, there was no point in locking people up for crimes associated with feeding their addiction without trying to cure them. Treatments were needed to attack the root problem, which Leshner claimed was biochemical.

This hypothesis gradually became more complex as it was

discovered that although dopamine levels are increased in alcoholism, this is not the case for all addictions.[38] Many addictive recreational drugs, such as nicotine, cocaine and amphetamines, alter dopamine concentrations in the same part of the brain, but they do this in different neurons, by different routes and in different ways – for example, opiates suppress dopamine, while benzodiazepines enhance the firing of dopaminergic neurons.[39] Nevertheless, leading US physicians continue to argue not only that substance addiction can be explained by a biochemical 'brain disease' model, but also that this can be extended to other supposed addictions, such as the internet, food and sex.[40] Confusingly, the main implications of the model relate to behavioural treatments and policy changes, rather than the need for drugs directed against a supposed common biochemical basis.

These scientific approaches have crept into popular culture, and it is now commonly claimed that the supposedly addictive power of everything from pornography to social media is due to the activation of our dopamine system. In 2017 Sean Parker, one of the founders of Facebook (he resigned in 2005), claimed that they deliberately designed the website to be addictive – 'we … give you a little dopamine hit', he boasted.[41] This is nonsense. Although one study has reported that dopamine was released in the brain while subjects (eight of them) played a video game, the investigation had no connection with addiction, nor did it present any proof that the reported effects had anything to do with interacting with a computer (the control measure was a blank screen rather than, say, reading a book).[42] There is no evidence that Twitter has hacked your dopamine system. According to Wolfram Schultz, who should know, it is not even clear that the activation of dopaminergic neurons produces a pleasurable sensation. The claim that all addictive behaviour can be blamed on dopamine is an example of what is commonly called neurobollocks. Despite the enthusiasm for the dopamine brain disease model of addiction, it seems certain that although different addictive behaviours may look – and feel – the same, they probably have different underlying mechanisms.

Part of the problem in seeking a link between mental illness and physiology is that psychiatric diagnosis is not very precise. In the USA it is based on the *Diagnostic and Statistical Manual of Mental Disorders* (commonly known as DSM), a collective production of the American Psychiatric Association that effectively defines what is considered to be a mental health disorder.[43] These views change not least because the boundaries of mental health are partly socially determined – in the 1980s homosexuality was removed from the drafts of an earlier version of DSM only after a huge battle. In most cases, the causes of mental health problems are hard to explain in terms of brain function or chemistry. One partial exception is Alzheimer's disease, which is linked with the appearance of abnormal forms of protein that disrupt the structure of the brain. But even here it is difficult to untangle cause, effect and contributory factors, and even harder to come up with an effective treatment. Our understanding of the origins of mental health problems, and how to treat them, remains profoundly unsatisfactory.

*

The most well-known attempt to fuse pharmacological approaches and the physiological basis of mental health disorders has come from work on the role of serotonin in depression. When neurotransmitters are released into the synapse, they bind with the receptors on the postsynaptic cell; the neuronal signal comes to an end as the neurotransmitter is reabsorbed into the presynaptic cell. The discovery of this 'reuptake' led to the creation of drugs known as selective serotonin reuptake inhibitors (SSRIs) that could thus increase serotonin levels. It is claimed that these drugs increase the levels of serotonin in the brain, thereby alleviating the symptoms of depression. Often known by the US name of one of their most successful versions, Prozac, SSRIs have become widely prescribed around the world, and many patients consider that their lives have been transformed as a result.

And yet our understanding of what happens when you take an SSRI remains virtually non-existent. It is not known if someone who is depressed actually has low serotonin, or how SSRIs then affect this – at a cellular level serotonin reuptake is altered very quickly by SSRIs, yet the effects on mood take weeks to be felt, if at all.[44] There is no physiological marker of depression (or of any other mental health disorder), and a recent genome-wide analysis of genetic factors associated with depression in over 800,000 people noted that 'an intriguing omission among the depression-associated genes identified in our study are genes linked with the serotonergic system' (they replicated their study with a further 1.2 million subjects, and again found no link with serotonin).[45] This was not the first failure to find any connection between depression and genetic factors involved in serotonin metabolism. To put it bluntly, there is no decisive evidence that low mood is caused by low serotonin levels, nor of what SSRIs actually do to serotonin levels in the brains of patients.

Many patients report no improvement while taking SSRIs, and a combination of scientific disputes over data, suspicion of the motivation of the drug companies and desperation on the part of some patients who suffer serious side effects have all led to a fractious debate over whether SSRIs are effective.[46] This may not be the best way of framing the issue: the key question seems to be what proportion of patients are helped, and to what extent, and how – if – such patients could be detected before being prescribed the drugs.[47]*

Perhaps the most intriguing aspect of the entry of SSRIs into our culture has been that the public has embraced the explanation of depression that has been offered by scientists, even though it remains unproven. Two men are generally given the credit for developing the hypothesis that low serotonin levels cause depression; in reality, neither of them said any such thing. In 1965 Joseph Schildkraut summarised various ways in which the class of chemicals known as monoamines – noradrenaline, dopamine and serotonin – might explain depression and other disorders, without pointing the finger at low serotonin levels. Two years later, the role of monoamines in depression was explored by Alec Coppen, a psychiatrist working for the Medical Research Council in the UK, but even he would go no further than implicating all three substances in a range of disorders: 'monoamine deficiency is not the sole cause of the disorder', he wrote.[48]

Nevertheless, the idea took hold in psychiatric circles and in

---

*Please, if you have been prescribed SSRIs, or any other medication for mental health problems, do not stop taking them without first consulting your physician.

1974 two researchers from Philadelphia reviewed a large number of studies 'to evaluate the hypothesis that clinical depression is associated with reduced biogenic amine activity'. They paid particular attention to research in healthy subjects of the effects of PCPA, a drug that depletes brain serotonin levels. They noted that although there were reports of increased agitation and confusion, there was no tendency for subjects to become depressed. In more extensive animal studies, the observed behavioural changes included insomnia and hyper-aggressive behaviours, which were, 'if anything, reminiscent of mania'. Like Coppen in the previous decade, the researchers concluded that the depletion of monoamines 'is in itself not sufficient to account for the development of the clinical syndrome of depression'.[49]

Five years later, researchers reported that depressive patients with persistent serotonin disturbances had higher frequencies of depression than depressive patients who did not have such problems, and they concluded that this indicated a predisposing factor to depression.[50] This nuanced view was quickly transformed into something far more definite and by the 1980s the idea that low serotonin levels might directly cause depression had taken root, becoming known as the chemical imbalance theory of depression.[51] This concept was soon extended to explain other mental health problems, such as bipolar disorder, ADHD and anxiety, and it is now deeply rooted in popular conceptions, in pharmaceutical advertising and in the minds of journalists, although some psychiatrists now claim they never really embraced the theory.[52] At one level, the idea of a 'chemical imbalance' is merely shorthand for the empirical truth that a drug that alters brain chemistry might bring relief from distressing symptoms. But it is notable that the way patients – and physicians – think about this explanation for faulty brain function is, at root, not so different from Galen's 'four humours' explanation of disease that dominated European culture and what passed for medicine for well over a thousand years.*

---

*In his 1621 book *The Anatomy of Melancholy*, Richard Burton described how the humours and the mind interact in producing the symptoms of melancholy: 'For as the Body workes upon the Mind, by his bad humors, disturbing the Spirits, sending grosse fumes into the Braine; and so per consequens disturbing the Soule, and all the faculties of it, with feare, sorrow

One probable reason for the widespread acceptance of the chemical imbalance theory is that this is what it feels like. Depressed people often describe their symptoms as overwhelming, and say that the feelings of hopelessness, their inability to feel joy, are like some vast grey blanket covering the mind. Similarly, people who are addicted feel impelled by some force beyond their control – the 'monkey on my back'. Just because an explanation feels right does not make it true, but it might tell us why we accept inadequate and potentially false accounts.

It seems unlikely that there is a single explanation and a single best treatment for depression, or for any other mental health problem. This may explain the current lack of interest shown by major pharmaceutical companies in developing new drugs for mental health. Big pharma rode an astonishing wave of luck in the 1950s, but that is far behind us. In 2012 H. Christian Fibiger, a psychiatrist who has played a leading role in the global pharmaceutical industry, reported gloomily: 'Psychopharmacology is in crisis. The data are in, and it is clear that a massive experiment has failed: despite decades of research and billions of dollars invested, not a single mechanistically novel drug has reached the psychiatric market in more than 30 years.'[53] And that is not about to change any time soon – in 2010, Glaxo-Smith-Kline and AstraZeneca, two of the world's major pharmaceutical companies, both announced they were withdrawing from the development of new drugs designed to treat mental illness. The explanation was simple – follow the money. Both companies felt the probability of failure was too great to justify the risk to their shareholders. We cannot expect new treatments in the foreseeable future. As the British sociologist Nikolas Rose has put it: 'The pipeline is empty!'[54]

etc. which are ordinary symptomes of this Disease: so on the other side, the Minde most effectually workes upon the Body, producing by his passions and perturbations, miraculous alterations, as Melancholy, Despaire, cruell diseases, and sometimes death it selfe.' If you replace 'bad humors' by 'serotonin' and update the spelling and the grammar, it does not seem so different from something you might read today. Burton, R. (1621), *The Anatomy of Melancholy, What it Is. With All the Kindes, Causes, Symptomes, Prognostickes, and Severall Causes of It* (Oxford: Cripps), p. 119.

*

The other framework for understanding the origin of mental health problems that strikes a chord with the public is the role of genes in determining our behaviour. While genetics has become a major tool in investigating brain function in laboratory animals, it has been much less successful in explaining human brain function and mal-function. Nevertheless, many people embrace the suggestion that our genes lie at the root of our mental health problems. Again, the power of these apparent explanations appears to lie in our subjective experience. For many patients, their mental health problems feel con-stitutional – it is the way they are. But just because we are the way we are, that does not mean there is a strong or clear genetic component in some or all of those essential aspects of our character. Handedness is strongly constitutional, it feels very much 'the way we are', but the contribution of genetic factors remains obscure, and those that are involved are presumably very intricate.[55]

In reality, there are no examples of precise and identifiable major genetic components that explain mental health problems. Schizo-phrenia and autism both have strong heritable elements, but there is no gene for either of these conditions, any more than there is a gene for depression. Instead, dozens or hundreds of genes, each with a very small effect, may contribute to a predisposition to a given condition. The pursuit of the genetic bases of mental health dis-orders has, in one case at least, led to a dead end. From the late 1990s researchers became interested in a gene that codes for a serotonin transporter called SLC6A4. Variants in the gene seemed to be linked to depression, fitting in with the SSRI model. Hundreds of papers were published, virtually all of them contributing to a scientific con-sensus in which SLC6A4, along with a number of other genes, held the key to understanding depression, and in particular the link with anxiety. In 2019 researchers studied the role of all these genes using massive data sets (up to 443,264 individuals) and rigorous statistical techniques that required them to describe expected results before doing the study, rather than endlessly fishing for statistical signifi-cance afterwards. Their conclusion was that all that time and effort had been wasted; there was no evidence that the eighteen genes that were thought to play a role in depression, including SLC6A4, actually do so.[56]

According to Kevin Mitchell, a geneticist at Trinity College Dublin, because our diagnostic tools are so weak when it comes to mental health, we may be unable to identify the genes that are truly involved in a subset of these conditions.[57] If we were to start by identifying those genes that are consistently implicated in some patients with a particular diagnosis, we might be able to improve both our diagnostic techniques and also our understanding of underlying causes, potentially leading to more effective treatments.

Whatever the case, genes are not magic forces affecting our brains. In one way or another, they simply determine the proteins that our bodies produce; no matter how much a particular phenomenon may feel deeply rooted and simply the way we are, if it has a strong genetic component, that will ultimately be expressed in terms of the kinds of protein that are produced at particular times and places in our brain and that in turn will be affected by a myriad of environmental factors. Given our poor understanding of even very simple nervous systems, unravelling the genetic architecture of the human brain and how it interacts with the environment will be the work of centuries.

A huge project launched by the NIH in the USA called Psych-ENCODE involved fifteen research institutes and had the ambitious aim of identifying all the genetic factors that are involved in the human brain, characterising their roles in its evolution and development, and above all in neuropsychiatric disorders.[58] At the end of 2018 the first slew of papers from this project was published, but there were no massive reveals, partly because the approach assumed that categories associated with mental health are reliable and valid (e.g. 'schizophrenia' is a single thing that can be confidently identified) and that the ultimate cause is molecular. Neither of these things is known to be true. While the immense database produced by the consortium is a useful starting point, the underlying assumption – that there are reliable biomarkers of mental health problems that are tightly linked to genetic variants – is almost certainly flawed.

*

In the absence of any clear solutions to mental health problems, once-fashionable treatments such as electroconvulsive therapy (ECT) have been making a comeback. ECT, which induces a fit in the patient, was pioneered in the 1930s and by the 1940s became widely used in the USA as a treatment for depression.[59] But it fell out of fashion as pharmacological approaches appeared to offer a better alternative. There were also repeated claims of memory loss, and the public became horrified by a version of what happens to the patient during treatment, as seen in Miloš Forman's 1975 film *One Flew Over the Cuckoo's Nest* or, perhaps more movingly, in Sylvia Plath's 1963 book *The Bell Jar*:

> Something bent down and took a hold of me and shook me like the end of the world. Whee-ee-ee-ee-ee, it shrilled, through an air crackling with blue light, and with each flash a great jolt drubbed me till I thought my bones would break and the sap fly out of me like a split plant.[60]

Thanks to the use of muscle relaxants ECT is generally now much less horrific than Plath's account, but it retains a whiff of brimstone. Part of the reason for this is that we do not know how, or if, it works. Some patients find it a godsend, others are implacably hostile to it. Every year about 1 million people around the globe receive ECT.[61]

In another throwback to the 1950s, there is growing scientific and medical interest in LSD.[62] The drug that seemed to offer a window into brain chemistry, and – according to some users – into another reality, may have more use than mere recreation. Researchers are trying to understand how LSD produces its effects, in particular by altering connectivity within the brain, with the aim of producing a brain-wide model of neuromodulator activity.[63] The researchers claim this approach could 'lead to fundamental insights into human brain function in health and disease and be used for drug discovery and design in neuropsychiatric disorders'.[64] This may turn out to be a pious hope, but recreational drugs, once divested of the media-concocted fear that surrounds them, can lead to new treatments. Ketamine, a powerful anaesthetic that became popular as a recreational drug in the club culture of the 1990s, provoking shrieking tabloid headlines, has now been adapted and approved for use as an antidepressant in the USA – these therapeutic effects were first noticed by physicians in 2000.[65] Joshua Gordon, the head of the NIMH, described this as 'amazing news … the first truly novel antidepressant in decades and first to target treatment-resistant patients.'[66] Given the repeated cycles of boom and bust that have characterised the pharmaceutical treatment of mental health problems, we may have a less optimistic view in a few years' time.

It is striking that despite the massive shift in public awareness of mental health, the large amount of funding devoted to such research, and the growing numbers of scientists and physicians who are focusing on understanding the causes of mental health problems and finding potential solutions, the overall impact on the distress felt by patients has apparently been minimal. Thomas Insel, who led NIMH from 2002 to 2015, recently acknowledged this:

> I spent 13 years at NIMH really pushing on the neuroscience and genetics of mental disorders, and when I look back on that I realize that while I think I succeeded at getting lots of really cool papers published by cool scientists at fairly large costs – I think $20 billion – I don't think we moved the needle in reducing suicide, reducing hospitalizations, improving recovery for the tens of millions of people who have mental illness.[67]

It is hard to know what to say. We do not understand how a healthy brain and mind work, so it is hardly surprising that we do not know how to fix things when problems arise. Researchers like myself, working on systems far removed from mental health disorders, can recognise this immense gap between our fundamental knowledge of how the human brain works and the distant prospect of any effective treatment, and then turn back to the day job (in my case, studying the maggot nose). Things are not so easy for physicians faced with someone in desperate need, or, above all, for patients or their families (I am a member of such a family). Effective and safe therapies are urgently required. Ultimately, it would not matter if there is no deep understanding of how those therapies work, as long as they do.

FOURTEEN

LOCALISATION

A repeated theme in our long quest to understand the brain has been the claim that certain functions are localised to particular bits of our nervous system. At first, all this was speculation – the earliest written traces of these ideas were the various versions of ventricular localisation that were popular in Europe for a thousand years or so up until the sixteenth to seventeenth centuries. The early nineteenth-century phrenologists located dozens of vague psychological or behavioural constructs in lumps on the skull, and thence to fanciful organs in the brain. From the mid-nineteenth century onwards, firm evidence was found for localisation of some functions – for example, speech and motor control – but there was little evidence for the localisation of higher mental activities. Scientists lacked ways of measuring brain activity during a psychological or behavioural task.

The development of techniques for recording the electrophysiological activity first from the whole brain – the electroencephalogram (EEG), as initially developed by Berger in the 1920s – and then from regions and ultimately from cells, did not resolve the problem. These techniques were either incredibly general, like EEG, or were extremely specific, reporting only how a particular small area was responding. To be more confident about the unique role of a particular region in a given process, researchers needed a method that combined an overall measure of activity with the simultaneous detection of localised changes. This happened at the beginning of the 1990s, when a new brain-imaging technique led to an explosion of research which has transformed how we view the brain.[1]

The earliest developments in brain imaging were focused on anatomy. Computer-assisted X-ray computed tomography (the CT scan) became widespread in the 1970s and involved taking multiple X-ray images from around the patient's head. The CT scanner was developed in the 1960s by the British electrical engineer Godfrey Hounsfield (unknown to Hounsfield, Allan Cormack in South Africa had been doing theoretical work along similar lines). The scanner was first used in 1971 to image the brain of a patient with a suspected frontal lobe tumour – when the surgeon eventually operated on her, he said that what he found looked 'exactly like the picture'.[2] This new approach, relying upon the growing availability of computers to make the necessary calculations to produce the image, rapidly transformed diagnosis of physical brain diseases and in 1979 Hounsfield and Cormack were jointly awarded the Nobel Prize for their breakthrough.

CT scans, like simple X-rays, reveal structure at a fairly coarse level, and provide no direct information about function. That changed with the development of positron emission tomography (PET) scans, which appeared in the mid-1970s through the work of Marcus Raichle, Michael Phelps and Michel Ter-Pogossian. This technique measured the metabolic activity in a particular brain area by injecting a weak radioactive tracer, such as water made with a radioactive oxygen isotope. The rapidly decaying isotopes used in PET lead to the emission of gamma rays, which can then be detected.[3] Because these radioisotopes are rapidly integrated into the normal metabolic activity of the brain, in 1988 Raichle and his colleagues were able to use PET to show changes in localised brain activity when subjects were listening to words.[4] One of their 1988 papers had a title that laid out the new approach heralded by their work – 'Localization of Cognitive Operations in the Human Brain'.

Nevertheless, PET scans remained too slow and imprecise for researchers to be able to establish clear links between brain structure and subtle psychological functions, and above all they involved the injection of radioisotopes, limiting their attractiveness. The decisive breakthrough came about with the appearance of the most influential brain-imaging technique, functional Magnetic Resonance Imaging (fMRI), which measures the behaviour of atoms in a strong magnetic field and is now the dominant form of brain imaging. In 1991 Jack Belliveau and his colleagues published an article in *Science* showing changes to blood flow in the visual cortex during visual stimulation. The paper made the front cover of the journal (this is a big deal for a scientist), with a dramatic picture that showed a computer-generated greyscale image of a head from the rear. The back part of the head was sliced open, revealing a view of the surface of the brain, with small regions coloured in red and yellow, highlighting where the change in blood flow had been identified. The effect was as electrifying as the researchers had hoped; Nancy Kanwisher, now a leading fMRI scientist, was a young researcher at the time and recalled her excited reaction: 'these images changed everything ... Now, scientists could actually watch activity in the normal human brain change over time as it sees, thinks, and remembers.'[5]

Belliveau's method still relied on the injection of a contrast agent. The next step, taken within a year and more or less simultaneously by three groups, was to measure the level of oxygenation in the blood in a particular brain region by observing the behaviour of the iron atoms in oxygenated and deoxygenated haemoglobin (this measure is said to be Blood Oxygen Level Dependent – BOLD) while the patient was in the magnetic scanner carrying out simple psychological tasks.

fMRI detects differences in the magnetic responses of oxygenated and deoxygenated haemoglobin in the blood in different regions, which are expressed as bright colours on a picture of the brain. These images are said to show the brain 'lighting up' when the subject is engaged in a particular mental activity.* fMRI therefore reports a simple measure of the brain's physiology, its function as an organ in our bodies – those images do not directly describe anything like the actual activity of the brain's neurons. The brain in an fMRI scan is not a computer, nor a neural network, it is a gland.

---

*In case you are in any doubt, this is simply how differences in blood oxygenation levels are represented on a screen. Brains do not glow.

Kanwisher recalled her excitement when she first observed the recordings from an fMRI scan, in 1995 (the brain was her own):

Most thrillingly you could see in the raw time course of the fMRI response of individual voxels [the smallest unit of the image] that the signal was higher during the periods when I was looking at faces than the periods when I was looking at objects.

It seemed possible that cognitive processes could indeed be localised to very specific areas of the brain, as Raichle had proclaimed a few years earlier. The fMRI revolution had begun.

The final step was to convince sceptical scientists that fMRI does directly reflect the neuronal activity of the brain, by simultaneously recording from single neurons and measuring fMRI responses. This was an immense technical challenge, not least because placing electrodes in the magnetic field of the scanner induces electrical activity that can make it hard to identify the responses of the neurons the electrodes are recording from. Eventually, in 2001 – ten years after the BOLD breakthrough – Nikos Logothetis and his colleagues published an article in which they showed that fMRI is indeed tightly linked to neuronal activity.[6]

The impact of fMRI has been extraordinary. In less than thirty years, over 100,000 scientific articles have been published on the topic, currently at the rate of about 8,000 articles each year. These studies are much loved by the media, because of the striking images they involve, and the relatively simple story they appear to tell about how the brain works. For example, journalists tell us that fMRI can explain individual differences ('gambling addict's brain is built differently') or even read our minds ('fMRI knows your secrets'). In a weird twist, these images are sometimes used to confirm our subjective experiences, as though coloured blobs make our feelings more real ('brain imaging provides visual proof that acupuncture alleviates pain' or 'fat really does bring pleasure').[7] There are also spurious claims that brain scans can tell if you are lying, although no court has yet accepted its use, while one research groups says that the brains of murderers show a distinctive organisation.[8] The dystopian potential of this new technology is obvious, and is a growing focus of interest for sociologists and ethicists.[9]

*

The seductive power of fMRI, for both researchers and the general public, flows from its apparent ability to identify the precise areas of the brain that are activated when a particular mental activity is being performed. But these intuitive images are far less simple than they appear. The brain is a living organ, performing all sorts of actions, so there is a high level of activity against which the researchers have to pick out the change they are interested in. Calculating the often very small differences in BOLD levels between regions – differences that are expressed as bright colours, thereby implying a precision and intensity that the data may not justify – involves complex software packages that can produce important errors. In 2016 a group of fMRI researchers surveyed over 3 million recordings from published studies and discovered that 'the most common software packages for fMRI analysis … can result in false-positive rates of up to 70%. These results question the validity of some 40,000 fMRI studies and may have a large impact on the interpretation of neuroimaging results.'[10] Some researchers contested this alarming conclusion (the 40,000 figure was eventually downgraded to a few thousand), but many welcomed it as an opportunity to clarify the scientific underpinning of a method that has become widespread over the last three decades.[11]

This was not the first time that doubts have been raised about the interpretation of fMRI data. In 2008 Nikos Logothetis warned of major problems with the methods used in many fMRI studies. In particular, he focused on the fact that fMRI provides a surrogate measure of brain activity – blood flow in large regions, rather than the actual activity of cells, even if the two are clearly linked. Researchers had to keep this in mind, he argued:

> The limitations of fMRI are not related to physics or poor engineering, and are unlikely to be resolved by increasing the sophistication and power of the scanners; they are instead

due to the circuitry and functional organisation of the brain,
as well as to inappropriate experimental protocols that ignore
this organisation.[12]

This kind of criticism is particularly valid when fMRI studies
claim a particular part of the brain 'lights up' when, say, a given
emotion is felt. As fMRI researcher Russell Poldrack has pointed out,
'there is rarely such a one-to-one correspondence; most brain regions
are activated in many different contexts'.[13] Even where there is such
a tight link, it is merely a correlation – to prove that an area really
is the sole site of a particular thought or experience requires studying patients with lesions in that location, or stimulating the region
in some way. There are few such studies, and those that have been
carried out have sometimes failed to prove the predicted causal
connection.[14]

An example of the problems associated with interpreting fMRI
data sprang to global attention in 2009, following the appearance
of an article provocatively entitled 'Voodoo Correlations in Social
Neuroscience' in an obscure psychology journal. The article highlighted the 'puzzlingly high' correlations observed in many studies
between the activation of precisely identified parts of the brain and
particular behaviours or feelings. The authors went so far as to say
that the results of some of these studies were 'impossibly high' and
challenged the authors to reanalyse their data.[15]

To everyone's surprise, this argument, which rested on a rather
arcane statistical niggle, went viral.[16] (Sadly, at this point the editors
of the journal to which the article had been submitted lost their
nerve, and insisted that the paper be given a less flamboyant title.)
The outcome of the kerfuffle was a general agreement that improved
tests and greater rigour were needed. Given that the same kind of
things were said again five years later when the glitch in the imaging
software was discovered, it seems that such lessons are learned for a
while but are soon forgotten.

Some fMRI researchers have demonstrated their recognition of
these problems with a twinkle in their eye. A couple of months after
the voodoo row erupted, the chair of the Organization for Human
Brain Mapping spoke at their annual meeting and described one of

the presentations at the conference that he thought was of particular interest. This was a study by Craig Bennett and colleagues, and it was carried out on a dead salmon. Bennett's description of their procedure is worth a read:

> One mature Atlantic Salmon (*Salmo salar*) participated in the fMRI study. The salmon was approximately 18 inches long, weighed 3.8 lbs, and was not alive at the time of scanning. The task administered to the salmon involved completing an open-ended mentalizing task. The salmon was shown a series of photographs depicting individuals in social situations with a specified emotional valence. The salmon was asked to determine what emotion the individual in the photo must have been experiencing.[17]

While the ex-fish was questioned about the series of photos, the fMRI scan picked up several significant responses ($p < .001$, to use the jargon) in a small 27 mm^3 area of the defunct animal's brain. According to the traditional interpretation of fMRI results, the dead fish was processing the photos in a very particular region of its brain. The well-made point of this satirical study was that more rigorous and complex statistical methods were needed – there were random variations in the BOLD measurements in the dead animal's brain, which could be mistakenly interpreted as being significant. The unstated implication was that the same thing might be happening in investigations of living subjects. A fuller version was soon published in the *Journal of Serendipitous and Unexpected Results,* and in 2012 the study was awarded the Ig Nobel Prize.

A more serious approach to the same problem appeared in 2017, when a group of researchers led by Russell Poldrack, whose work was among the studies that had been skewered in the voodoo correlations article, made a series of recommendations for how the validity of fMRI studies could be improved.[18] In particular they suggested researchers should increase the number of subjects, and state clearly before the experiments begin which brain areas are expected to be involved (too many studies trawl through the data looking for some effect, which increases the likelihood of making a mistake). This will go a long way towards removing some of the methodological doubts about experimental design that are voiced by critics of fMRI.

*

But even if experiments are designed more rigorously, the more fundamental problem about what exactly fMRI tells us will remain. Few imaging papers use their findings to provide insight into how the brain works – this would involve describing what kind of representational and computational processes are being carried out in different regions – because their data have little to say on the matter. In this respect there is a gulf between the public perception of what such studies tell us and how the scientists involved interpret their data. Many fMRI researchers know that their results need to be integrated into a comprehensive framework of brain function, but at the moment they cannot do so – neither the appropriate data nor a suitable theoretical framework exist.

Despite these existential problems, fMRI researchers are understandably proud of their experiments and some are robust in their defence of the localisation that their results imply. For example, in 1997 Nancy Kanwisher used fMRI to identify a region of the brain that appeared to be involved in the processing of faces – the fusiform face area. Despite a long-running argument with researchers such as Jim Haxby, who has advocated a view of face processing that is more distributed,[19] as well as criticism over the method she used to identify the area in the first place, Kanwisher claims her results provide clear evidence of highly localised functions. And indeed, the electrophysiological work of Doris Tsao has shown that in the macaque the same brain area is also devoted to processing faces.

Although Kanwisher defends herself against the suggestion that she is some latter-day phrenologist ('no complex cognitive process is accomplished in a single brain area, and arguments for the specificity of these regions by no means imply that other brain regions play no role', she has explained[20]), and she points to stimuli that do not have specific processing areas (for example, flowers, spiders and snakes), she insists that there are clear anatomically based functional modules in the brain. She has even claimed that 'functional imaging of the brain has begun to reveal, in a very concrete way, the functional organization of the human mind'. And that functional organisation, she says, is modular – different bits of the brain do different things. As Kanwisher put it in 2017, in an article surveying the two decades since her discovery, 'I would argue that, for an understanding of the human mind, functionally specific brain regions do in fact carve nature at its joints, capturing structure inherent in both cognitive and neural data.'[21] Many neuroscientists would not agree with her.

Irritation at the occasionally facile interpretation of fMRI data, coupled with the recognised methodological problems, have created a gulf between those who use the technique and many other neuroscientists. For example, commenting on Twitter about the problems associated with fMRI, Albert Cardona, who is leading the attempt to establish the connectome of the *Drosophila* maggot brain, said: 'I have only once been to a neuroscience talk featuring fMRI where I did not feel I was being sold snake oil.' A few months later Daniel MacArthur, a leading human geneticist, tweeted that he had become 'conditioned to disbelieve anything involving the word "fMRI"', while in 2019 Dublin-based neurogeneticist Kevin Mitchell highlighted the inherent lack of resolution in fMRI studies of brain structure, bluntly stating that it was 'a general issue with neuroimaging – it's just a bit crap'.[22]

These critics are unimpressed because they are used to exploring very precise effects in individual cells or those exerted by particular genes, whereas fMRI cannot measure what is truly important for the brain – action potentials, the actual signal in the neuron.[23] The brain is so dense that in 2008 Nikos Logothetis estimated that in each pixel ('voxel' in fMRI jargon) of an image of the brain there are a staggering 5.5 million neurons, between 2.2 and 5.5×10^{10} synapses, 22 km of dendrites and 220 km of axons.[24*] The scale at which the real action is taking place – in individual cells and synapses and in networks of cells – is hopelessly blurred out by the coarseness of fMRI. Furthermore, fMRI measures activity changes in seconds, whereas neurons send information in the millisecond range. Even more strikingly, fMRI is unable to reveal one of the key aspects of how the

---

* Yes, those figures are correct.

brain works – the difference between activation and inhibition. fMRI cannot tell us what single cells, or networks of cells, are up to. Even at the level of neural tracts, it cannot tell us meaningfully what is happening, merely where, at an extremely coarse level, something is happening relatively more or less than elsewhere. Perhaps.

In 2015 Doris Tsao and her colleagues showed that the coarse-grained nature of fMRI means that even negative results cannot be relied upon – when an fMRI study says that a given region does not 'light up', we cannot draw any reliable conclusions. Comparing the results from fMRI and from single-cell recordings in the face-processing region of the macaque's visual cortex, Tsao's team found that fMRI suggested that this region did not process the identity of human faces. However, the far more precise single-cell recordings showed that this information was indeed present in the activity of cells in the region; it just could not be detected because of the lack of precision in fMRI. The cells that were involved in recognising faces were too few, and too scattered, to be identified by imaging.[25]

Some fMRI researchers have kicked back against this kind of criticism. In 2017 Olivia Guest and Bradley Love of UCL's Alan Turing Institute used neural networks to explore fMRI data, looking at how similarities and differences between visual objects were represented in the data.[26] The deep learning network identified signals in fMRI data from the initial levels of the visual processing pathway, but at higher levels of the brain it was less well able to pick out clear responses to precise objects. The representation, argued Guest and Love, tended to become more diffuse and symbolic at these higher levels. Surprisingly, Guest and Love advanced an explanation of the material basis of perception that appeared to eschew any focus on anything like the cellular level:

> fMRI's success might mean that when one is interested in the nature of computations carried out by the brain, the level of analysis where fMRI applies may be preferred. To draw an analogy, one could construct a theory of macroeconomics based on quantum physics, but it would be incredibly cumbersome and no more predictive nor explanatory than a theory that contained abstract concepts such as money and

supply. Reductionism, while seductive, is not always the best path forward.

Guest and Love could be right, but there is one very good reason to think they are not. As Barlow argued, the functional unit of the brain is the neuron, organised as a node in a network. What the brain does, however mysterious it might be, is in the end reducible to the firing of neurons. Those neurons combine to show coordinated functional activity that produces psychological phenomena, but that does not mean we can simply ignore that such population-level activity is composed of many individual neurons, or that any calculations carried out by ensembles of neurons will be based on the activity of cells. The success of reductionism in explaining how the human brain works will not mean producing a theory based on the individual activity of 80 billion neurons, but rather showing how the processing of sensory phenomena, and the mental life of humans and animals, can be explained by the patterns of activity of populations of neurons. And that will require knowing what individual cells are doing, even if they are subsequently analysed at a population level.

This highlights the fundamental weakness of fMRI – it is too coarse to allow a real understanding of the computational activity of the brain. Far more precise imaging methods will need to be developed – precise in terms of time, space and function – and explained in terms of more detailed connectomes.[27] There are hints that such developments may be possible, with the advent of ultra-high-field MRI giving access to submillimetre scale precision. However, this technique is in its infancy, and it is still far from enabling us to distinguish the activity of even hundreds of thousands of cells.[28]

One repeated claim is that imaging techniques such as fMRI reveal anatomical and functional differences between human male and female brains, and that these differences can explain differences in behaviour. On one level the existence of differences between brains is a truism – you and I have anatomical and functional differences between our brains simply because we are not the same individual. There are plenty of reasons to assume that overall, taken as two groups, the brains of men and women will have different characteristics – men and women generally have different roles in modern society and tend to behave in different ways (overall, men are more aggressive, for example). In terms of our evolution, sexual selection, operating within and between the sexes, was a feature of our past (and perhaps our present), while our different roles in reproduction – in particular through maternal behaviour – played a decisive role in shaping human society. All these factors will have shaped the anatomical, functional and behavioural differences between the sexes. The key questions are what those anatomical differences are, whether we can detect them and, above all, how determining they are for our behaviour.[29]

There is one clear difference between the brains of men and women: men's brains are, on average, larger. This is partly because, on average, men are larger than women. But this is a population-level difference and cannot be used to tell whether a given brain is a man's or a woman's. There is no archetypal 'male brain' or 'female brain' that can be identified from a scan or a dissection.[30] The population-level differences that have been identified are hard to interpret – the relative size of the corpus callosum shows population-level differences between the sexes, and the connectivity of the brains of men and women differs when they are put in various testing procedures.[31] The existence of differences in brain structure between the sexes in newborns reinforces the suggestion that there are innate biological differences between us, but some of these differences disappear later in life, while others appear.[32] Above all, the existence of a difference tells us nothing about its significance or consequences.

No imaging technique has been able to reveal structural differences that can explain differences in behaviour between the sexes, because our measures are too crude. For the moment, the nature of the differences in brain function that underlie differences between the sexes remain unclear. That such differences exist seems certain – one set of behaviours that has been sustained throughout our evolution are those involved in mating and parenting; no matter how fluid our past sexual activity may or may not have been, selective pressure on these behaviours will have left traces in our genes, and in our brains. As to how constraining such differences might be – historical and recent changes in society would appear to suggest that many functional constraints, if they exist, are weak and that much of human behaviour is highly plastic.

*

One response to the dramatic claims of some fMRI studies, articulated by Fernando Vidal and Francisco Ortega in their book *Being Brains*, a boisterous critique of all things neuro, is simply to ask 'So what?', much as Marr had suggested we respond to the discovery of a grandmother cell.[33] Vidal and Ortega propose that we take fMRI results at face value and simply ask what it really means for a particular behaviour, idea or emotion to be located to a particular brain area, how that discovery sheds light on the organisation or evolution of thought and behaviour, what it tells us about brain function. And often the answer is, remarkably little – beyond a partial localisation that may or may not be true.

Sometimes there is not even much evidence of localisation. In 2016 a group of researchers at Berkeley described an experiment in which seven subjects lay in a scanner and listened to two hours of stories from *The Moth Radio Hour*, involving a total of 10,470 words.[34] The aim was to see how different meanings corresponded to activity in different parts of the cortex. The researchers grouped the words into twelve categories (tactile, visual, emotional, social and so on) and discovered that the responses of the brain to these groups covered the whole of the cortex, with virtually no localisation. Although some of the categories consistently excited the same area in all subjects, relatively little difference was found in the responses of the two hemispheres, even though, going back to Wernicke, studies of patients with brain lesions suggest that the left hemisphere is where words are processed.[35] This implies that assumptions about highly localised bits of our brain representing particular words or concepts are wrong. To the extent that fMRI can truly provide an answer to this kind of question, those functions seem to be distributed across different parts of the brain.

Some researchers have dismissed fMRI interpretations as 'the new phrenology' or 'internal phrenology'. Despite the occasionally overblown claims about some fMRI studies, such criticisms are wrong and unfair – fMRI represents a powerful non-invasive technique for identifying regional variation in activity and correlating those changes with behaviour or psychological state. It can be argued that imaging studies have done much to emphasise the dynamic role of the brain, and to highlight the significance of connections between regions during mental processing.[36] Furthermore, Nancy Kanwisher's fMRI work on face recognition regions was the foundation of Doris Tsao's research on the activity of single cells in this area. There is indeed a face-processing area in the human brain, and fMRI helped to identify it.

Nevertheless, despite all this immense ingenuity and astonishing technological prowess, fMRI studies have not significantly contributed to our understanding of how the brain works in terms of creating a model of its overall activity – with one potential exception. In 2001 Marcus Raichle's group used PET scans to identify a set of areas, symmetrically arranged on both sides of the brain and scattered across the cortex, that decreased their activity during attention-demanding tasks compared to their level of activation when the subject was lying still.[37] These regions have become known as the default mode network – they seem to relate to the intrinsic activity of the human brain when nothing else is happening.

For the last two decades brain-imaging researchers have become increasingly interested in this enigmatic phenomenon, which has also been observed in other mammals. It appears to be involved in large-scale base-level functional coordination of the brain's activity, and there is growing evidence that it is actively involved in cognition and has some effects on memory.[38] But although over 4,500 papers have been published on the topic, it has so far defied simple explanation. The way the network sprawls over the brain, and its involvement in base-level functions are extremely intriguing, but despite the recent identification of electrophysiological correlates of changes in the default network when an individual performs certain tasks, for the moment it can appear to non-specialists to be a network in search of a function.[39]

Locating functions to particular structures, or even suggesting that particular concepts are somehow represented by the activity of a particular part of the brain, has not explained how the ensemble of neurons interact to produce either perception or behaviour, nor will it be able to do so. A map – and at their best that is what fMRI data are – does not tell you how something works. Where is not how. The next time you read a claim that a particular ability, or emotion, or concept has been localised to a particular region of the human brain using fMRI, ask yourself, 'So what?'

*

There is a deeper issue at stake with approaches that seek to localise function to particular structures. There is clear evidence from neuroanatomy that different regions of the brain are discrete and specialised – that they are linked with a particular sensory modality, or they possess a particular type of cell. Lesion studies of patients and of animals often indicate that certain areas play a significant role in a particular ability or function and are often taken as key evidence in favour of localisation. The problem here is that not only can brains recover from damage, in particular in the young, but also that much of the logic behind the tendency to identify location and function is faulty. Just because a particular cell is activated by a photo of Jennifer Aniston does not mean that is all the cell does – other faces, or other stimuli entirely, may also be represented – or that cells from no other network are involved.

Virtually all of our limited understanding of brain function flows from results obtained through our adoption of something like Steno's 1665 suggestion that we should aim 'to dismantle it piece by piece and to consider what these can do separately and together'. The dismantling has been done in a variety of ways – surgically, genetically or with the point of an electrode – but all these methods use essentially the same approach. In 2017 the underlying problem was explained in fancy language by French neuroscientist Yves Frégnac, who argued that because of the complexity of nervous systems, 'causal-mechanistic explanations are qualitatively different from understanding how a combination of component modules performing the computations at a lower level produces emergent behaviour at a higher level'.[40] In other words, we are using a relatively crude model of cause and effect, based on some major and mistaken assumptions, to investigate how complexity emerges from interacting units in the brain.

For well over a century, scientists and philosophers have repeatedly highlighted the same issue, which relates to how we can logically locate function to a particular structure. In 1877 the German philosopher Friedrich Lange used a simple analogy:

> If any one shows me that a slight injury to some portion of the brain makes an otherwise healthy cat give up mousing, I will believe that we are in the right path of psychological discoveries. But even then I will not assume that the point has been found in which the ideas of mouse-hunting have their exclusive seat. If a clock strikes the hours wrongly because a wheel is injured, it does not follow from this that it was this wheel that struck the hours.[41]

As my friend US neuroscientist Mike Nitabach put it on reading a draft of this book, claims of localisation of function are generally a gross overextrapolation – at best we have identified a location necessary for function, and often merely shown a correlation between location and function.

Throughout the second half of the twentieth century the British psychologist Richard Gregory repeatedly raised this problem. At the 1958 symposium where Selfridge unveiled his Pandemonium program, Gregory argued that identifying function by ablating or lesioning a particular structure was not only logically flawed but also failed to provide real insight – you might be focusing on the output of a damaged, misfiring system. To properly understand the role of the component, you need a theoretical model of how the system works. And therein lies the difficulty, argued Gregory: 'The biologist has no "Maker's Manuals" or any clear idea of what many of the "devices" he studies may be. He must guess the purpose, and put up for testing likely looking hypotheses for how it may function.'[42] During the discussion of Gregory's paper, McCulloch expressed his agreement, pointing out that 'the argument from lesions, unless we are careful, leads us into utter rubbish'.

Over the years Gregory extended this critique, presenting all sorts of analogies to undermine researchers' confidence that if a particular structure was ablated, the altered behaviour had to be localised to that part of the brain. These analogies often involved what at the time was cutting-edge technology, but which now looks quaint or even mysterious to young people – valves in television sets, spark plugs in car engines, and so on – but they all focused on the same problem of interpreting what appear to be simple experiments that remove a key component.[43]

One of Gregory's most thought-provoking arguments took issue with what had been Steno's starting point. Steno suggested that we could understand a brain in the same way we understand a machine – by taking it apart and isolating functions to each part. In his sprawling 1981 masterwork, *Mind in Science*, Gregory questioned whether this was actually true, pointing out that particular functions could rarely be revealed by removing parts one by one:

> One finds, rather, that bizarre things happen when parts are removed; or nothing may happen, except under special conditions such as extreme demands or loading. For example, spokes of a bicycle wheel can be removed, one by one with little effect until there is a sudden collapse. Removing parts from an electrical circuit may produce output characteristics that were not present – such as whistles for a radio or complex patterns for a television ... In truth the relations between parts, and their causal interactions, and the functions that they achieve, are highly complex and subtle beyond common understanding. It is particularly difficult to say *where* functions are located. This is a most serious problem for brain research.[44]

Although these arguments all hit the mark, they had little effect on the wave of ablation and stimulation studies that marked the 1960s and 1970s, nor did they affect the subsequent tide of investigations that used the same method when it was applied to studies of gene function in the nervous system and elsewhere. The twenty-first century has seen a massive growth in optogenetic, imaging and single-cell methods, many of which are based on the assumptions that Gregory was criticising. Even where direct manipulation of a particular cell or network alters or restores a given function, that still does not mean that the function is located in that structure. What it does show is that the structure is required for the function, which generally involves a large network of neurons. Those grandmother cells that respond to an actor's face or an equation are not strictly grandmother cells, they are merely one part of a vast network that is active when presented with the stimulus, one cell that scientists happen to have recorded from.

To incorporate both the apparent specificity of function suggested by many fMRI studies, and the recognition that aspects of function are distributed across the brain, Karl Friston has explored what he has described as 'the dialectic between functionalism and connectionism',[45] seeking correlations between patterns of activity shown by different brain areas during a given behaviour. He calls this 'functional connectivity', and it is an approach that has attracted a great deal of interest from fMRI researchers. However, this mathematical description of broad-scale correlations between the activity of different brain areas has yet to prove itself on smaller, more precisely understood animal brains.

Repeatedly, when a function has been claimed to be localised to a particular structure, scientists eventually discover that the situation is more complex. For example, for over thirty years researchers studying fear in mammals have focused on the role of the amygdala – paired structures deep in the brain. There is a rare human condition called Urbach–Wiethe syndrome, which leads, among other things, to the degeneration of the amygdala. Patients with this condition often show reduced levels of fear. Awareness of this research has become so widespread that internet forums have recently been the focus of debates over whether it is possible to get amygdalas removed, in order to conquer fear. Fear is located in the amygdala, it is suggested.

In reality things are not so simple: in rodents (the main animals to have been studied) these structures are now associated with defensive behaviour – specifically, freezing – rather than the emotion of fear itself, which is distributed across various areas of the brain.[46] In humans, the effects of Urbach–Wiethe syndrome are not limited to the amygdala, nor is fear completely abolished (patients with the

condition are still afraid of being suffocated). And the amygdala is
not only involved in fear, it seems to play a role in emotional and
autonomic responses linked to pain and other negative stimuli, as
well as integrating various sensory stimuli that are not associated
with punishment and reward, and even (in mice) sex differences in
parental behaviour.[47] Localisation of function has become blurred
and more complex than originally claimed. Fear is not specifically
located in the amygdala, nor does the amygdala only have a role in
fear. And, in case you had any doubts, you should not have them
removed.

Not everything about localisation of function in the brain is confusing. At the sensory level, some networks of cells carry out precise
and limited actions and there is clear localisation of function, at least
in the initial stages. Sensory stimuli are initially processed separately – there is no evidence of olfactory signals being represented
in primate visual cortex V1, nor are visual signals found in the vertebrate olfactory bulb. This also applies to the equivalent structures
in an insect brain. However, in mice, Terry Sejnowski's group has
shown that top-down inputs from the hippocampus and the entorhinal cortex project to the area of the olfactory bulb where odours
are identified.[48] This implies that memory or stress may influence
how we perceive odours; similar pathways may be found relating
to vision, suggesting that the function of regions of the brain that
process sensory stimuli may be more complex than was originally
thought. And just a few synapses away from these areas, in the
higher regions of the brain where, presumably, thinking is done in
both you and an ant, things get interesting. The signals from the different sensory modalities are integrated and our understanding of
what is localised becomes messy.

The same is true of some of the most intensively studied bits
of the mammalian brain. Individual place cells in the hippocampus
encode location, but they also respond to particular sensory modalities corresponding to events that have occurred at that place, such
as touch and smell and light. The cerebellum is indeed associated
with motor control, as shown by Flourens in the nineteenth century,
but it is now known to be involved in a wide range of psychological
functions and receives input from cortical and sensory regions and

projects forward to the reward regions of the brain. It also plays a role in visual attention and social behaviour.[49]

Function is both localised and distributed – or rather, to be clearer, both terms are misleading: localisation is rarely precise, and distributed functions are also localised to particular networks and cells, even if these may sprawl over the brain. Brain function therefore involves both segregation and integration.[50] Even the simplest animal brain is not homogeneous but has a highly developed internal structure – but in most cases a single function cannot be precisely localised to a given area; for the function to exist, that region needs to be integrated into a functional whole.

*

Such is the power of the idea that different bits of the brain carry out specific tasks, much like a machine, that we repeatedly find ourselves pulled back by the claims of the highly specific localisation of some exciting psychological capacity. For example, a few decades ago the most important idea to influence popular perceptions of what science has to say about the human brain was the entirely erroneous idea that we have, deep in our skulls, a 'reptilian brain' that is responsible for our basest behaviours.[51] This view – which is still doing the rounds – was based on the work of neurologist Paul MacLean, who claimed that we have three brains:

> One of these brains is basically reptilian; the second has been inherited from lower mammals; and the third is a late evolutionary development that has made man peculiarly man ...
> The reptilian brain is filled with ancestral lore and ancestral memories and is faithful in doing what its ancestors say, but it is not a very good brain for facing up to new situations.[52]

MacLean's ideas, which neuroscientists never took seriously, surged into popular culture in the 1960s and 1970s when they were adopted by two of the period's most influential popular science writers. In 1967 Arthur Koestler summarised MacLean's work in his best-seller *The Ghost in the Machine*, chucking in everything from the Christian doctrine of original sin to Freud's theory of infantile sexuality to bolster his odd claim that conflicts between the three brains 'provide a physiological basis for the paranoid streak running through human history'.[53]

Back in the 1960s, this kind of guff sold like hotcakes, and MacLean shot to fame, speaking to packed lecture theatres. One of the people who flocked to hear MacLean was the astronomer Carl Sagan, who subsequently published a Pulitzer Prize winning book, *The Dragons of Eden*, based on MacLean's ideas.[54] Like Koestler, Sagan mixed small pieces of scientific fact with a huge dollop of psychoanalytic nonsense and a thick wodge of poorly understood anthropology, and then washed the whole thing down with a strong dose of speculation. For example, as well as wondering if childhood nightmares of monsters might be some relic of our ancestors' encounters with dinosaurs and owls, Sagan suggested the story of the serpent in the Garden of Eden might be a metaphor for our reptilian brains.[55] This would explain, he claimed, why myths about dragons are so pervasive – the dragons of Eden were us, man …*

In 1990, aged seventy-seven, MacLean summarised his ideas in *The Triune Brain in Evolution*.[56] *Science* magazine reviewed the book respectfully but savagely, pointing out that his basic hypothesis was 'not consistent with current knowledge', which explained why neuroscientists 'came to ignore the idea'.[57] While MacLean's desire to put the functioning of the human brain into an evolutionary context was laudable, his fundamental idea never had a leg to stand on.† As the Oxford anatomist Ray Guillery suggested in *Nature*, it should have been classified as neuromythology.[58]

More recently, similar excitement was caused by a more reliable and far more interesting finding. In 1992 researchers at the University

---

*I can hear you saying 'All right, clever-clogs, how much of your book will stand up to rereading in forty years' time?' I hope the vast majority of it, even if some parts will seem quaint, naive and overoptimistic.

† Three simple examples reveal the depth of MacLean's errors. The 'reptilian brain' is not reptilian, it is also found in fish. The early mammalian brain is not solely responsible for parenting behaviour – birds, which are excellent parents, do not have this structure. The neocortex is not a mammalian adaptation – elements of it can also be seen in birds and fish.

of Parma described their fortuitous discovery of a set of neurons in the ventral motor precortex of the monkey brain that fired not only when the monkey performed an action but also when the monkey saw that action performed by another individual.[59] Cleverly named 'mirror neurons' shortly afterwards (it always helps to have a snappy description), these cells soon attracted a great deal of attention, not to say hype. Some researchers speculated these cells might have been involved in the evolution of language, while others suggested that the deficits in social interactions shown in autism may be due to dysfunctional mirror neurons.[60] In 2006 the *New York Times* proclaimed that mirror neurons are 'cells that read minds', while one neuroscientist described them as 'the neurons that shaped civilization', because of their alleged role in enabling humans to feel empathy.[61] None of these claims is true.

When neurons with a mirror function were finally identified in the human brain in 2010, they revealed something unexpected and very interesting. The cells fired when patients either observed an action or carried it out (some of the neurons produced an inhibitory response, suggesting their role was to prevent imitation), but unexpectedly, these cells were not restricted to the areas where they had been identified in monkeys – 11 per cent of the human mirror neurons were found in the hippocampus.[62] The presence of mirror neurons in part of the motor cortex, apparently carrying out a cognitive function, and their presence in the hippocampus, where they are involved in an apparently motor function, shows that the sensory/motor division is not as absolute as it is often presented. Something similar to mirror neurons – 'simulation neurons', which appear to represent the behaviour of another animal when it is making a decision – have been identified in the amygdala of rhesus macaques.[63] This highlights not only that the amygdala is involved in responses other than fear, but also that representations of others and their behaviour can be found in many different regions. Mirror neurons – if indeed they all truly share a common identity and are not simply united by us giving them the same fancy name – are distributed across the brain, along with their mixed function.

This complex reality of exceptions to the identification of function to a particular structure is reinforced by some recent clinical examples of surprising plasticity in the human brain. In Marseilles there is a middle-aged man who has a cortex that is compressed into a tiny, thin layer of cells, yet he is of nearly average intelligence and holds down a job as a civil servant, while a number of women in Israel have been identified who have no detectable olfactory lobes in their brains, but have a normal sense of smell.[64] A young Chinese woman completely lacks a cerebellum; although her voice is slurred, and she has slight mental retardation and poor coordination, these symptoms are nowhere near as severe as would be expected from what happens to an animal when this structure is removed.[65] Finally, an Argentinian woman who suffered two catastrophic strokes that extensively damaged brain areas involved in sensorimotor skills and high-level psychological functions, has made a near-complete and unexplained recovery.[66]

Recent studies of animals have revealed further problems. In rats and songbirds, it has been claimed that certain learned behaviours are controlled by very specific brain areas, because they can be disrupted by briefly inactivating that part of the brain. Paradoxically, if those structures are *permanently* damaged, the animal can recover its learned ability. The explanation of this surprising plasticity is that structures that are dependent on the altered area cannot rapidly change their activity over a short period of time to respond to the new situation, leading to an absence of the relevant behaviour during the experiment. However, over longer periods, such as those seen after recovery from surgery, they are able to do so, much as some stroke patients can recover aspects of their former abilities given time.[67]

The underlying explanation of these kind of reports is that structures in the brain are not modules that are isolated from one another – they are not like the self-contained components of a machine. Because brains are composed of living matter, neurons and networks of neurons are both interconnected and able to affect adjoining regions by changing not only the activity of neighbouring structures but also the patterns of gene expression. Function can be spread, and even induced, by both synapses and neuromodulators, which can act in complex ways.[68] This presumably underlies some examples of plasticity and emphasises the difficulty of precisely identifying function with a given location.

Even something as apparently simple as thirst turns out to be remarkably complex: in 2019 researchers reported a study of the activity of 24,000 neurons in thirty-four brain regions of the mouse brain as the mice drank and became sated. Over half of these neurons were involved in various ways in this extremely simple behaviour – thirst, and the behavioural response to this sensation, appears to be broadly distributed across the mouse brain.[69] Furthermore, parts of the brain not usually considered to be involved in motor control are activated when mice are running or moving their whiskers, affecting the activity of neurons in the visual cortex. A similar mouse study of 30,000 neurons from 42 brain regions, carried out by researchers at University College London, also found that when the mouse began doing something, neurons from all regions became active. However, when the mouse was given a choice, very specific cells in particular parts of the brain responded. These data give us a glimpse into the complexity of the brain, and how it uses both localised and distributed functions, even though the actual circuits involved remain a mystery.[70]

Finally, the way that the mammalian brain does things may not be the best or the only way, indicating that the identification of structure and function may not necessarily be tightly determined. The role of the cortex in higher psychological functions has been repeatedly demonstrated by stimulation, ablation and comparative studies. Humans, with our complex cortical folds, are seen as showing the highest levels of both cortical complexity and psychological richness. And yet birds, which do not have a layered cortex like that found in mammals, are capable of some highly complex psychological processes, matching mammals in many respects. New Caledonian crows can not only make tools but also make tools to make tools. Magpies have even passed the 'mirror self-recognition test', which is generally seen as a proxy for an animal having a concept of self.[71] Even though avian and mammalian ways of organising the brain may have a common developmental root, the key point is that different structures can apparently give rise to the same function.[72]

This raises the biggest problem of all in understanding how the brain works – how it produces consciousness, and in which animals. For centuries this issue has been the domain of philosophers; in the last half century or so, scientists have begun to seriously address this question of questions.

FIFTEEN

CONSCIOUSNESS

1950 TO TODAY

In 2005 the editors of the journal *Science* highlighted 125 unsolved scientific questions that they thought we had a good possibility of answering in the coming decades. Second on the list, after 'What is the universe made of?', was 'What is the biological basis of consciousness?'[1] Just sixteen years earlier, the British psychologist Stuart Sutherland had been much less optimistic: 'Consciousness is a fascinating but elusive phenomenon; it is impossible to specify what it is, what it does, or why it evolved. Nothing worth reading has been written about it.'[2]*

The profound shift in attitude that occurred in that brief space of time reflected a renewed interest in consciousness – there are now hundreds of books devoted to the question, while TED talks on the topic garner millions of views. Since Sutherland wrote, scientists have published over 16,000 articles with the word 'consciousness' in the title. And yet there is no agreement about how – and in some quarters even if – the brain produces consciousness.[3]

---

*In the second edition of *The Hitchhiker's Guide to the Galaxy* the one-word entry on Earth was altered from 'harmless' to 'mostly harmless'. Sutherland's quip about consciousness could perhaps be amended to: 'Nothing *much* worth reading has been written about it.'

Francis Crick is often blamed for this explosion of interest. At the time that Sutherland was writing, Crick was arguing that researchers should look for what he called the neural correlates of consciousness – patterns of neuronal activity correlated with consciousness-related phenomena. But although Crick's intellectual drive helped shape the modern scientific study of consciousness, the topic had never really gone away.[4] One of the earliest collective explorations of the question took place in August 1953, when twenty scientists, including Edgar Adrian, Donald Hebb, Karl Lashley and Wilder Penfield, met in a chalet in Quebec for a five-day symposium on Brain Mechanisms and Consciousness.[5] The meeting was dominated by a breakthrough made four years earlier by Horace 'Tid' Magoun, in which electrical stimulation of the brainstem of an anaesthetised cat could induce the kind of EEG changes seen when an animal wakes up.[6] Now that EEG could be manipulated, it seemed as though scientists had a way of investigating the nature and location of consciousness.

However, in a prophetic opening presentation, Magoun warned his colleagues of 'the head-shaking sympathy with which future investigators will probably look back upon the groping efforts of the mid-twentieth century, for there is every indication that the neural basis of consciousness is a problem that will not be solved quickly'.[7] Tid would probably have been amused to learn that nearly seventy years later the neural basis of consciousness is still not understood, nor, the optimism of *Science* magazine notwithstanding, is there any sign of an answer on the horizon. Despite massive technological innovations, the two key questions debated in Quebec – localised versus distributed activity, and the significance of physical correlates of consciousness – are still the focus of our long quest to understand the brain.

The Quebec meeting heard some compelling evidence for localisation of function from Penfield, who described his work showing that electrical stimulation of the cortex could evoke both dream-like states and motor activity. But as Penfield explained, although the patient's body moved if the motor cortex was stimulated, the subjects always said that this occurred 'independent of, or in spite of, their own volition'. Similarly, the very precise experiences he was able to evoke never resembled 'things seen or felt in ordinary experience' but were more like dreams. This was not what would be expected if the part of the brain directly involved in consciousness was being stimulated.[8]

In a way, this conclusion was not surprising. Most of the speakers at the meeting felt that consciousness was a function of some kind of integration of nervous activity across the brain. As Stanley Cobb (no relation) had explained the year before:

> It is the integration itself, the relationship of one functioning part to another, which is mind and which causes the phenomenon of consciousness. There can be no centre. There is no one seat of consciousness. It is the streaming of impulses in a complex series of circuits that makes mind feasible.[9]

With new EEG technology and new surgical interventions, it now seemed possible to identify a focus for that integration. But a fundamental problem remained – as the French physiologist Alfred Fessard put it, the key issue was how localised that integration might be, whether it should be seen as 'concentrated or diffuse, as specific to a narrowly limited region of the brain, or as capable of being identified with variously situated nervous structures'.[10] As the Quebec discussion continued, even the excitement over EEG as a measure of states of consciousness and their localisation began to seem overblown. The ever-sceptical Lashley pointed out that it was not clear how EEG was related to patterns of neuronal activity or states of consciousness, forcing the EEG enthusiast Richard Jung to admit that 'no absolute correlation is possible between the form of the EEG and the state of consciousness or perception'.[11] Penfield concluded the meeting by admitting his complete ignorance of how neural activity is turned into thought: 'Here is the fundamental problem. Here physiology and psychology come face to face. We are far from this final understanding and life is short!'[12]

There was still hope that the problem might be solved, however. Hebb outlined a scientific approach that was to prove highly influential when it was independently propounded by Crick four decades later:

We should not try to devise a theory that will be completely adequate to account for all of what we know that human beings know, feel and do. We should try to account for those aspects of the problem we might have some chance of accounting for and not worry if the theorising appears to be inadequate in some respects to cover all known features of the system.[13]

Not everyone agreed that consciousness arises from neuronal activity. Earlier in 1953, John Eccles had published *The Neurophysiological Basis of Mind*, in which, following his PhD supervisor Sherrington, he suggested that the mind was a non-material substance that merely interacted with the brain in some way, effectively repeating Descartes's dualist ideas from three centuries earlier.[14] Eccles's views had been rehearsed in the pages of *Nature* in 1951, when he outlined his hypothesis that the density of neurons in the cortex somehow turn it into a detector of non-physical reality: 'mind achieves liaison with the brain by exerting spatio-temporal "fields of influence" that become effective through this unique detector function of the active cerebral cortex'.[15] Eccles argued that psychokinesis and other supposed extrasensory abilities were of 'particular significance' in providing support for his idea.

Eccles's approach was not viewed with enthusiasm by other scientists – his *Nature* article, not that journal's finest hour, has been cited only ten times, mostly by historians. At the conclusion of the 1953 Quebec meeting, Horace Jasper accepted the difficulty of linking mind with brain activity before concluding waspishly: 'Dr Eccles has attempted to meet this problem by leaving the physical and by going to the spiritual world for an explanation.'[16] Eccles, who was a devout Catholic and worked for some time with the philosopher Karl Popper, remained a dualist throughout his life, although he changed the detail of his views several times, propounding each successive version with equally combative certainty.[17]

One of Eccles's greatest contemporaries eventually took a similar view. Wilder Penfield's whole career was based upon the assumption that 'activities of the highest centres and mental states are one and the same thing, or are different sides of the same thing'.[18] However, in 1975, shortly before his death, Penfield explained that 'after years of striving to explain the mind on the basis of brain-action alone, I have come to the conclusion that it is simpler (and far easier to be logical) if one adopts the hypothesis that our being does consist of two fundamental elements'. Penfield's justification was that 'there is no good evidence, in spite of new methods, such as the employment of stimulating electrodes, the study of conscious patients and the analysis of epileptic attacks, that the brain alone can carry out the work that the mind does'.[19] But just because there was as yet no explanation of consciousness, that did not mean that the prevailing materialist framework was wrong.

*

In the 1950s, following the publication of Gilbert Ryle's *The Philosophy of Mind*, philosophically minded psychologists (and psychologically minded philosophers) began to focus on the nature of consciousness. In 1956 the Oxford psychologist Ullin Place argued that a 'reasonable scientific hypothesis' was that consciousness is a process in the brain, and insisted that it was not good enough to dismiss this suggestion on philosophical grounds alone.[20] Three years later, the University of Adelaide philosopher Jack Smart developed Place's arguments, helping to establish the philosophical basis of the long-held working hypothesis that consciousness and brain processes are different aspects of the same thing.[21]

Given the difficulties of attacking the question head-on, scientific interest in consciousness began to wane, to the extent that in 1962 the US psychologist George Miller argued, 'We should ban the word consciousness for a decade or two.'[22] Whatever the seriousness of Miller's suggestion – he used the word over eighty times in his book, including devoting a whole chapter to the question – in the following year a major review on brain function was published that surveyed over 1,000 scientific articles and yet managed to avoid using the term consciousness.[23] But despite the absence of the word, the fundamental question was still there – the review highlighted an astonishing set of results that radically challenged our view of how the human brain works and are still profoundly disturbing.

The fashion for psychosurgery that dominated US psychiatry in the middle of the twentieth century did not only have catastrophic effects, as in the case of poor Henry Molaison. In some patients, debilitating epilepsy could be relieved by separating the left and right hemispheres of the brain, simply by cutting through the structure that connects them – the corpus callosum. Patients often improved substantially without any apparent ill effects, leading to the assumption that the corpus callosum was merely some kind of structural element.[24] However, during the 1950s, animal experiments by Roger Sperry showed that if this part of the brain is cut, something extremely odd takes place.

In 1956 Sperry's student Ronald Myers studied visual learning in cats and explored the well-known fact that retinal signals pertaining to the left side of the visual field go to the right half of the brain, while those pertaining to the right side go to the left half of the brain, making it possible to present a visual stimulus to only one side of the brain. Myers showed that if a cat had its corpus callosum cut, it apparently behaved quite normally until it was tested using a very specific procedure. If it was trained to perform a task based on stimuli in the left visual field and was then tested in the right visual field, it behaved as though no training had ever occurred. Unlike in a normal cat, the left side of its brain did not know what the right side had learned. The corpus callosum enabled the transfer of learning of all kinds between the two hemispheres; in animals in which the structure was cut – Sperry dramatically called them 'split-brain' animals – that transfer could not occur. In 1961 Sperry summarised his findings: 'The split-brain cat or monkey is thus in many respects an animal with two separate brains that may be used either together or in alternation.'[25]

This was truly remarkable. Brain activity associated with perception and learning was neither specifically localised to one particular place, nor did it rely upon the activity of the whole brain. The capacity to perceive and learn could be equally and separately present in each half of the brain – the brain could act as one or as two distinct neurological centres.

Broader, disturbing implications lurked unstated in Sperry's 1961 article. He made no mention of what might happen to a human whose corpus callosum had been severed.

Australian newspaper cartoon from 1961 explaining split-brain experiments in animals. The use of eye-patches is not accurate.

Within a year, there were astounding answers to this question, thanks to the work of another of Sperry's students, Mike Gazzaniga, and the willing help of a forty-eight-year-old man, WJ, who was stricken with debilitating epilepsy.[26] Partly because of this work, in 1981 Sperry received the Nobel Prize, along with Hubel and Wiesel.

In February 1962 WJ – whose name was Bill Jenkins – had his corpus callosum sectioned in order to relieve his terrible bouts of epilepsy. It was also an opportunity for the scientists involved to discover something fundamental about how the brain worked. WJ fully appreciated the dual nature of the intervention and generously said before the operation: 'You know, even if it doesn't help my seizures, if you learn something it will be more worthwhile than anything I've been able to do for years.'[27]

Six weeks after surgery, Gazzaniga visited WJ at home – he had apparently made a full recovery and had substantial relief from his fits – and began a decades-long study of what it means to live with a split brain. In the early days, Gazzaniga used simple tests that involved flashing images on screens either in the patient's left or right visual field and therefore could be seen by only one side of his brain.[28] The transcripts of these early experiments are amazing. On the very first test, Gazzaniga briefly presented an image of a box to the right visual field, which could be detected only by the left half of WJ's brain, which also controlled his speech:

> MG: What did you see?
> WJ: A box.
> MG: Good, let's do it again.

This time, Gazzaniga presented a different image that could be seen only by the right half of the brain:

> MG: What did you see?
> WJ: Nothing.
> MG: Nothing? You saw nothing?
> WJ: Nothing.

Gazzaniga recalls that at this point his pulse was racing with excitement, and he broke into a sweat. Like the experiments with animals, it looked as though one side of WJ's head was not aware of what the other side had seen.

But there was a catch – the left side of the brain controls speech, so only the left side would be able to answer Gazzaniga's question. To reveal if there was anything at all happening in the right side of the brain, when WJ said he had seen nothing Gazzaniga showed him a set of cards with objects on them and asked him to take a guess at the image that had been projected, by pointing at one of the cards. Using his left hand (controlled by the right half of his brain, which had seen the image), WJ unerringly pointed to the correct card. This astonishing experiment suggested that each half of WJ's brain now had its own separate presence (Gazzaniga used the less emotive term 'mental control system').[29] One could talk, the other could not, but both could hear, see, recognise objects and respond to questions. 'Oh, the sweetness of discovery,' recalled Gazzaniga. Without knowing it, he had found support for the theory that Gustav Fechner had come up with in 1860, over a century earlier: if you split a brain, you get two minds instead of one.

It was not all plain sailing. In the early months WJ would some-times experience a conflict between the two sides of his brain – his hands would work in different ways when he was pulling on his trousers or doing up his belt.[30] This initially caused trouble in experi-ments, with the two hands competing to complete a task, often preventing an answer from being given.[31] These conflicts gradually died down, apparently as each version of himself got used to sharing a body (although neither mind was aware of the other's existence). Eventually, WJ lived a normal life, although it is mind-boggling to imagine what was going on in his head.

These results were simply extraordinary. It was not only the brain of an animal that could be harmlessly split in two – the same seemed to be true of a human brain, and the human mind. Each half of the brain was, on its own, sufficient to produce a mind, albeit with slightly different abilities and outlooks in each half. From one mind, you had two. Try that with a computer.

The initial assumption that the right hemisphere had no access to language turned out to be oversimplistic: it was sometimes able to recognise written words and even to control speech to a limited degree.[32] In one patient, the right hemisphere could respond verbally to simple questions, even though the left side of the subject's brain was expected to have sole access to speech production. How this information transfer occurred was not clear, but it may have been through intact subcortical structures, indicating that connections exist between the two sides of our brains that are not involved in consciousness.[33]

In one experiment, Gazzaniga flashed a picture of a female nude to the right hemisphere of a woman patient called NG. Although her left hemisphere reported that she had seen nothing, she smirked, looked embarrassed and eventually began to giggle:

MG: Why are you laughing?
NG: Oh, I don't know. That is a funny machine you have there.

Her left hemisphere did not know what the joke was, but the right side of her brain did, and was laughing.[34] The right hemisphere had produced emotional responses which the left side of the brain experienced but did not comprehend.

Gazzaniga and Joseph Ledoux studied another patient, a young man called PS. If they presented images with instructions on them (stand, stretch, laugh and so on) to the right hemisphere, PS would obey. When they asked him why he was behaving this way, his left hemisphere responded with some invented justification – he needed to stretch, or he thought the experimenters were funny. If they simultaneously presented different images to each hemisphere, and then asked PS to pick out a card with a linked image on it, each hand (controlled by the opposite side of the brain) would pick out the appropriate card. But when PS was asked why the left hand had picked the card it had, the left side of PS's brain, which had not seen the image that was projected to the right hemisphere, came up with a circuitous explanation, finding a spurious link between the two cards.

Gazzaniga and Ledoux realised that the examples of NG and PS showed that the mind in the left side of the brain, unaware of the reasoning going on in the right side, would try to explain away behaviours that it could not understand. It would just make stuff up that seemed to fit the situation. As Gazzaniga recalled:

> Though the left hemisphere had no clue, it would not be satisfied to state it did not know. It would guess, prevaricate, rationalise, and look for cause and effect, but it would always come up with an answer that fit the circumstances … This is what our brain does all day long. It takes input from various areas of our brain and from the environment and synthesizes it into a story that makes sense.[35]

One problem with this interpretation is that you can only have conversations with the left side of the brain in a split-brain patient – in general, the right side cannot articulate its feelings using words. Maybe the right hemisphere is also trying to come up with explanations for what on earth is going on, but it simply cannot express its perplexed solutions because it has no control over speech production. Although there have been some recent suggestions that some of these patients can in fact retain a degree of integration between the two sides of the brain, there is no reason to dispute the well-established fundamental finding – if you split the brain, you split the mind.[36]

In everyday life, the patients were able to function quite normally, partly because our sensory world is not strictly divided up into areas controlled solely by each side of our brains. Auditory stimuli are, in the jargon, ipsilateral (the left ear goes to the left side of the brain), while touch information is processed in a complex way by both sides. And, like you and me, in normal circumstances the patients would continuously move their heads and eyes, getting visual information to both sides of the brain. The real world is not like a psychology experiment.

Gazzaniga and his colleagues also studied more complex, moral responses. They presented split-brain patients with stories involving a character behaving – either deliberately or accidentally – in a way that harmed another character in the story, and then asked the patients to make a moral judgement about what they had been told.[37] People with intact brains considered deliberate harm to be more reprehensible, but when questioned orally the patients viewed deliberate and accidental harm as being similar. They ignored what the characters in the stories believed and based their negative moral judgement solely on the harmful outcome. The implication is that both halves of our brain are necessary to make appropriate moral judgements. But again, there was a catch – these studies involved verbal expression, which is largely the domain of the left side of the brain.

To understand how the right side of the brain saw moral judgements, a split-brain patient was presented with a series of visual, non-linguistic morality plays. The right hemisphere on its own responded like intact subjects, whereas the left hemisphere produced the same outcome-based judgements as with the stories, showing 'a propensity ... to create false hypotheses in order to explain events'.[38] It seems that in split-brain patients the left hemisphere makes inappropriate moral judgements, while the right side of the brain takes a more usual moral standpoint.

As to what it is like to live with a split brain, that is probably the strangest thing of all. The verbal left hemisphere appears to have no understanding of how things were different before the operation, no sense of what it has lost. It is hard to say what the right hemisphere thinks, because it generally has no control over speech, but it is clear that there are truly two minds in these patients, with different outlooks and abilities, and each seems happy with its lot, with no feeling of oddness.[39]

These studies, together with psychological investigations of intact subjects, have led to a popular but wrong view that the two sides of our brains are strongly differentiated, with different abilities and thinking styles – this is often expressed in the inaccurate terms 'left brain' and 'right brain'. This is profoundly mistaken. The existence of two minds in a split-brain patient is a consequence of the operation; it does not imply that everyone has two minds in their head, nor two different brains. Nevertheless, some popular accounts suggest that we all have a dominant hemisphere, related to our personality, with the 'right brain' being more 'creative' and the 'left brain' more 'logical'; some people claim these are linked to sexual preference. None of this is true.* In reality, with the exception of language in the left hemisphere and a tendency for emotional responses to be the responsibility of the right hemisphere (this is also seen in our primate relatives), there are no clear fundamental differences in the functions of the two sides of the brain. This is further demonstrated by a small number of patients, each of whom, as children, had one of their brain hemispheres completely removed to alleviate crippling epilepsy. Remarkably, these people, now adults, show normal levels of cognition and behaviour, and fMRI measures of brain connectivity in the remaining hemisphere also appear normal.[40]

The brain does not work as two separate halves, but as an integrated whole. In some way we do not understand, consciousness is unitary by nature, but it can be divided in split-brain patients, with the weird results that Gazzaniga and his colleagues have revealed. These

---

*For a robust debunking of many popular brain myths, see Christian Jarrett's *Great Myths of the Brain* (Chichester: Wiley-Blackwell, 2015). As Anne Harrington pointed out in *Medicine, Mind, and the Double Brain* (Princeton: Princeton University Press, 1987), the idea of two competing brains goes back to Dr Arthur Wigan, who put it forward in his 1844 work A *New View of Insanity: Duality of Mind*.

differences between the two hemispheres strongly support the general working hypothesis that the mind emerges from the structure of the brain. Any non-materialist explanation of the mind–brain link, like Eccles's suggestion that the brain somehow 'detects' the non-material mind, has to explain how, when separated, the two hemispheres enable such different minds to appear.

The powerful and provocative studies on split-brain patients are now reaching their natural end. The patients are all quite aged, and new anti-epileptic drugs mean that few operations to sever the corpus callosum are now carried out. Through their suffering and generosity, split-brain patients have provided us with a window on the brain, but that will soon be closing. The full significance of what happens inside their heads is not clear – Gazzaniga, who spent over half a century studying them, still does not fully understand. As he put it in 2014: 'Today, it is still haunting and challenging to ponder the question: what does it mean that one can split the mind?'[41]

Gazzaniga was on the receiving end of a less awestruck view when he first presented his astonishing split-brain results – the veteran psychologist William Estes told him: 'Great, now we have two things we don't understand.'[42]

*

In Spring 1977, aged sixty, Francis Crick left the Laboratory of Molecular Biology at the University of Cambridge to study neuroscience at the Salk Institute in California. Through his keen intelligence and his unprecedented access to leading journals for publication of his ideas, Crick had a massive impact on the field.[43] The neurologist and writer Oliver Sacks said that meeting Crick was 'a little like sitting next to an intellectual nuclear reactor … I never had a feeling of such *incandescence*.'

Within a year of making this switch, Crick was writing an invited article for a special issue of *Scientific American*, alongside Hubel, Wiesel, Kandel and other prominent neuroscientists.[44] Crick, who had never done any research on the brain, wrote the closing piece in the collection in which he told his readers (and his fellow-authors) that a new approach was needed. Avoiding any explicit mention of consciousness, Crick argued that researchers should focus on developing 'theories dealing directly with the processing of information in large and complex systems'.

As good as his word, Crick soon turned to the visual system, and in particular to visual attention, breaking the subject into tractable aspects. His first research article on the question appeared in 1984. Inspired by the work of Anne Treisman at Princeton, it was a firecracker of a paper, full of speculative hypotheses that were boldly predicted 'to apply to all mammals and also to other systems, such as the language system in man'.[45] Following Treisman, Crick seized on the idea that attention could be seen as a spotlight – the brain would serially focus on different elements of a visual scene, a function that should be detectable in the activity of neurons. Crick proposed that the searchlight is controlled by the reticular complex of the thalamus and that it should be possible to detect its attention-related activity in this area. The key point is not whether Crick was right or wrong (he was wrong), but that he was developing a way of approaching consciousness – choosing a clearly defined aspect and then looking for its neuronal basis. If something relatively simple could be understood, there was hope that the larger problems could eventually be dealt with, too.

This idea was made explicit in 1990, when Crick published the first fruits of his collaboration with a young German theoretical neuroscientist, Christof Koch, which lasted until Crick's death in 2004. In an article entitled with typical Crick flair 'Towards a Neurobiological Theory of Consciousness', the pair laid out an approach that could lead to the location of what they called 'the neural correlates of consciousness'.[46] For most of us, 'consciousness' refers to awareness, the magical way in which we experience the world. Crick, however, was not immediately interested in this question, because he thought it was impossible to answer with current knowledge. Instead, he felt that by looking for correlates of waking arousal – something that would be present in any animal – it would be possible to identify the conditions that might enable consciousness to emerge. Nailing down how exactly changes in membrane potentials and synaptic strengths turn into our sensations of the world and our impression of individuality would come later. Much later.

This approach does not involve a strong hypothesis of how consciousness works – it merely treats physiology and mind as being tightly correlated. This does not satisfy philosophers, who point out that Crick's starting point and working hypothesis – that mind is somehow identical with neuronal activity – is not consistent with the search for correlations; after all, a correlation is not an identity. But that underlines the difference in approach between a philosopher and a scientist. For Crick, as for Hebb before him, understanding how and why we experience the world the way we do was not something he could immediately address, whereas it was in principle possible to establish a correlation between neuronal activity and arousal. And science is about the doable, about experiments that can test hypotheses, rather than necessarily establishing a logically watertight framework that can resist all potential counter-arguments, as favoured by philosophers. As Crick and Koch explained:

> No neural theory of consciousness will explain everything about consciousness, at least not initially. We will first attempt to construct a rough scaffold, explaining some of the dominant features and hope that such an attempt will lead to more inclusive and refined models.[47]

Although Crick and Koch's intention was to avoid any suggestion that there is some kind of homunculus inside our heads observing what is going on, in 1991 their article was criticised for exactly this reason by the philosopher Daniel Dennett in *Consciousness Explained*. Dennett's book was part of a renewed wave of philosophical interest in consciousness that began in the late 1980s and which continues to the present day.* Crick and Koch wrote that 'one of the functions of consciousness is to present the result of various underlying computations'. With a glint in his beady philosophical eye, Dennett swooped:

---

*For clear summaries of the various philosophical arguments, which are outside the scope of this book, see Susan Blackmore, *Consciousness: A Very Short Introduction* (Oxford: Oxford University Press, 2017); Andrea Cavanna and Andrea Nani, *Consciousness: Theories in Neuroscience and Philosophers of Mind* (Berlin: Springer-Verlag, 2014); Josh Weisberg, *Consciousness* (Cambridge: Polity, 2014).

'present the result', he repeated, 'but to whom? The Queen? ... And then what happens?'[48] Crick and Koch, he claimed, were typical neuroscientists, focusing on minor issues and avoiding the fundamental question of what consciousness is. Dennett's textual criticism is valid, but it flows from that difference in approach between the philosopher and the scientist.

Crick's starting point was a materialist assumption that everything we feel and perceive is 'in fact no more than the behaviour of a vast assembly of nerve cells and their associated molecules'.[49] As he emphasised, there was no absolute proof of this hypothesis (there still is not), but there is a lot more evidence in support of it than there is for any of the competing views, all of which view mind as non-material.

Crick expected that by hard scientific investigation and careful experimentation we would eventually be able to explain 'all aspects of the behaviour of our brains'. Although he was under no illusion that the citadel would fall under the first assault, his timescale may still have been optimistic: 'I do not contend this will happen quickly. I do believe that, if we press the attack this understanding is likely to be reached some day, perhaps some time during the twenty-first century.'[50] We are one-fifth of the way through.

None of Crick's specific proposals for the identity and location of the neural correlates of arousal have stood the test of time.[51] In his final paper, written with Koch and completed on his deathbed in 2004, Crick argued that part of the site of these correlates was the claustrum, a thin layer of poorly understood cells that lie underneath the cerebral cortex with intricate connections to both the cortex and to neighbouring areas such as the hippocampus. Because of its complexity, Crick and Koch suggested that the claustrum might be the focus of the integration that apparently underlies consciousness. The article concluded with what were literally Crick's last words on the topic:

> The neuroanatomy of the claustrum is compatible with a global role in integrating information at the fast time-scale. This should be further experimentally investigated, in particular if this structure plays a key role in consciousness. What could be more important? So why wait?[52]

Although there is evidence that the claustrum is involved in aspects of conscious state, Koch now accepts that this structure is not the site of neural correlates of consciousness.[53]

A wave of research was unleashed by Crick's work, but the general region that produces arousal, and the form that this takes in terms of neuronal activity, both remain elusive. Probably the most precise agreed localisation is that the level of consciousness is largely determined by the brainstem and the basal forebrain, while its content – what is being perceived – is processed by the cortex, hypothalamus and so on. This in itself is both informative and perplexing. The cerebellum is a denser structure than the cortex, with many more neurons, and yet it is generally considered not to be involved in the processes of consciousness. This enigma highlights the fact that no one can yet explain why the activity of one set of neurons produces consciousness, whereas that of another does not.

For some researchers, early interest in frontal regions of the cortex as being the focus for consciousness has now been replaced by a focus on a posterior cortical 'hot zone'. Others disagree.[54] History shows that many sites have been proposed but none has so far resisted the gnawing criticism of experimental study – it would be rash to assume that either the posterior or the frontal cortical area is truly where the action is, if it indeed is in one particular place.[55]

One persistent problem in the field is that of designing reliable measures of conscious activity that are not clouded by irrelevant aspects of the experiment (such as speaking or pressing a button). The ideal approach is to use what are called 'no-report' measures, but this is difficult, and discussion of the significance of experimental results often focuses on tedious alternative interpretations of very precise methodological details. Given that there are so many popular articles about the brain lighting up during conscious activity, it is perhaps surprising that there are no consistent measures – neither EEG nor fMRI – that can reliably distinguish conscious from unconscious individuals. The search is still on for the perfect no-report paradigm.

This research is not simply of academic interest. There have been descriptions of fMRI or EEG responses in patients with locked-in syndrome or in a comatose state who are unable to communicate verbally but whose brains clearly respond when asked, say, to imagine playing tennis.[56] Despite recent progress using a complex mathematical model of EEG function, and claims that fMRI measures can distinguish the brains of unresponsive patients from those of healthy and minimally conscious individuals, for the moment there are no generally accepted correlates of consciousness.[57] Such a measure will eventually be found using these techniques, because consciousness is a physical phenomenon. However, because neither fMRI nor EEG can tell us directly what neurons are doing, it will at best be a correlate of neural correlates of consciousness. That would satisfy the clinician, but not the neuroscientist.

*

The search for neuronal correlates of consciousness has shown most promise by following Crick's focus on the visual system. Crick's aim – to identify a subset of neurons that embody the correlation with consciousness – has not been achieved, but only a tiny fraction of the neurons and of the possible kinds of visual stimulation have been studied. As Crick and Koch pointed out in 1998: 'it will not be enough to show that certain neurons embody the NCC (neural correlations of consciousness) in certain – limited – visual situations. Rather, we need to locate the NCC for all types of visual inputs, or at least in a sufficiently large and representative sample of them.'[58] We are still far from this goal.

In 2008 Itzhak Fried's group described the responses of medial temporal lobe cells in awake patients as they were presented with very brief presentations of images such that in some cases the images could not be consciously identified.[59] The responses of these cells were tightly correlated with the patient's ability to recognise the image. For example, one cell in one patient responded strongly to a picture of Elvis Presley if the image was presented long enough to be recognised, but did not produce a single spike if the presentation was so short as to preclude recognition. In a more recent study, Fried and Koch looked at neurons that are involved in creating a binocular perception from images presented to each eye.[60] By alternating normal binocular images, say either of the actress Annette Bening or of snakes, with simultaneous presentation of both Bening and snakes, which could not be resolved into a binocular image, they found neuronal correlates of this non-conscious process, which is taking place in your head as you read these words. Some cells responded up to two seconds before the patients reported what they saw.

As this work indicates, one of the significant implications of Crick's work was that there may be aspects of what our brain does when we perceive that are essential for the overall process, but which are *not* part of consciousness. This insight – first made by Helmholtz – saw the return of the unconscious as a respectable term in neuroscience, not in its mythical Freudian sense, but rather in terms of processes that are not accessible to conscious experience. The main focus of this research has been the primate visual cortex, and in particular attempts to determine which elements of the activity of the earliest stage of visual processing are part of consciousness, and which are not.

In 1995 Crick and Koch claimed that we are not aware of activity in the part of the primate visual cortex known as V1, which processes the earliest signals. It is generally accepted that activity in this region corresponds to the identity of the physical stimulus, rather than the full perception, which involves structures higher up in the brain. This would tend to suggest that, overall, V1 is not part of the neural correlates of consciousness. This provoked criticism from the American philosopher Ned Block, who thought that Crick and Koch were using the term consciousness a bit too freely.[61] Block's contribution was significant because he distinguished between what he called phenomenal consciousness (because it deals with phenomena, not because it is amazing, although it is) and access consciousness (the use of consciousness to guide action). This distinction is not universally accepted by philosophers (what is?) but some scientists have embraced Block's view, hoping to find a difference that they can explore experimentally, as a way of further understanding the nature of consciousness.[62] Rigorous experimental evidence – psychological and neurobiological – will be required before these two supposed aspects of consciousness are widely accepted. Crick's view of the intervention of philosophers into the discussions over consciousness – a field they had to themselves for several millennia

– was typically frank: 'Listen to their questions, but don't listen to their answers.'[63]*

A radical challenge to our everyday experience of consciousness appeared in a series of studies by the veteran neuroscientist Benjamin Libet, contributing to the philosophical excitement that began in the 1980s and 1990s.[64] Libet's work is generally taken to undermine the notion of free will – our feeling that we can choose how to behave. In a very complicated experiment that has since been replicated many times in various forms, Libet found that EEG traces which revealed subjects' intentions to move a finger slightly preceded their conscious decision to do so. For many scientists and some philosophers, this finding suggested that consciousness and free will, in the form of a mental homunculus, are an illusion. The conscious sensation of deciding to move your finger, they claim, is a rationalisation of a decision that has already been taken by your nervous system.

The strong interpretation is that we have no free will, being controlled instead by neuronal activity that is not immediately available to consciousness, but which is 'made sense of' immediately afterwards. Although the results of Libet's experiment are not in question, this interpretation and its implications are still disputed.[65] One recent study has shown that Libet's basic finding holds only if subjects are making arbitrary choices, not if they are making important, deliberate decisions.[66] This issue is far from being resolved.[67]

Many people find the belief that they have free will and can decide what to do in any given circumstance too compelling to be able to accept any alternative; others are deeply hostile to the strict interpretation of Libet's work because it implies that we cannot make moral choices, and that the punitive framework of much legislation is ill-founded – punishing people for doing things they had no control over seems unfair and pointless. Even if this interpretation is correct, and free will is an illusion, it does not explain how and why

---

*The philosophers can give as good as they get. According to the US philosopher Jonathan Westphal: 'It is instructive to see some of the philosophers mixing it up with the scientists on their own ground and offering what are undeniably testable scientific theories, even if the reverse phenomenon has not been very edifying.' Westphal, J. (2016), *The Mind-Body Problem* (Cambridge, MA: MIT Press), p. 137.

we perceive that illusion – what exactly is happening in the brain to produce this impression – nor does it tell us anything about when in the evolutionary past this illusion first emerged.

Towards the end of his life, Libet explored these more profound problems, suggesting that there was what he called a cerebral mental field, a non-physical expression of neuronal activity that was not separable from the brain – 'it is a non-physical phenomenon, like the subjective experience that it represents'.[68] As to how this field arises, Libet dismissed the question, saying it was just one of the givens in the universe, like gravity or magnetism. Libet had little to say about how many neurons, of which type, showing what kind of activity, would be necessary to generate the field. The question of the neural correlates of consciousness did not feature in his considerations.

*

At one level, neural correlates of consciousness have been established – those Jennifer Aniston cells that fired when a patient saw her photo. But this correlated activity gives no insight into why seeing a picture of this particular actor produces this particular response in this particular individual. (Almost certainly a different response would be seen in the same cell in the brain of another individual looking at the same photo.) Most significantly, it tells us nothing general about consciousness or perception. It is a partial neural correlate of what happens in one person's head when they see a picture of a certain individual, and nothing more. To avoid this kind of problem, researchers have refined their objectives slightly. They now generally agree that they are looking for 'the minimum neuronal mechanisms jointly sufficient for any one specific conscious percept'.[69] The Jennifer Aniston cells would not count, because they are only one component of the hundreds of thousands of neurons that are required to be aware of a photo of her.

The ultimate test of the causal link that is assumed to lie behind these correlations will occur after we have identified the relevant neural networks and activated them with an appropriate pattern of stimulation – for example by a transcranial magnetic pulse, by implanted electrodes or, in the case of an experimental animal, by

optogenetics. If there is a causal link between the identified neuronal activity and consciousness, then the subject should perceive the
relevant thing (or, if the neurons involved are blocked, should be
unable to perceive it).

Some steps have been made in this direction. In 2014 researchers described the consequences of stimulating the regions of the
human cortex that are responsible for detecting faces, again using
electrodes that were implanted for the treatment of epilepsy. When
the face-detecting regions on the right side of the brain were stimulated, the patients reported weird perceptual effects, specifically to
do with the perception of faces – 'You just turned into somebody else,
your face metamorphosed. Your nose got saggy and went to the left,'
reported one. Another said, 'The middle of the eyes twist … chin
looks droopy,' while yet another told the researcher, 'It was almost
like you were a cat.'[70]

In 2018 French researchers reported that stimulation of this
region in a patient produced very precise hallucinations as she
looked at various photos.[71] As she reported in a series of different
tests: 'the photograph of Sarkozy was transposed onto the other face'
and 'they were not your eyes, they were the eyes of someone I had
already seen'. Only some parts of the face were affected, and unlike
the 2014 study, the individual facial elements were both undistorted
and were hallucinated in the appropriate position. Although these
eerie results of very precise stimulation of the face-detecting regions
succeeded in recreating elements of perception, the consequences of
that activation on other parts of the brain's activity are unknown –
only one small part of some of the neural correlates of face detection
has been explored.

In 2013, again working on patients who were being prepared
for treatment, researchers at Stanford University led by Josef Parvizi
stimulated a very specific part of the midregion of the anterior cingulate cortex (mACC), a region deep in the front of the brain. Both
patients reported the same astonishingly specific response: they
started to feel both the bodily and the mental symptoms associated
with being prepared to meet a great physical challenge. As one of
them reported (each of these phrases corresponded to a different
period of stimulation):

My chest and respiratory system started getting shaky ... I started getting this feeling like I was driving into a storm ... like this thing of trying to figure your way out of, how you're going to get through something ... it was more of a positive thing like push harder, push harder, push harder, to try and get through this.[72]

These feelings were reported only if and when this particular area was stimulated (so no response in regions only a tiny distance away, or if no current was applied), and the feelings grew in intensity and precision with increasing current and ceased as soon as the current was turned off. The authors summarised their findings in the title of their article: 'The Will to Persevere Induced by Electrical Stimulation of the Human Cingulate Gyrus.'

The precision of this effect, both in terms of the relatively small area that was stimulated and the clear feelings that stimulation induced, could lead us to imagine that we all have a tiny bit of our brain responsible for these feelings. As the philosopher Patricia Churchland put it somewhat facetiously, the result could be taken to imply that the researchers had identified 'a module for feeling-ominous-threats-and-mustering-courage'.[73] In reality, the same neurons will be involved in a wide variety of conscious states, but both their patterns of activity and their interconnections will vary with the particular state. These amazing results contribute to the growing mountain of evidence that our conscious experiences and the activity of our brains are the same thing and suggest that, eventually, the great mystery of how it all works will be solved. As Churchland pointed out:

Even in this century, some philosophers have grandly announced that consciousness, for example, cannot possibly be a property of the human brain. For all the philosophical finger wagging, however, it is more than modestly auspicious that a few milliamps of current applied to the human mACC can spawn a complex cascade of feelings, feelings that vanish with cessation of the current ... So far as anyone knows, non-physical souls do not respond to milliamps of current.[74]

For the moment, no artificial stimulation has been able to consistently create what would in effect be a complete hallucination, altering every aspect of an individual's perception. Psychedelic drugs may induce altered states, including seeing things that are not there, but their impact is brain-wide, and their outcome is highly unpredictable. In the real world – not the world of thought experiments – we still cannot consistently induce a conscious experience using artificial means. That time is coming, though.

*

In 1995 the philosopher David Chalmers focused everyone's minds on the various questions associated with consciousness, distinguishing what he called the 'easy problems' – those that involved explaining phenomena such as attention, control, categorisation and so on (neuroscientists might cavil at the suggestion that explaining any of these is 'easy') – from the 'hard problems', which is why we experience anything at all: 'It is widely agreed that experience arises from a physical basis, but we have no good explanation of why and how it so arises. Why should physical processing give rise to a rich inner life at all? It seems objectively unreasonable that it should, and yet it does.'[75]

On the one hand, this cunning exercise in rebranding – Chalmers was not emphasising anything that had not been recognised for over 300 years – had the advantage of separating the problem into discrete elements. On the other, as Crick warned, philosophers play by different rules from scientists. Chalmers is one of several modern philosophers who have embraced non-materialist explanations of consciousness, claiming that it does not obey the physical laws of the universe, and that new laws of physics will be necessary if we are ever to understand it. This possibility cannot be logically excluded, but at the moment there is no reason to support this view, beyond frustration at our current perplexity and the desire for something novel. For scientists to abandon the materialist approach that has got us so far, and which provides experimental tools for investigating mysterious phenomena such as consciousness, we would need far stronger motivations, such as inexplicable experimental results that contradict the materialist working hypothesis. No such data have been forthcoming.

Another philosophical contribution that has influenced scientific approaches to the question of consciousness was made in 1974 by Thomas Nagel, in his article 'What Is It Like to be a Bat?' (he did not invent the question).[76] Nagel emphasised that vivid subjective experiences (the philosophical term for these sensations is 'qualia' – such as the experience of seeing a red berry) are intrinsic to what it is like to be you – or a bat – and knowing what that is like in another – be it bat or human – is, Nagel argued, impossible. As striking as the question is, the scientific implications of Nagel's argument are not clear, beyond throwing one's hands up in horror at the complexity of it all.* More recently, Nagel has predicted that making progress will require 'a major conceptual revolution at least as radical as relativity theory', a revolution that will be non-materialist.[77] Without some kind of pointers as to where we should look for this new theory, and above all clear experimental evidence of its necessity, this is not much help.

These views are really a confession of despair, for we know even less about hypothetical immaterial substances or speculative exotic states of matter and how they might or might not interact with the physical world than we do about how brain activity produces consciousness. Not one piece of experimental evidence directly points to a non-material explanation of mind. And above all, the materialist scientific approach contains within it an investigative programme that can in principle resolve the question through experimentation. This is not the case for any of the alternatives.

Over the last thirty years, scientists have intensified their attempts to understand the problems of consciousness. The hard question in particular remains hard, and largely unaddressed, apart from views like Libet's which treat it as a given, and therefore a non-question

---

* At a Salk Institute discussion of the suggestion that 'we'll never be able to figure out consciousness because we'll never know what it is like to be a bat', Joseph Bogen quipped: 'Of course you'll never know. That doesn't mean we won't understand consciousness. I don't have the slightest idea what it's like to be my wife!' It was a good gag, but that was kind of Nagel's point. Bogen, J. (2006), *The History of Neuroscience in Autobiography* 5:46–122.

(this is also the position of some philosophers).[78] For those trying to investigate the question from a strictly materialist point of view, the gulf between physical and mental phenomena remains as yawning as it was to Leibniz in the eighteenth century or to du Bois-Reymond and Tyndall a hundred and fifty years later. But the fact that there is a gap does not mean it cannot be bridged.

*

In the last decade or so, the insight provided first by Hebb and then by Crick into how to study consciousness scientifically by focusing on precise, solvable problems seems to have been somewhat forgotten. Much of the theoretical work done in this field veers into the realm of speculation – the theories seek to describe many or most things about consciousness, rather than explaining one tractable aspect. Although there are many different ways of theorising consciousness, there are currently two main scientific approaches, neither of which is widely accepted.

The French neuroscientists Stanislas Dehaene and Jean-Pierre Changeux, following on from the ideas of Bernard Baars, have developed global neuronal workspace theory, according to which consciousness arises when information is made available to multiple brain systems, in particular through the activity of neurons with axons that spread across brain regions.[79] As Dehaene has put it, inadvertently using an old metaphor: 'consciousness is nothing but the flexible circulation of information within a dense switchboard of cortical neurons'.[80] 'Nothing but' is doing a lot of work in that sentence, and the theory does not explain why flexible and dense circulation of information causes consciousness to pop up. Ultimately, it appears as a given. This may be true, but it seems unsatisfying as an explanation.

The other approach is called integrated information theory, and was developed by Giulio Tononi, Gerald Edelman and a number of collaborators, including Christof Koch.[81] This is a complex mathematical approach that involves a series of mathematically expressed axioms relating to essential properties of experience, together with a set of postulates regarding the organisation of the physical substrates of those axioms.[82] According to this theory, consciousness is simply the integration of information involved in these networks, and can be measured by the degree of connectivity, which in turn can be given a quantity indicating the degree of consciousness.* Again, the link between consciousness and the chosen focus of the theory – in this case integration of information – is unclear. It simply is.

Although few scientists currently working in this area have followed Sherrington, Eccles and Penfield and openly adopted dualist positions, some have been happy to embrace other solutions to the mind–brain problem that were first clearly proposed in the seventeenth century, in particular the suggestion that all matter might be somehow conscious – panpsychism (Tononi claims that his theory vindicates some of the 'intuitions' of panpsychism; other researchers prefer to hypothesise that only living matter, from single-celled organisms onwards, is conscious[83]). This has the great advantage of not requiring any specific explanation of the existence of the human or animal mind, but it explains nothing, and often leads to untestable mystical beliefs as seen in Koch's claim that integrated information theory has teleological implications – he suggests there is some kind of urge in matter to become conscious, and enthusiastically references the Jesuit mystic Teilhard de Chardin.[84] It is hard to imagine Crick appreciating such company.

There are a number of psychological theories of consciousness, linked to the way the brain interprets and acts on the world, which tend to be more focused on the function of consciousness than on fundamental mechanistic questions.[85] The theories of consciousness that the general public finds particularly fascinating are those that invoke the quantum realm, such as the mathematician Roger

---

*I confess I do not fully understand integrated information theory. As I suspected, I am not alone – Matthias Michel, a French philosopher of science, has surveyed scientists' attitudes to this theory and found that many non-expert researchers do not really grasp it, but are somehow impressed: 'In a sense, the apparent complexity of the theory is used as a proxy for its probability of being true. They don't really understand it, but they come to believe that if they understood it, they would likely consider it as the right theory of consciousness.' Sohn, E. (2019), *Nature* 571:S2-S5; Michel, M., et al. (2018), *Frontiers in Psychology* 9:2134.

Penrose's suggestion that quantum effects in neural microtubules in the brain are at the heart of conscious experience (why human microtubules might show different quantum effects from those in a worm is not clear).[86] Recently, Gazzaniga has also taken the quantum route, although his theory is more like a general framework, in which he sees consciousness as just the most complex example of the deeper problem of determining what is alive and what is not, with the quantum concept of complementarity playing a key, if vague, role.[87] Quantum approaches to unexplained biological phenomena are attractive to some people (generally physicists and mathematicians), partly because of the assumption that if two things are mysterious then they may be linked, but there is no evidence that quantum mechanics can explain consciousness.[88]

Frustratingly, many of the theoreticians do not engage with rival theories, even when these are apparently quite similar.[89] This surprising situation in which different theories largely go their own separate ways is not only possible but also widespread, because there are many ideas but few decisive experimental findings.

This is the key point. To convince scientists that any of these theories are right or, more likely, that they need to be discarded or adapted, will require clear experimental results. This will probably only become possible when the neural correlates of consciousness are finally discovered, and the theoreticians can focus on more precise and localised predictions. For the moment, many of the predictions made by these theories are so vague as to be uninformative in terms of designing experiments. In October 2019, the proponents of integrated information theory and global workspace theory agreed to a series of tests that might be able to demonstrate that one of the theories is more accurate. Whether this will show that either theory is actually true is another matter.[90]

There is one other possible – but highly unlikely – route forward: global neural workspace theory and integrated information theory have contrasting implications with regard to the possibility that machines could become conscious. The clear implication of global neuronal workspace is that a machine could be conscious if it had circuitry that replicated the global distribution of information that is at the heart of the theory (Dehaene has argued this is extremely unlikely).[91] On the other hand, one interpretation of integrated information theory suggests that only something that is as organisationally complex as the brain could possibly contain the degree of integrated information to allow consciousness. The appearance of truly conscious machines (but how could we tell?) would perhaps resolve this issue, although it would raise many more significant questions. For the moment, all this is speculation. There is no sign that consciousness is about to pop out of integrated circuits.

Major scientific progress in our understanding of consciousness and its origins in brain function will likely require a re-engagement with the resolutely experimental focus proposed first by Hebb and then by Crick. A subsidiary suggestion would be that scientists should perhaps leave the philosophising to the philosophers. Studying doable parts of the problem, rather than worrying about theoretical explanations of the most complex aspects of consciousness would appear to be the most fruitful approach.

That does not mean that researchers should avoid either the eerie results of the split-brain patients or the routine but deeply perplexing experience of losing consciousness and then regaining it, as happens after waking from sleep or recovering from a general anaesthetic. These findings are telling us something highly significant about the nature of the link between mind and parts of the brain.[92] But, for the moment, trying to integrate these profoundly challenging facts into an explanation of how the brain works is probably a mistake. Clarity will come when we have a more solid basis to build upon. That is not an approach that would satisfy a philosopher, but it is a method that every scientist will recognise.

As Patricia Churchland has astutely pointed out, it is very unlikely that there will be a single experiment and a single theory that demonstrates how brain activity becomes consciousness.* Between the fifteenth and eighteenth centuries European thinkers gradually accepted that the site of thought is the brain and not the heart; there

---

*Churchland's writings are limpid pools of good sense that should be plunged into when, in her words, you are feeling 'bamboozled by philosophical flimflam'. Churchland, P. (2017), in K. Almqvist and A. Haag (eds.), *The Return of Consciousness: A New Science on Old Questions* (Stockholm: Axel and Margaret Ax:son Johnson Foundation), pp. 39–58, p. 59.

was no brain-centric moment then, and it seems unlikely that there will be a neural-network-centric moment in the future. Instead, the slow accumulation of evidence will gradually shed light. Whatever the case, there is no reason to retreat into the kind of pessimism that affected thinkers in the 1870s. We will resolve this thorny problem, eventually.

In Autumn 2020, two studies of animals provided some exciting insights into what is generally seen as a uniquely human problem.[93] Andreas Nieder's group at the University of Tübingen recorded the activity of brain cells in crows while they were carrying out a visual detection task. Birds do not have a layered cortex and yet the activity of these cells was correlated with the birds' perception about the presence or absence of a stimulus. If this result is indeed due to conscious perception, rather than reflecting memory or computation, it implies that either consciousness involves structures that are common to many animals and were present before the evolution of the mammalian cortex, or that consciousness appeared separately in birds and does not necessarily require a cortex. Shortly afterwards, Doris Tsao's group reported that the activity of single cells in the macaque cortex corresponded to what the animal was actually seeing. The population activity of cells in the inferotemporal cortex, at the base of the brain, appears to represent a neural correlate of consciousness in the monkey.

Extending these insights to humans, where the existence of consciousness is more readily accepted, will be challenging. In 1998, after a late-night drinking session at a conference in Bremen, Koch made a bet with Chalmers that within a quarter of a century (so, by 2023), we will have identified the neural correlates of consciousness – not necessarily its cause – in terms of the activity of 'a small set of neurons characterised by a small list of intrinsic properties'.[94] The winner will get a case of fine wine. At the moment it looks like Koch will be paying out.

FUTURE

It is hard to predict how we will eventually come to understand the brain, and what that understanding will consist of. It is also foolhardy – many readers (especially the neuroscientists among you) will undoubtedly disagree with some of what follows, and prediction is a mug's game, in particular when it concerns the future. Nevertheless, here goes.

Astonishing new techniques now provide a degree of control in brain experiments that would have been dismissed as science fiction only a few years ago, while our ability to image what is happening where in brains of all kinds is becoming increasingly precise. And yet scientists have repeatedly argued not only that all these data have not enabled us to understand the brain, but also that we are not even on the road to achieving that goal.[1] As Olaf Sporns has put it, 'neuroscience still largely lacks organising principles or a theoretical framework for converting brain data into fundamental knowledge and understanding'.[2] Our understanding of the brain appears to be approaching an impasse.

In 2017 *Science* magazine explored this question through a series of articles under the general title 'Neuroscience: In Search of New Concepts'.[3] French neuroscientist Yves Frégnac focused on the current fashion of collecting massive amounts of data in expensive large-scale projects. For Frégnac, this represents the industrialisation of brain research, where funding agencies (and researchers) believe that 'using the fanciest tools and exploiting the power of numbers could bring about some epiphany'.[4] There are projects like this all over the world, from the US (BRAIN initiative, the Human Connectome Project and others), through China (Brain Project) to Europe (Human Brain Project and many others), as well as in Australia and Japan. Paradoxically, the tsunami of data they are producing is leading to major bottlenecks in progress, partly because, as Frégnac put it pithily, 'big data is not knowledge':

> Only 20 to 30 years ago, neuroanatomical and neurophysiological information was relatively scarce while understanding mind-related processes seemed within reach. Nowadays, we are drowning in a flood of information. Paradoxically all sense of global understanding is in acute danger of getting washed away. Each overcoming of technological barriers opens a Pandora's box by revealing hidden variables, mechanisms, and nonlinearities, adding new levels of complexity.

Frégnac had no direct answer to the problem, beyond a series of suggestions that would tame and enrich the big-data projects by encouraging greater interdisciplinary collaboration and focusing on hypothesis testing rather than simply collecting vast amounts of information.

Although the size of the mountain of data that is being produced is new, the problem is not. In 1992 Patricia Churchland and Terry Sejnowski published *The Computational Brain*, in which they described the latest models of sensation, plasticity and sensorimotor integration, but nevertheless argued that there had been little theoretical progress – 'almost everything remains to be done, and on all sides major puzzles loom'.[5] Nearly a quarter of a century later, Churchland's daughter, the neuroscientist Anne Churchland, made a similar diagnosis. Writing with Larry Abbott, Churchland *fille* emphasised our difficulties in interpreting the massive amount of data that is being produced by laboratories all over the world: 'Obtaining deep understanding from this onslaught will require, in addition to the skilful and creative application of experimental technologies, substantial advances in data analysis methods and intense application of theoretic concepts and models.'[6]

These repeated calls for more theory may be a pious hope. It can be argued that there is no possible single theory of brain function, not even in a worm, because a brain is not a single thing (scientists are even hard put to come up with a precise definition of what a brain is[7]). As Crick observed, the brain is an integrated, evolved structure with different bits of it appearing at different moments in evolution and adapted to solve different problems. Our current comprehension of how it all works is extremely partial – for example, most neuroscience sensory research has been focused on sight, not smell, which is conceptually and technically more challenging. But the way that olfaction and vision work are different, both computationally and structurally. By focusing on vision, we have developed a very limited understanding of what the brain does and how it does it.[8]

The nature of the brain – simultaneously integrated and composite – may mean that our future understanding will inevitably be fragmented and composed of different explanations for different parts. After all, as Marr put it, the brain is composed of 'a whole lot' of information-processing devices. Churchland and Abbott spelt out the implication: 'Global understanding, when it comes, will likely take the form of highly diverse panels loosely stitched together into a patchwork quilt.'[9]

*

For more than half a century, all those highly diverse panels of patchwork we have been working on have been framed by thinking that brain processes involve something like those carried out in a computer. But that does not mean this metaphor will continue to be useful in the future. At the very beginning of the digital age, in 1951, Karl Lashley argued against the use of any machine-based metaphor:

> Descartes was impressed by the hydraulic figures in the royal
> gardens and developed a hydraulic theory of the action of the
> brain. We have since had telephone theories, electrical field

theories, and now theories based on computing machines and automatic rudders. I suggest we are more likely to find out about how the brain works by studying the brain itself and the phenomena of behaviour than by indulging in far-fetched physical analogies.[10]

This dismissal of metaphor has recently been taken even further by the French neuroscientist Romain Brette, who has challenged the most fundamental metaphor of brain function: coding.[11] Since its inception by Adrian in the 1920s, and above all its enthusiastic adoption by Horace Barlow in the 1960s, the idea of a neural code has come to dominate neuroscientific thinking – over 11,000 papers on the topic have been published in the last ten years.[12] Brette's fundamental criticism was that, in thinking about 'code', researchers inadvertently drift from a technical sense, in which there is a link between a stimulus and the activity of the neuron, to a representational sense, according to which neuronal codes represent that stimulus. This question was raised back in 1990 by Walter Freeman and Christine Skarda, when they published an article entitled 'Representations: Who Needs Them?'[13] Freeman, who had studied electrophysiological responses to odours for decades, explained that by ceasing to worry about how nervous systems represent the environment, he was able 'to focus less on the outside world that is being put into the brain and more on what brains are doing'. The idea that nervous systems represent or encode information contains an even more fundamental implication. As Dennett said to Crick and Koch – represent to whom?

The unstated implication in most descriptions of neural coding is that the activity of neural networks is presented to an ideal observer or reader within the brain, often described as 'downstream structures' that have access to the optimal way to decode the signal. But the ways in which such structures actually process the activity of peripheral neurons is unknown and is rarely explicitly hypothesised, even in simple models of neural network function. The processing of neural codes is generally seen as a series of linear steps – like a line of dominoes falling one after another, much as in a reflex. The brain, however, consists of highly complex neural networks that are interconnected, and which are linked to the outside world to effect action. Focusing on sets of sensory and processing neurons without linking these networks to the behaviour of the animal misses the point of all that processing. 'Action potentials are potentials that produce actions,' concluded Brette, 'not hieroglyphs to be deciphered'.

A similar view of the brain has been presented by György Buzsáki in his recent book *The Brain from Inside Out*.[14] According to Buzsáki, the brain is not simply passively absorbing stimuli and representing them through a neural code, it is actively searching through alternative possibilities to test various options. His conclusion, building on the insights of Helmholtz and Marr, is that the brain does not represent information: it constructs it.

The metaphors of neuroscience – computers, coding, wiring diagrams and so on – are inevitably partial. That is the nature of metaphors, which have been intensely studied by philosophers of science and by scientists, as they seem to be so central to the way scientists think.[15] But metaphors are also rich and allow insight and discovery. There will come a point when the understanding they allow will be outweighed by the limits they impose, but in the case of computational and representational metaphors of the brain there is no agreement that such a moment has arrived.[16] From a historical point of view, the very fact that this debate is taking place suggests that we may indeed be approaching the end of the computational metaphor. What is not clear, however, is what would replace it.

Scientists often get excited when they realise how their views have been shaped by the use of metaphor and grasp that new analogies could alter how they understand their work, or even enable them to devise new experiments. Coming up with those new metaphors is challenging – most of those used in the past with regard to the brain have been related to new kinds of technology. This could imply that the appearance of new and insightful metaphors for the brain and how it functions hinges on future technological breakthroughs, on a par with hydraulic power, the telephone exchange or the computer. There is no sign of such a development; despite the latest techno buzzwords that zip about – blockchain, quantum supremacy (or quantum anything), nano-tech and so on – it is unlikely that these fields will transform either technology or our view of what brains do.

The advent of the internet and of cloud computing led to a brief vogue for thinking of the brain as some kind of distributed computer system. And indeed, a recent study has shown that our neurons are not like the simple components in a computer; instead, through their myriad dendritic connections, many of which involve multiple neurotransmitters and which nuance the output of the cell, they are able to carry out highly complex processes that are the equivalent to what are called linearly non-separable functions. Each dendrite can respond to local stimulation from other neurons by sending a spike towards the cell body, not in a one-to-one linear fashion but by increasing their spike frequency disproportionately. One of the researchers involved in this study, UK neuroscientist Mark Humphries, has said that this implies that each cell can act in a way that resembles a complex minicomputer.[17]

However, that does not mean that the cloud/internet analogy helps us very much. One of the key points about the internet, which was built into its structure from its earliest incarnations, was that it would continue to function even if some key parts of it were removed, for example by a nuclear strike. No matter how distributed our view of brain activity, and despite the very real evidence for plasticity, some aspects of brain function can indeed be decisively disrupted if particular areas are damaged.

*

One sign that our metaphors may be losing their explanatory power is the widespread assumption that much of what nervous systems do, from simple systems like the rhythmic grinding of the lobster's stomach right up to the appearance of consciousness in humans, can only be explained as emergent properties – things that you cannot predict from an analysis of the components, but which emerge as the system functions.

In 1981 Richard Gregory argued that the reliance on emergence as a way of explaining brain function indicated a problem with the theoretical framework: 'the appearance of "Emergence" may well be a sign that a more general (or at least different) conceptual scheme is needed … It is the role of good theories to remove the appearance of Emergence. (So explanations in terms of Emergence are bogus.)'[18] This overlooks the fact that there are different kinds of emergence, weak and strong. Weak emergent features, such as the movement of a shoal of tiny fish in response to a shark, can be understood in terms of the rules that govern the behaviour of their component parts. In such cases, apparently mysterious group behaviours are based on the behaviour of individual animals, each of which is responding to factors such as the movement of a neighbour, or external stimuli such as the approach of a predator.

This kind of weak emergence cannot explain the churning of the lobster's stomach, never mind the working of your brain, so we fall back on strong emergence, where the phenomenon that emerges cannot be explained by the activity of the individual components; it obeys its own lawfulness. You and the page you are reading this on are both made of atoms, but your ability to read and understand comes from features that emerge through atoms in your body forming higher-level structures, such as neurons and their patterns of firing, not simply from atoms interacting. Strong emergence has recently been criticised by some neuroscientists as risking 'metaphysical implausibility', because there is no evident causal mechanism nor any single explanation of how emergence occurs. Like Gregory, these critics claim that the reliance on emergence to explain complex phenomena suggests that neuroscience is at a key historical juncture, similar to that which saw the slow transformation of alchemy into chemistry.[19] But faced with many of the mysteries of neuroscience, emergence is often our only resort. And it is not so daft – the amazing properties of deep learning programs, which at root cannot be explained by the people who design them, are essentially emergent properties.

Interestingly, while some neuroscientists are discombobulated by the metaphysics of emergence, researchers in artificial intelligence revel in the idea, believing that the sheer complexity of modern computers, or of their interconnectedness through the internet, will lead to what is dramatically known as the singularity. Machines will become conscious. There are plenty of fictional explorations of this possibility (things often end badly for all concerned), and the subject certainly excites the public's imagination, but there is no reason, beyond our ignorance of how consciousness works, to suppose that it will happen in the near future. In principle, it must be possible, because the working hypothesis is that mind is a product of matter, which we should therefore be able to mimic in a device. The scale of complexity of even the simplest of brains dwarfs any machine we can currently envisage. For decades – centuries – to come, the singularity will be the stuff of science fiction, not science.

A related view of the nature of consciousness turns the brain-as-computer metaphor into a strict analogy. Some researchers view the mind as a kind of operating system that is implemented on neural hardware, with the implication that our minds, seen as a particular computational state, could be uploaded onto some device or into another brain. In the way this is generally presented, this is wrong, or at best hopelessly naive. The materialist working hypothesis is that brains and minds, in humans and maggots and everything else, are identical. Neurons and the processes they support – which somehow include consciousness – are the same thing. In a computer, software and hardware are separate; however, our brains and our minds consist of what can best be described as wetware in which what is happening and where it is happening are completely intertwined.

Imagining that we can repurpose our nervous system to run different programs, or upload our mind to a server, might sound scientific but lurking behind this idea is a non-materialist view going back to Descartes and beyond. It implies that our minds are somehow floating about in our brains and could be transferred into a different head or replaced by another mind. It would be possible to give this idea a veneer of scientific respectability by posing it in terms of reading the state of a set of neurons and writing that to a new substrate, organic or artificial. But to even begin to imagine how that might work in practice, we would need both an understanding of neuronal function that is far beyond anything we can currently envisage and would require unimaginably vast computational power and a simulation that precisely mimicked the structure of the brain in question. For this to be possible even in principle, we would first need to be able to satisfactorily model the activity of a nervous system capable of holding a single state, never mind a thought. Once again, the lobster's stomach shows our ignorance, and the distance we have to travel.

*

For the moment, the brain-as-computer metaphor retains its dominance, although there is disagreement about how strong a metaphor it is.[20] In 2015 the roboticist Rodney Brooks chose the computational metaphor of the brain as his pet hate in his contribution to a collection of essays entitled *This Idea Must Die*. Less dramatically, but drawing similar conclusions, over two decades earlier the historian S. Ryan Johansson argued that 'endlessly debating the truth or falsity of a metaphor like "the brain is a computer" is a waste of time. The relationship proposed is metaphorical, and it is ordering us to do something, not trying to tell us the truth.'[21] In a similar vein, the neuroscientist Matteo Carandini has recognised that parallels with what is seen as cutting-edge technology can soon appear quaint and outdated,* but he nevertheless emphasised that the computer metaphor has some value: 'the brain is undeniably an information-processing organ, so it may pay to compare it to our best information-processing devices'. Gary Marcus has made a far more robust defence of the computer metaphor:

> Computers are, in a nutshell, systematic architectures that take inputs, encode and manipulate information, and transform their inputs into outputs. Brains are, so far as we can tell, exactly that. The real question isn't whether the brain is an information processor, per se, but rather how do brains store and encode information, and what operations do they perform over that information, once it is encoded.[22]

Marcus went on to argue that the task of neuroscience is to 'reverse engineer' the brain, much as one might study a computer, examining its components and their interconnections to decipher how it works. This suggestion has been around for some time. In

---

*The pitfalls of an overenthusiastic application of the latest technology can be seen in the ideas of Karl Pribram, who in the 1960s and 1970s argued in a series of papers that 'the brain may exploit, among other things, the most sophisticated principle of information storage yet known: the principle of the hologram'. It does not. Pribram, K. (1969), *Scientific American* 220(1):73–86.

1989 Crick recognised its attractiveness, but felt it would fail, because of the brain's complex and messy evolutionary history – he dramatically claimed it would be like trying to reverse engineer a piece of 'alien technology'.[23] Attempts to find an overall explanation of how the brain works that flow logically from its structure would be doomed to failure, he argued, because the starting point is almost certainly wrong – there is no overall logic.

Reverse engineering a computer is often used as a thought experiment to show how, in principle, we might understand the brain. Inevitably, these thought experiments are successful, encouraging us to pursue this way of understanding the squishy organs in our heads. But in 2017 a pair of neuroscientists decided to actually do the experiment on a real computer chip, which had a real logic and real components with clearly designed functions. Things did not go as expected.

The duo – Eric Jonas and Konrad Paul Kording – employed the very techniques they normally used to analyse the brain and applied them to the MOS 6507 processor found in computers from the late 1970s and early 1980s that enabled those machines to run video games such as *Donkey Kong*, *Space Invaders* or *Pitfall*. First, they obtained the connectome of the chip by scanning the 3510 enhancement-mode transistors it contained and simulating the device on a modern computer (including running the games programs for ten seconds). They then used the full range of neuroscientific techniques, such as 'lesions' (removing transistors from the simulation), analysing the 'spiking' activity of the virtual transistors and studying their connectivity, observing the effect of various manipulations on the behaviour of the system, as measured by its ability to launch each of the games.

Deleting transistors – this is the equivalent of lesioning a brain area – produced some seductively clear results. For example, there were ninety-eight transistors, each of which, if removed on their own, prevented the system from booting up *Donkey Kong*, but had no effect on *Space Invaders* or *Pitfall*. But as the authors recognised, this did not mean that there was anything like a *Donkey Kong* transistor – such an interpretation would be 'grossly misleading', they said. In reality, each of these components merely carried out a simple, basic function, which was required for *Donkey Kong* but not for the other two games.

Despite deploying this powerful analytical armoury, and despite the fact that there is a clear explanation for how the chip works (it has ground truth, in technospeak), the study failed to detect the hierarchy of information processing that occurs inside the chip. As Jonas and Kording put it, the techniques fell short of producing 'a meaningful understanding'. Their conclusion was bleak: 'Ultimately, the problem is not that neuroscientists could not understand a microprocessor, the problem is that they would not understand it given the approaches they are currently taking.'[24]

This sobering outcome suggests that, despite the attractiveness of the computer metaphor and the fact that brains do indeed process information and somehow represent the external world, we still need to make significant theoretical breakthroughs. Even if our brains were designed along logical lines, which they are not, our present conceptual and analytical tools would be completely inadequate for the task of explaining them. This does not mean that simulation projects are pointless – by modelling (or simulating) we can test hypotheses and, by linking the model with well-established systems that can be precisely manipulated, we can gain insight into how real brains function.[25] This is an extremely powerful tool, but a degree of modesty is required when it comes to the claims that are made for such studies, and realism is needed with regard to the difficulties of drawing parallels between brains and artificial systems.

Even something as apparently straightforward as working out the storage capacity of the human brain falls apart when it is attempted. Terry Sejnowski's group has made a careful anatomical study of the number and size of dendritic spines, together with the number of neurotransmitter vesicles at a synapse, and calculated that, on average, each synapse can store at least 4.7 bits of information.[26] This suggests that the human brain can hold at least a petabyte (1 million gigabytes) of information. No matter how dramatic this might sound, nor how appealing it might be to those who like the idea that maths and engineering can tell us how the brain works, the starting point of such a calculation is skew-whiff. Neurons are not digital (the basis of a bit of information), while brains – even the not-really-a-brain of the worm *C. elegans* – are not hard-wired. Each brain is continually changing the number and strength of its synapses, and above all it does not work by synapses alone. Neuromodulators and neurohormones are also responsible for the way that the brain functions, but because the way they work, and the timescales upon which they operate, do not fit with the computer metaphor, they do not feature in such studies.

Calculating the storage capacity of a brain is fraught with conceptual and practical difficulties. Brains are natural, evolved phenomena, not digital devices. They cannot be fully understood using crude – or even sophisticated – notions of information.

Even more fundamentally, the very structures of a brain and a computer are completely different. In 2006, Larry Abbott contributed to a book by twenty-three leading neuroscientists, focusing on unsolved problems (most of the questions have still not been satisfactorily answered).[27] In his essay 'Where Are the Switches on This Thing?', Abbott explored the potential biophysical bases of that most elementary component of an electronic device – a switch. Although inhibitory synapses can change the flow of activity by rendering a downstream neuron unresponsive, such interactions are relatively rare in the brain.

Every cell is not like a binary switch that can be turned on or off, forming a wiring diagram. Instead, the main way that the nervous system alters its working is by changes in the pattern of activity in networks of cells composed of large numbers of units; it is these networks that channel, shift and shunt activity. Unlike any device we have yet envisaged, the nodes of these networks are not stable points like transistors or valves, but sets of neurons – hundreds, thousands, tens of thousands strong – that can respond consistently as a network over time, even if the component cells show inconsistent behaviour.

This is the great problem we have to solve. On the one hand, brains are made of neurons and other cells, which interact together in networks the activity of which is influenced not only by synaptic activity but also by various factors such as neuromodulators. On the other hand, it is clear that brain function – in any animal – involves complex dynamic patterns of neuronal activity at a population level. Finding the link between these two levels of analysis will be a challenge for much of the rest of the century, I suspect.

*

There is an even bigger problem with theories that are intended to understand brain function on the basis of structure – connectomes, or whatever – which can be seen if we imagine that the MOS 6507 chip and its associated components were an example of Crick's alien technology, a device found on a Martian spaceship that fell to earth. A full analysis of its components would reveal that inputs from the exterior could alter its function, but it seems unlikely that we would realise that a Martian would use the device to play a game. And without observing a Martian interacting with the machine, we would never fully grasp how it worked. In the absence of that decisive external element both the meaning and the mode of functioning of the device would be obscure.

When we extend this insight to understanding the brain, the key implication is that, as the title of a 1997 article strikingly put it, 'The Brain Has a Body'. And the body has an environment, and both affect how the brain does what it does. This might seem trivially obvious, but neither the body nor the environment feature in modelling approaches that seek to understand the brain. The physiological reality of all brains is that they interact with the body and the external environment from the moment they begin to develop. Excluding these aspects from the model, or from the experimental set-up, will lead at best to an inadequate understanding. This issue goes even further. As Alex Gomez-Marin and Asif Ghazanfar have recently argued, 'the behaviour of animals is not the behaviour of their brains'. Animals are not robots piloted by brains, we are all, whether maggots or humans, individuals with agency and a developmental and evolutionary history. Those factors are all involved with how our brains work and need to be integrated into our models.[28]

Simulating a brain in a vat – essentially what the Human Brain Project is doing, except it involves only part of a tiny sliver of rat brain – deprives the system of an essential component it needs, which is input from the outside world. In the words of Olaf Sporns: 'Neurons don't just passively respond to inputs – by contributing to motor activity and behaviour they actively determine what the inputs are.'[29] It is possible that what we observe in simulations or

in isolated networks of neurons is a system that is not functioning
normally. Comparison of simulation studies with the brain activity
of behaving animals, as has been done for the zebrafish, will be necessary to clarify this.[30]

This viewpoint also undermines some of the recent excitement
about brain organoids – blobs of stem-cell-derived brain tissue grown
in a dish. Researchers have found that appropriate brain cell types
– including microglia – appear consistently and reproducibly within
brain organoids. Neurons in organoids can show rhythmic behaviour just like those primitive computer simulations from the 1950s
and which have even been claimed to resemble those of the preterm
infant. In other experiments one organoid responded to light in a
region where retinal tissue was growing, while others have even
been observed to join up with bits of mouse spinal cord and induce
muscle contractions.[31] Eerie as this might seem, organoids will never
grow beyond a few millimetres across and around 3 million cells – a
tiny proportion of a human brain – because they are isolated from the
myriad factors produced by the body in interaction with the environment, which guide the brain's development.

These lentil-sized blobs are poised to give us important insights
into how simple brain structures develop in both health and disease,
and how brains have evolved.[32] But they have already been lined
up for some less savoury stunts – one smart-alec researcher aims to
create brain organoids using the Neanderthal genome, wire them
up to 'robots that resemble crabs' and then race them against robots
controlled by human brain organoids.[33] Such a spectacle would tell
us nothing. Faced with this kind of thoughtlessness, scientists and
bioethicists have argued for an ethical framework for organoids that
would weed out trivial or potentially damaging experimentation on
these latest scientific playthings.[34] Although there is only a remote
possibility that an organoid might become conscious, it is hard to
know how we could tell; thinking back to what was done to poor
Mary Rafferty in 1874, caution should dominate over curiosity or
amusement.

The significance of remembering that brains are in bodies can
be seen from the way that brains interact with the gut microbiome.
'Germ-free' mice have no microbes living in their gut, and as a result

show altered levels of serotonin in the brain and lower levels of anxious behaviour. The unlikely causal link between microbes and behaviour was shown when the introduction of a normal micro-biome into the mice reversed both these effects: fundamental aspects of brain biochemistry can be affected by the microbes that live in the gut.[35]

Many scientists do adopt an integrative approach to under-standing the brain. For example, in their 2018 book *The Neuroscience of Emotion*, Ralph Adolphs and David Anderson concentrate on one of the trickiest but most powerful areas of mental life that has barely been touched upon in these pages – emotion. Using studies from across the animal kingdom, including octopuses and flies as well as mammals, Adolphs and Anderson have explored how physiologi-cal and mental states interact, even in supposedly simple organisms. Whatever the validity of their particular theory, the lesson is that, to understand emotion fully, it needs to be studied in the context of a whole organism, interacting with the outside world.[36] A similar position is argued by neuroscientist Alan Jasanoff, in his book *The Biological Mind*, in which he criticises what he calls 'the cerebral mys-tique' – the view that reduces human mental life to the activity of our brains, often with the implication that our minds are spirits floating in a complex mass of neurons.[37] By situating the brain in its anatomical, physiological and evolutionary context, we get a richer understand-ing of how the various bits of our bodies interact to produce our behaviour and, ultimately, our minds. This even extends to the way that neurons function – in their academic book *Principles of Neural Design*, Peter Sterling and Simon Laughlin emphasise the importance of understanding the basic rules of brain construction, stemming from physiology and bioenergetics, even in the simplest brains.[38]

The significance of our bodies in our mental experience also sug-gests that those old views about our minds not being in our heads but elsewhere in our bodies may not have been so far off. Finnish researchers asked subjects with different cultures and native tongues to describe bodily sensations associated with emotions, and the physical location of various feelings.[39] Perhaps unsurprisingly, the torso, and more specifically the heart region, seemed to be associ-ated with many of these emotions, in particular anxiety, pride, fear and anger, while all of the cognitive feelings – thinking, reasoning, remembering and so on – were focused in the head. My guess is that this brain-centred location of thought is a consequence of modern knowledge, while the localisation of certain emotions in bits of our bodies may be a direct product of our biology.

*

My own preference for how best to proceed in understanding the brain would be to pour resources into discrete, doable projects able to provide insight that can subsequently be integrated into a more global approach. Crick's approach to studying consciousness applies to the brain as a whole, it seems to me. As some parts of theoretical physics demonstrate, high-flying ideas that are not rooted in experimental reality can generate vast amounts of excitement and occupy whole academic careers, without necessarily advancing understanding. By developing analytical techniques and theoretical frameworks to understand what a fly thinks, we will lay the ground for understanding more complex brains; trying to understand simple animal brains will keep us busy for the rest of the century, at least. If you feel that any study of the brain must involve a vertebrate to be truly interesting, the brain of the tiny zebrafish larva consists of only 100,000 neurons, and easily falls into the small-brain category.

Imaging studies of the human brain, along with future, more precise, brain-wide measures of neuronal activity and interconnection, may indeed provide some insight, but it seems more likely that conceptual advances will come from simpler systems. This does not imply that all studies of the brain and its function should be reductionist, but rather that where there are similarities or even identities in structure and function, in different species, it is easier to develop methodological and analytical techniques on simpler systems. This was the approach used for the massive Human Genome Project, which began by obtaining and analysing the genomes of simple organisms – bacteria, worms and flies – before applying those lessons to humans. This was a far simpler problem, both technologically and conceptually, than understanding the brain of any animal.

Small brains also enable us to investigate how the structure of the brain is a function of two kinds of history. There is the individual history of the animal, of the internal and external stimuli that affected it during its embryonic and pre-adult development and which continue to change its activity, and there is the evolutionary history of the species. Developmental effects help explain inter-individual differences, while comparative studies between species provide insight into some fundamental questions. For example, there are many related *Drosophila* species, which show differences in their sensory structures and differences in behaviour, depending on their ecological niche. These differences will be reflected in brain structure and function, just as Darwin predicted. Comparing such species provides the possibility of exploring the significance of both individual and evolutionary history in understanding brain function.[40] It will also help answer the vexed question of whether all brains are homologous, that is, if the common ancestor of you and a fly and an octopus had a brain. If so, common genes, structures and processes involved in brain function would be expected across the animal kingdom; if not, we would expect to find important differences between brains from different animal lineages when we investigate them more closely.

A focus on brains in insects, worms, larval zebrafish and other organisms does not mean that we cannot study complex behaviours. In 2007, when the results of the first genomic study of a large number of related species were announced (these were eleven species of *Drosophila*), my friend the US neuroscientist Leslie Vosshall published an article in *Nature* provocatively entitled 'Into the Mind of a Fly'. She predicted that we stood on the threshold of a whole new area of study, enabled by comparative genomics:

> It now seems possible to approach in the fly more complex behaviours and even emotions, the neurobiological basis of which is not well understood at the genetic or functional level in any animal: sociality, common sense, altruism, empathy, frustration, motivation, hatred, jealousy, peer pressure, and so on. The only a priori limitation to studying any of these traits is the belief that flies can show such emotions and the design of a plausible behavioural paradigm to measure them.[41]

Although at the time I was dubious, the intervening years have only confirmed her bold prediction, and the advent of CRISPR now makes it possible to alter a gene in pretty much any animal that can be reared in a lab, providing powerful new tools for the study of the brains of what are quaintly called 'non-model organisms' (in other words, neither the mouse, nor *Drosophila*, nor *C. elegans*). As the developmental biologist Nipam Patel put it recently: 'Evolution has solved all the problems we're interested in, we just have to find those organisms and figure out how to ask them how they did it.'[42]

Small brains are now known to produce behaviours that look very similar to those produced by our own, from perception and learning through excitement, indecision, prediction, foresight, aggression, personality and responding to pain.[43] They could even shed insight on a key aspect of our being that is represented by the twin senses of proprioception and interoception. Proprioception is our sense of where our limbs are (this is what enables you to touch your nose with your finger while your eyes are closed), while interoception is our sense of being in our body. In humans, introspection suggests that our sense of self is at least partly intertwined with these feelings. *Drosophila* flies know how big they are and will avoid trying to cross a gap that is too large for their tiny legs to straddle. This knowledge is learned – shortly after hatching, young flies tend to overreach, apparently thinking they are the same size as when they were maggots with an extendable body; visual feedback leads to rapid improvement in their ability to gauge their reach. The flies encode their body size memory, through the activity of an identified set of neurons in the centre of the brain.[44] If we can think of the right experiments to do, the processes that underlie this phenomenon may shed light on more complex examples of how brains represent the body and its relation to the outside world.

Craik's suggestion that the human brain is 'a calculating machine capable of modelling or paralleling external events' applies equally to small brains – they enable animals to interpret events in the environment and, no matter how crudely, to predict outcomes. If we can understand what Darwin called these most marvellous atoms of matter – that is, predict how they behave, both globally and in terms of their components and their interactions, under a

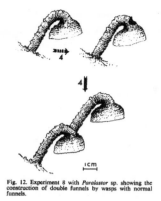

Fig. 12. Experiment 8 with *Paralastor* sp. showing the construction of double funnels by wasps with normal funnels.

Figure from Andrew Smith showing the sequences with which a mud wasp builds its nest.

range of circumstances – we will have taken a massive step towards understanding our own brains. Some scientists argue that such an approach may even unveil the ancient origins of consciousness, but for the moment even the control of movement in the worm *C. elegans* is proving more complex than expected.[45] Whether we will be able to understand the neurobiological basis of the glimmerings of consciousness in animals before we have understood how human consciousness works is unclear.[46]

As well as exploring more complex, conditional behaviours, one fruitful approach might be to take a behaviour that is apparently entirely determined by external factors impinging on internal sensory templates, showing little or no variation between individuals, and seek to understand the underlying neural networks that control this behaviour. For example, in 1978 in one of my favourite scientific papers, Andrew Smith described how an Australian solitary mud wasp constructs its nest entrance – a kind of bendy umbrella-shaped funnel that projects above the ground.[47] The wasp builds the structure in phases, and Smith broke bits off the structure, or raised the ground around it, to reveal the key sensory stimuli that lead the wasp to behave in different ways.

For example, if the wasp saw a hole, it started building a vertical funnel; while it was away collecting mud, Smith pierced a hole at the top of the near-complete structure, the hapless insect simply started building a new vertical funnel, producing a double structure. The wasp had no overall image of the final structure in its brain, merely the next step that had to be performed given a particular stimulus. The behavioural pathways producing this invariant behaviour can be described in terms of a simple flow diagram. Somewhere in the wasp's brain there are neuronal equivalents to those steps in the pathway; it must be possible to identify what they are and how they interact to produce this behaviour.

Although solitary wasps are not ideal laboratory animals, the advent of CRISPR means that, in principle, it should be possible to manipulate these organisms genetically and understand how their brains work. Already there are detailed anatomical descriptions of the brain of the widely studied parasitic wasp *Nasonia vitripennis*, providing a basis for comparative investigations of animals such as the mud wasp.[48] If the mud wasp is too difficult to rear, then exploring how something like navigation is represented in more friendly insect brains may prove an alternative – this is the approach currently being explored by Barbara Webb at the University of Edinburgh and by Marc Gershow at New York University. But as Adam Calhoun and his colleagues have recently shown, even something that seems completely hard-wired, such as *Drosophila* courtship behaviour, is in fact continuously modulated by feedback signals, which vary as the animal switches between various states.[49] The implication is that by revealing the neuronal basis of what is apparently a tightly controlled behaviour, it may be possible to gain insights into how more flexible and complex behaviours emerge.

As to the complexity of the mammalian brain, studies of mice, including increasingly complex connectomes, with the ability to manipulate individual cells, will create a framework that will eventually be able to explain the functioning of the human brain. As we understand more, localisation of function will become increasingly blurred and imprecise, and brains will be understood primarily in terms of circuits and their interaction rather than on the basis of anatomical regions, viewed as modules. The application of models developed on small brains, models that see the brain as active, responding to incoming sensory information and exploring and selecting future possibilities, rather than simply processing and transmitting signals, will provide a dynamic view of brain function.

There is a growing interest in using brain-wide data to explore how neurons respond as populations and correlating these complex responses with the rich behavioural repertoire shown by even the simplest animals. For example, the zebrafish larva alternates between two behavioural states, either limiting its movement and hunting locally or roaming for food and inhibiting hunting. Researchers have identified a small number of cells in the brain that encode hunting; their activity apparently represents the zebrafish's motivational state. Manipulation of such networks will provide insight into the link between network activity, behaviour and motivation. We may even find ourselves revisiting the ideas of McCulloch and Pitts with regard to the immanent logic of nervous system organisation – as this book was going to press, an article appeared describing single neurons in the human brain that compute what is known as an XOR function, that is, they responded if two inputs to the cell were different. This function was previously thought to be a property of networks of cells, with individual neurons able to calculate only AND or OR. The next challenge will be to investigate how this surprisingly rich activity of individual cells affects network function.[50] Small changes in the activity of single cells in the mouse primary visual cortex can have rippling effects on the activity of surrounding cells.[51]

Not everyone agrees with this emphasis on understanding brain activity at the level of single cells. Many researchers working on mammalian brains would share the view of David Robinson of Johns Hopkins University, who in 1992 argued that 'trying to explain how any real neural network works on a cell-by-cell, reductionist basis is futile and we may have to be content with trying to understand the brain at higher levels of organization.'[52] But despite our inability to understand the lobster's stomach, and the undoubted power of the growing number of population-level studies of brains of all kinds, population activity is affected by individual components. Out of, or despite, this complexity, the mystery of consciousness will eventually be solved, in ways I cannot begin to guess.

This is just what I think should happen. There are many

alternative scenarios about how the future of our understanding of the brain could play out:

Perhaps the various computational projects will come good and theoreticians will crack the functioning of all brains, or the connectomes will reveal principles of brain function that are currently hidden from us. Or a theory will somehow pop out of the vast amounts of imaging data we are generating. Or we will slowly piece together a theory (or theories) out of a series of separate but satisfactory explanations. Or by focusing on simple neural network principles we will understand higher-level organisation. Or some radical new approach integrating physiology and biochemistry and anatomy will shed decisive light on what is going on. Or new comparative evolutionary studies will show how other animals are conscious and provide insight into the functioning of our own brains. Or models developed to explain simple brains will turn out to be scalable and will explain ours too. Or the default mode network identified in humans will turn out to be applicable to other animals and to hold the key to overall function. Or unimagined new technology will change all our views by providing a radical new metaphor for the brain. Or our computer systems will provide us with alarming new insight by becoming conscious. Or a new framework will emerge from cybernetics, control theory, complexity and dynamical systems theory, semantics and semiotics. Or we will accept that there is no theory to be found because brains have no overall logic just adequate explanations of each tiny part and we will have to be satisfied with that. Or –

ACKNOWLEDGEMENTS

This book was first imagined in early 2015, in conjunction with my then publisher at Profile Books in London, John Davey, who sadly died in 2017. I think the book was John's idea; it certainly would not exist without his enthusiasm and friendship. Ed Lake took over as editor after John's death and has been exactly what you want of an editor – supportive when he needed to be, frank when he needed to be. The first draft I turned in was baggy and rambling. With Ed's encouragement I sliced 25,000 words off it without doing any harm to the ideas or the structure, and the final read is so much the better for it, even if I have repeatedly stuffed new things into it since. In New York, T. J. Kelleher of Basic Books came up with the title and encouraged me to set out my stall more clearly in the Introduction. Penny Daniel of Profile Books ably steered the manuscript through the various stages from completion to publication.

Although the book is dedicated to Kevin Connolly, all of Kevin's colleagues in the Department of Psychology at University of Sheffield in the mid-1970s deserve my thanks. Along with Barrie Burnet from the Department of Genetics, who with Kevin co-supervised my PhD, they played that precious, uniquely human role – they were my teachers. While writing this book I repeatedly found that many of the things they taught me back then are ideas and approaches that I have been thinking about for the last forty years, and which have found their way into these pages. So thanks to Graham Davey, Paul Dean, John Frisby, Margaret Martlew, John Mayhew, Rod Nicolson, Geoff Pilkington, Peter Redgrave, Terry Rick, David Shapiro, Adrian Simpson, Chris Smith, Max Westby and many others. You made me, and this book.

My friends and colleagues who read through various drafts or outlines deserve a particular mention: Ann-Sophie Barwich, Helen Beebee, Thony Christie, Jerry Coyne, Gabriel Finkelstein, Cathy McCrohan, Kevin Mitchell, Mike Nitabach, Damian Veal and Leslie Vosshall. They all made extremely useful comments, as did my agent Peter Tallack in the earliest planning stages. Special thanks goes to my uncle, Dr Gordon Langley. Many other people made suggestions or answered questions via email or Twitter or in person that steered me in the right direction, including Philip Ball, Sheree Cairney, Adam Calhoun, Albert Cardona, Dan Davis, Caspar Henderson, Andrew Hodges, Tom Holland (not Spider-Man, the other one), Brigitte Nerlich, Adam Rutherford, Sara Solla, Sophie Scott, Paul Summergrad, Josh Weisberg and whoever lurks behind the Neuroskeptic blog and the @ neuro_skeptic Twitter account. Students on Carsten Timmerman's first-year undergraduate course at Manchester – 'A History of Biology in 20 Objects' – have again been a sounding-board for my ideas. All these people helped in different ways, but of course any remaining errors and omissions are entirely the responsibility of my cats.

At lot of things happened in the fifty-odd months it took to produce the book, on both the global and personal scales, many of which have affected me. The growing climate emergency, Trump's presidency, the chaos of Brexit, the Manchester Arena bombing, repeatedly striking at my university in defence of jobs and pensions, illnesses and death in my close family, as well as working on two other books, all became intertwined with my feelings as I grappled with the writing. My loved ones kept me going through all this, so, to Tina, Lauren and Eve, and in particular to Tina – sorry, and thank you.

NOTES

To save space, the titles of journal articles and book chapters have been omitted; it should nonetheless be straightforward to locate all the material cited here using the information provided. A website, theideaofthebrain. com, contains the full bibliography (archived for all time at the Wayback Machine https://tinyurl.com/Cobb-bibliography), along with various videos and other forms of supplementary information. Most of the articles listed here are available on the internet, although some are behind paywalls. There are various means for people without a subscription to obtain these articles, including contacting the author of an article by email and asking for a copy, or using Sci-Hub. Most of the pre-1945 books can be read for free either on Google Books, or at archive.org; the remainder can be ordered from any library. Where possible, website pages cited here have been saved to the Wayback Machine at archive.org to ensure that the relevant page is available long into the future. Two works not referred to below provide an excellent accompaniment to this book: Andrew Wickens's *A History of the Brain* (Psychology Press, 2014) is highly recommended for its more anatomical approach, while Jon Turney's *Cracking Neuroscience* (Cassell, 2018) is a highly accessible overview of the current state of neuroscience.

1. Heart: Prehistory to 17th century

1. Lind, R. (2007), *The Seat of Consciousness in Ancient Literature* (London: McFarland), pp. 57–8.
2. Wallis Budge, E. (1972), *From Fetish to God in Ancient Egypt* (New York: Blom), p. 15.
3. Alter, R. (2007), *The Book of Psalms: A Translation with Commentary* (New York: Norton), p. 19, note 10, and p. 255, note 21.

4. Hultkrantz, A. (1953), *Conceptions of the Soul Among North American Indians: A Study in Religious Ethnology* (Stockholm: Ethnographical Museum of Sweden), p. 178; Spier, L. (1928), *Anthropological Papers of the American Museum of Natural History 29*. The South Fore people of Papua New Guinea ceremonially ate the bodies of their dead, including the brain, inadvertently transmitting the lethal neurodegenerative prion disease kuru. However, the brain played no particularly significant role (Whitfield, J., et al. (2015), *Le Journal de la Société des Océanistes* 141:303–21).

5. Jung, C. (1983), *Memories, Dreams, Reflections* (London: Flamingo), p. 276.

6. Sheree Cairney of Flinders University helped me realise the complexity of what appears to be a simple question. Strikingly, in the whole of Horton, D. (ed.) (1994), *The Encyclopedia of Aboriginal Australia: Aboriginal and Torres Strait Islander History, Society and Culture* (Canberra: Aboriginal Studies Press), there are only two mentions of the brain, three of mind and eleven of the heart, none of which is relevant to this issue.

7. Shogimen, T. (2009), in C. Nederman and T. Shogimen (eds.), *Western Political Thought in Dialogue with Asia* (Plymouth: Lexington), pp. 279–300.

8. Sanchez, G. and Meltzer, E. (2012), *The Edwin Smith Papyrus: Updated Translation of the Trauma Treatise and Modern Medical Commentaries* (Atlanta: Lockwood Press); Finger, S. (2000), *Minds Behind the Brain: A History of the Pioneers and Their Discoveries* (Oxford: Oxford University Press).

9. In 1979 Julian Jaynes published *The Origin of Consciousness in the Breakdown of the Bicameral Mind* in which he used evidence from Homer's *Iliad* to suggest that consciousness appeared as a result of the complexity of society, about 4,000 years ago. Prior to that, he claimed, people experienced consciousness in terms of the voices of the gods. Neither *The Iliad* nor ancient Greek society represent the totality of human experience across the planet and down the ages, and they do not support his argument (Greenwood, 2015). The hunter-gatherer peoples who constitute virtually all of our history were – and are – just as conscious as you and me. Jaynes, J. (1979), *The Origin of Consciousness in the Breakdown of the Bicameral Mind* (Harmondsworth: Penguin); Greenwood, V. (2015), Nautilus https://tinyurl.com/Jaynes-Bicameral.

10. Lloyd, G. (1991), in G. Lloyd (ed.), *Methods and Problems in Greek Science: Selected Papers* (Cambridge: Cambridge University Press), pp. 164–98; Doty, R. (2007), *Neuroscience* 147:561–8.

11. Temkin, O. (1971), *The Falling Sickness: A History of Epilepsy from the*

Greeks to the Beginnings of Modern Neurology (Baltimore: Johns Hopkins University Press), pp. 5–10.

12. Gross, C. (2009), *A Hole in the Head: More Tales in the History of Neuroscience* (London: MIT Press), p. 26.

13. Temkin (1971).

14. Lisowski, F. (1967), in D. Brothwell and A. Sanderson (eds.), *Diseases in Antiquity: A Survey of the Diseases, Injuries and Surgery of Early Populations* (Springfield: Thomas), pp. 651–72; Gross, C. (2002a), in R. Arnott, et al. (eds.), *Trepanation: History, Discovery, Theory* (Lisse: Swets and Zeitlinger), pp. 307–22.

15. Gross, C. (1995), *The Neuroscientist* 1:245–50.

16. Von Staden, E. (1989), *Herophilus: The Art of Medicine in Early Alexandria* (Cambridge: Cambridge University Press), p. 26; Lang, P. (2013), *Medicine and Society in Ptolemaic Egypt* (Leiden: Brill), p. 258.

17. French, R. (2003), *Medicine Before Science: The Business of Medicine from the Middle Ages to the Enlightenment* (Cambridge: Cambridge University Press), pp. 30–31.

18. Boudon-Millot, V. (2012), *Galien de Pergame* (Paris: Les Belles Lettres).

19. Gill, C., et al. (2009), in C. Gill, et al. (eds.), *Galen and the World of Knowledge* (Cambridge: Cambridge University Press), pp. 1–18, p. 6.

20. Gleason, M. (2009), in C. Gill, et al. (eds.), *Galen and the World of Knowledge* (Cambridge: Cambridge University Press), pp. 85–114, p. 112; Rocca, J. (2003), *Galen on the Brain: Anatomical Knowledge and Physiological Speculation in the Second Century A.D.* (Leiden: Brill).

21. Following quotes from Gleason (2009), pp. 99–102.

22. Gleason (2009), p. 100.

23. Al-Khalili, J. (2010), *Pathfinders: The Golden Age of Arab Science* (London: Allen Lane).

24. Frampton, M. (2008), *Embodiments of Will: Anatomical and Physiological Theories of Voluntary Animal Motion from Greek Antiquity to the Latin Middle Ages, 400 BC – AD 1300* (Saarbruck: Verlag Dr. Müller), p. 370.

25. Micheau, F. (1994), in C. Burnett and D. Jacquart (eds.), *Constantine the African and 'Alī ibn al-'Abbās Maǧūsī: The Pantegni and Related Texts* (London: Brill), p. 15. See Kwakkel, E. and Newton, S. (2019) *Medicine at Monte Cassino: Constantine the African and the Oldest Manuscript of his Pantegni* (Turnhout: Brepols).

26. This and subsequent quote from Frampton (2008), pp. 335, 339.

27. Green, C. (2003), *Journal of the History of the Behavioral Sciences* 39:131–42.

28. Van der Eijk, P. (2008), *The Lancet* 372:440–41; Green (2003). Another alleged source of ventricular localisation was a writer called Posidonius, about whom little is known.

29. Manzoni, T. (1998), *Archives Italiennes de Biologie* 136:103–52.

30. Frampton (2008), p. 372.
31. Ibid., p. 381.
32. French (2003), p. 113.
33. Savage-Smith, E. (1995), *Journal of the History of Medicine and Allied Sciences* 50:67–110.
34. Frampton (2008), pp. 383–6.
35. Berengario da Carpi, J. (1521), *Commentaria cum amplissimis additionibus super Anatomia Mundini una cum textu ejusdem in pristinum & verum nitorem redacto* (Bologna: de Benedictis).
36. Dryander, J. (1536), *Anatomia capitis humani* (Marpurg: Cervicorni).
37. Catani, M. and Sandrone, S. (2015), *Brain Renaissance from Vesalius to Modern Neuroscience* (Oxford: Oxford University Press).
38. Fleck, L. (1979), *Genesis and Development of a Scientific Fact* (London: University of Chicago Press), p. 141.
39. Berengario da Carpi had reached a similar conclusion over twenty years earlier. Pranghofer, S. (2009), *Medical History* 53:561–86.
40. Catani and Sandrone (2015), pp. 153–4.
41. This and quotes in next paragraph from Catani and Sandrone (2015), pp. 49, 98, 48.
42. This and next quote from Du Laurens, A. (1599), *A Discourse of the Preservation of the Sight: of Melancholike Diseases; of Rheumes, and of Old Age* (London: Kingston, Iacson), pp. 3, 77.
43. For a rich discussion of how seventeenth century English authors explored the question of brain function, see Habinek, L. (2018), *The Subtle Knot: Early Modern English Literature and the Birth of Neuroscience* (London: McGill-Queen's University Press). Similar books could be written focusing on writers in other languages.

2. Forces: 17th to 18th century

1. Steno, N. (1669), *Discours de Monsieur Stenon, sur l'Anatomie du Cerveau* (Paris: de Ninville).
2. Martensen, R. (2004), *The Brain Takes Shape* (Oxford: Oxford University Press), pp. 52–5.
3. *The Passions of the Soul*, paragraph 33, in Descartes, R. (1985), *The Philosophical Writings of Descartes* (Cambridge, Cambridge University Press), p. 341.
4. There are philosophical arguments about exactly what Descartes meant about the bête-machine idea. See, for example: Newman, L. (2001), *Canadian Journal of Philosophy* 31:389–426.
5. *The Passions of the Soul*, paragraphs 32 and 34, in Descartes (1985), pp. 340–41.
6. Descartes (1985), pp. 100–101.

7. For a discussion, see Huxley, T. (1898), *Collected Essays*, vol. 1: *Method and Results* (London: Macmillan), pp. 211–12.

8. Steno, N. (1965), *Nicolaus Steno's Lecture on the Anatomy of the Brain* (Copenhagen: Nyt Nordisk Forlag Arnold Busck), p. 124.

9. All material from Swammerdam, J. (1758), *The Book of Nature* (London: Seyffert), vol. 2, can be found on pp. 122–32; Malpighi, M. (1666), *Philosophical Transactions of the Royal Society* 2:491–2; Cobb, M. (2002), *Nature Reviews Neuroscience* 3:395–400.

10. Dick, O. (ed.) (2016), *Aubrey's Brief Lives* (London: Vintage), p. cxx.

11. Frank, R. (1990), in G. Rousseau (ed.), *The Language of Psyche: Mind and Body in Enlightenment Thought* (Berkeley: University of California Press), pp. 107–47, p. 123; Zimmer, C. (2004), *Soul Made Flesh: Thomas Willis, the English Civil War and the Mapping of the Mind* (London: Heinemann).

12. Cole, F. (1944), *A History of Comparative Anatomy: From Aristotle to the Eighteenth Century* (London: Macmillan), p. 222.

13. Frank (1990), p. 126.

14. Material from Willis, T. (1684), *Dr Willis's Practice of Physick, Being the Whole Works of That Renowned and Famous Physician* (London: Dring, Harper and Lee), can be found on pp. 71, 75, 92–3, 96.

15. Cobb, M. (2006), *The Egg and Sperm Race: The Seventeenth Century Scientists Who Unravelled the Secrets of Sex, Life and Growth* (London: Free Press).

16. Kardel, T. and Maquet, P. (eds.) (2013), *Nicolaus Steno: Biography and Original Papers of a 17th Century Scientist* (London: Springer), p. 508.

17. This and next quote from Steno (1965), pp. 127, 136.

18. Kardel and Maquet (2013), p. 516.

19. Collingwood, R. (1945), *The Idea of Nature* (Oxford: Clarendon).

20. Hobbes, T. (1651), *Leviathan, or, The Matter, Forme, and Power of a Common Wealth, Ecclesiasticall and Civil* (London: Crooke), p. 1.

21. Whitaker, K. (2004), *Mad Madge: Margaret Cavendish, Duchess of Newcastle, Royalist, Writer and Romantic* (London: Vintage).

22. Cavendish, M. (1664), *Philosophical Letters* (London: n.p.), p. 185. See Cunning, D. (2006), *History of Philosophy Quarterly* 23:117–36.

23. https://tinyurl.com/Descartes-Elizabeth.

24. Spinoza, *Ethics*, part III, proposition 2.

25. Cunning (2006), p. 118. The quote can be found in Leibniz's *Monadology*. For a modern criticism of Leibniz's argument, see Churchland, P. (1995), *The Engine of Reason, the Seat of the Soul: A Philosophical Journey into the Brain* (London: MIT Press).

26. Yolton, J. (1983), *Thinking Matter: Materialism in Eighteenth-Century Britain* (Oxford: Blackwell); Hamou, P. (2007), in P. Anstey (ed.),

   John Locke: Critical Assessments of Leading Philosophers, series II, vol. 3
   (London: Routledge).
27. Locke, J. (1689), *An Essay Concerning Human Understanding*, 4.3.6.
28. Browne, P. (1728), *The Procedure, Extent, and Limits of Human
   Understanding* (London: Innys).
29. Bentley, R. (1692), *Matter and Motion Cannot Think, or, A Confutation
   of Atheism from the Faculties of Soul* (London: Parkhurst, Mortlock),
   pp. 14–15.
30. Giglioni, G. (2008), *Science in Context* 21:1–29.
31. Bentley (1692), p. 29.
32. Thomson, A. (2010), *Early Science and Medicine* 15:3–37, p. 20.
33. Uzgalis, W. (2008), in J. Perry (ed.), *Personal Identity* (London:
   University of California Press), pp. 283–314, p. 296.
34. Uzgalis (2008), p. 284.
35. Thomson (2010).
36. Uzgalis (2008).
37. Ditton, H. (1712), *A Discourse Concerning the Resurrection of Jesus Christ*
   (London: Bell and Lintott), p. 474; Ditton, H. (1714), *The New Law of
   Fluids* (London: Cowse), p. 9.
38. Ditton (1714), appendix, p. 24.
39. Vartanian, A. (1960), *La Mettrie's L'Homme Machine: A Study in the
   Origins of an Idea* (Princeton: Princeton University Press), p. 74;
   Niderst, A. (1969), *L'Ame matériel (ouvrage anonyme). Edition critique,
   avec une introduction et des notes* (Paris: Nizet).
40. Fearing, F. (1970), *Reflex Action: A Study in the History of Physiological
   Psychology* (Cambridge, MA: MIT Press).
41. Yolton (1983), p. 177.
42. These quotes and the subsequent one from Boerhaave, H. (1743), *Dr.
   Boerhaave's Academical Lectures on the Theory of Physic*, vol. 2 (London:
   Innys), pp. 290 and 312–13.
43. Koehler, P. (2007), in H. Whitaker, et al. (eds.), *Brain, Mind and
   Medicine: Neuroscience in the 18th Century* (New York: Springer),
   pp. 213–31, p. 219; Steinke, H. (2005), *Irritating Experiments: Haller's
   Concept and the European Controversy on Irritability and Sensitivity,
   1750–90* (Amsterdam: Rodopi), pp. 21–2.
44. This and next quotes from Temkin, O. (1936), *Bulletin of the History of
   Medicine* 4:651–99, pp. 675, 657, 661; Steinke (2005).
45. Koehler (2007), p. 223.
46. Munro, A. (1781), *The Works of Alexander Monro, M.D.* (Edinburgh:
   Elliot, Robinson), p. 324.
47. Smith, C. (2007), in H. Whitaker, et al. (eds.), *Brain, Mind and Medicine:
   Neuroscience in the 18th Century* (New York: Springer), pp. 15–28 and
   p. 27, note 4.

48. Anonymous (1747), *An Enquiry into the Origin of the Human Appetites and Affections, Shewing How Each Arises from Association* (Lincoln: Dodsley), p. 41.
49. Glassman, R. and Buckingham, H. (2007), in H. Whitaker, et al. (eds.), *Brain, Mind and Medicine: Essays in Eighteenth-Century Neuroscience* (Boston, MA: Springer), pp. 177–90.
50. Hartley, D. (1749), *Observations on Man, His Frame, His Duty, and His Expectations* (London: Hitch and Austen), part I, p. iv.
51. Whytt, R. (1751), *An Essay on the Vital and Other Involuntary Motions of Animals* (Edinburgh: Hamilton, Balfour and Neill), p. 239; French, R. (1969), *Robert Whytt, the Soul, and Medicine* (London: Wellcome), p. 69.
52. Temkin (1936), p. 683.
53. French (1969).
54. Quotes from Whytt (1751), pp. 2, 252.
55. Fearing (1970), p. 69.
56. French (1969), pp. 75, 91.
57. Fearing (1970), pp. 82–3.
58. Wellman, K. (1992), *La Mettrie: Medicine, Philosophy and Enlightenment* (London: Duke University Press); Thomson, A. (ed.) (1996), *La Mettrie: Machine Man and Other Writings* (Cambridge: Cambridge University Press).
59. Thomson (1996), p. 26.
60. La Mettrie, J. de (1748), *L'Homme machine* (Leiden: Luzac). For recent translations see Vartanian (1960); La Mettrie, J. de (1994), *Man a Machine; and, Man a Plant* (Indianapolis: Hackett); Thomson (1996). For a full bibliography of La Mettrie's writings see Stoddard, R. (1992), *The Papers of the Bibliographic Society of America* 86:411–59.
61. Quotes from Thomson (1996), pp. 13, 9, 25, 35, 6.
62. Makari, G. (2015), *Soul Machine: The Invention of the Modern Mind* (London: Norton).
63. This and next quote from Thomson (1996), pp. 28, 31, 33.
64. Vartanian (1960), p. 139.
65. Riskin, J. (2016), *The Restless Clock: A History of the Centuries-Long Argument Over What Makes Living Things Tick* (Chicago: University of Chicago Press), pp. 162–3.
66. De Saussure, R. (1949), *Journal of the History of Medicine and Allied Sciences* 4:431–49, p. 432.
67. Thomson (1996), p. x; Vartanian (1960), p. 116.
68. Riskin (2016), p. 156.
69. Morange, M. (2016), *Une Histoire de la biologie* (Paris: Seuil), p. 101.
70. Braudy, L. (1970), *Eighteenth-Century Studies* 4:21–40.
71. Riskin (2016), pp. 116–22.

72. Colliber, S. (1734), *Free Thoughts Concerning Souls* (London: Robinson), p. 8.

73. Priestley, J. (1777), *Disquisitions Relating to Matter and Spirit* (London: Johnson), p. 27.

74. Priestley, J. (1778), *A Free Discussion of the Doctrines of Materialism, and Philosophical Necessity* (London: Johnson, Cadell), p. 61.

75. Priestley (1777), p. 27.

76. Brown, T. (1974), *Journal of the History of Biology* 7:179–216.

77. Fearing (1970), p. 94.

3. Electricity: 18th to 19th century

1. Shelley, M. (2003), *Frankenstein* (London: Penguin), pp. 6–7.

2. Holmes, R. (2008), *The Age of Wonder: How the Romantic Generation Discovered the Beauty and Terror of Science* (London: Harper).

3. Fara, P. (2002), *An Entertainment for Angels* (London: Icon).

4. Ewald Georg von Kleist had invented a similar device a year earlier in Pomerania – Torlais, J. (1963), *Revue d'histoire des sciences et de leurs applications* 16:211–19.

5. Priestley, J. (1769), *The History and Present State of Electricity, with Original Experiments* (London: Dodsley, Johnson, Payne, Cadell), p. 98.

6. Bertucci, P. (2007), in H. Whitaker, et al. (eds.), *Brain, Mind and Medicine: Neuroscience in the 18th Century* (New York: Springer), pp. 271–83.

7. Beccaria, G. (1776), *A Treatise Upon Artificial Electricity* (London: Nourse), p. 270.

8. Haller, A. von (1762), *Mémoires sur les parties sensibles et irritables du corps animal, tôme troisième* (Lausanne: Grasset); Kaplan, P. (2002), *Journal of the Royal Society of Medicine* 95:577–8, p. 577.

9. Priestley (1769), p. 622.

10. Hartley (1749), part 1, p. 88.

11. Bonnet, C. (1755), *Essai de psychologie; ou considérations sur les opérations de l'âme, sur l'habitude et sur l'éducation* (London: n.p.), p. 268.

12. Bonnet, C. (1760), *Essai analytique sur les facultés de l'âme* (Copenhagen: Philibert), pp. 21–2.

13. Home, R. (1970), *Journal of the History of Biology* 3:235–51.

14. Koehler, P., et al. (2009), *Journal of the History of Biology* 42:715–63.

15. Material from: Piccolino, M. (2007), in H. Whitaker, et al. (eds.), *Brain, Mind and Medicine: Neuroscience in the 18th Century* (New York: Springer), pp. 125–43; Finger, S. (2013), *Progress in Brain Research* 205:3–17.

16. Home (1970), p. 250.

17. Bertholon, P. (1780), *De l'électricité du corps humain dans l'état de santé et de maladie* (Paris: Didot), pp. 70, 94.

18. Galvani, L. (1953), *Commentary on the Effect of Electricity on Muscular Motion* (Cambridge, MA: Elizabeth Licht); Bresadola, M. (2003), in F. Holmes, et al. (eds.), *Reworking the Bench: Research Notebooks in the History of Science* (Dordrecht: Kluwer), pp. 67–92.

19. Galvani (1953), p. 46.

20. Ibid., p. 97.

21. Ibid., pp. 60, 66, 67.

22. Ibid., p. 72.

23. Material in this paragraph from Valli, E. (1793), *Experiments on Animal Electricity, with Their Application to Physiology* (London: Johnson), pp. 5, 241–2

24. Fowler, R. (1793), *Experiments and Observations Relative to the Influence Lately Discovered by M. Galvani, and Commonly Called Animal Electricity* (Edinburgh: Duncan, Hill, Robertson & Berry, add Mudie).

25. Volta, A. (1816), *Collezione dell'opere del cavaliere Conte Alessandro Volta*, vol. 2, part I (Florence: Piatti), p. 111.

26. Finger, S., et al. (2013), *Journal of the History of the Neurosciences* 22:237–352.

27. Mauro, A. (1969), *Journal of the History of Medicine* 24:140–50.

28. Hoff, H. (1936), *Annals of Science* 1:157–72. This was first demonstrated by Matteucci, C. (1842), *Annales de chimie et de physique*, Série 3 6:301–39.

29. Darwin, E. (1801), *Zoonomia; Or, the Laws of Organic Life*, vol. 1 (London: Johnson), p. 83.

30. Pancaldi, G. (1990), *Historical Studies in the Physical and Biological Sciences* 21:123–60.

31. Volta, A. (1800), *Philosophical Transactions of the Royal Society of London* 90:403–31.

32. Holmes (2008), pp. 274, 325. There is no evidence that Godwin actually attended the lecture; however, the textual similarities between Davy's lectures and *Frankenstein* are striking – if Godwin did not attend the lecture, she clearly read the printed version.

33. I have spared you the details, some of which can be found in Aldini, J. (1803), *An Account of the Late Improvements in Galvanism* (London: Murray), pp. 68–80.

34. Aldini (1803); Aldini, J. (1804), *Essai théorique et expérimental sur le galvinisme* (Bologna: Piranesi).

35. Aldini (1803), p. 193.

36. *The Times*, 22 January 1803, p. 3.

37. Aldini (1804), p. 216.

38. This and next quote from Aldini (1803), pp. 57, 63–4.

39. Quotes from Aldini (1804), pp. 116–20; Bolwig, T. and Fink, M. (2009),

Journal of Electro-Convulsive Therapy 25:15–18. Aldini also used this procedure on Charles Bellini, a labourer, with similar results.

40. Finger, S. and Law, M. (1998), *Journal of the History of Medicine* 53:161–80, p. 167.

41. Neuburger, M. (1981), *The Historical Development of Experimental Brain and Spinal Cord Physiology Before Flourens* (London: Johns Hopkins University Press), p. 199. For even more detail, if you can stomach it (remember that while the results are certainly untrue, the procedure is probably accurately described), see Finger and Law (1998).

42. Finger and Law (1998), p. 169.

43. Neuburger (1981), pp. 199, 220.

44. Finger and Law (1998), p. 165.

45. Roget, P. (1824a), *Supplement to the Fourth, Fifth, and Sixth Editions of the Encyclopaedia Britannica*, vol. 6 (Edinburgh: Constable), p. 187 – entry on Physiology.

46. Rogers, J. (1998), in E. Yeo (ed.), *Radical Femininity: Women's Representation in the Public Sphere* (Manchester: Manchester University Press), pp. 52–78.

47. Sharples, E. (1832), *The Isis* 6:81–5, p. 85.

48. Anonymous (1844), *Vestiges of the Natural History of Creation* (London: Churchill); Secord, J. (2000), *Victorian Sensation: The Extraordinary Publication, Reception, and Secret Authorship of Vestiges of the Natural History of Creation* (Chicago: University of Chicago Press).

49. Anonymous (1844), p. 334.

50. Ibid., p. 335.

51. Longet, F.-A. (1842), *Anatomie et physiologie du système nerveux de l'homme et des animaux vertebrés*, vol. 1 (Paris: Fortin, Masson et Cie), pp. 138–9.

52. Matteucci, C. (1844), *Traité des phénomènes électro-physiologiques des animaux* (Paris: Fortin, Masson et Cie).

53. Matteucci, C. (1845), *Philosophical Transactions of the Royal Society of London* 135:303–17, p. 317.

54. Matteucci, C. (1850), *Philosophical Transactions of the Royal Society of London* 140:645–9, p. 648.

55. Finger, S. and Wade, N. (2002a), *Journal of the History of the Neurosciences* 11:136–55; Finger, S. and Wade, N. (2002b), *Journal of the History of the Neurosciences* 11:234–54.

56. This and next quote from Müller, J. (1843), *Elements of Physiology* (Philadelphia: Lea and Blanchard), pp. 513, 515, 532.

57. Otis, L. (2007), *Müller's Lab* (Oxford: Oxford University Press).

58. Finkelstein, G. (2014), in C. Smith and H. Whitaker (eds.), *Brain, Mind and Consciousness in the History of Neuroscience* (New York: Springer), pp. 163–84, p. 164.

59. Clarke, E. and Jacyna, L. (1987), *Nineteenth-Century Origins of Neuroscientific Concepts* (London: University of California Press), p. 211.

60. Bowditch, H. (1886), *Science* 8:196–8, pp. 196–7.

61. Meulders, M. (2010), *Helmholtz: From Enlightenment to Neuroscience* (Cambridge, MA: MIT Press).

62. Finger and Wade (2002a), p. 152.

63. Lenoir, T. (1994), *Osiris* 9:184–207.

64. Helmholtz, H. (1875), *On the Sensations of Tone as a Physiological Basis for the Theory of Music* (London: Longmans, Green), p. 224.

65. Odling, E. (1878), *Memoir of the Late Alfred Smee, FRS, by his Daughter* (London: Bell and Sons). I had not heard of Smee until I encountered him in the work of the historian of science, Iwan Rhys Morus, while researching this chapter.

66. Smee, A. (1849), *Elements of Electro-Biology, or, the Voltaic Mechanism of Man* (London: Longman, Brown, Green, and Longmans), p. 39.

67. Ibid., p. 45.

68. This and quotes in subsequent paragraphs from Smee, A. (1850), *Instinct and Reason Deduced from Electro-Biology* (London: Reeve, Benham and Reeve), pp. 29, 211, 98. The Figure is Plate VIII, opposite p. 210.

69. Morus, I. (1998), *Frankenstein's Children: Electricity, Exhibition, and Experiment in Early-Nineteenth-Century London* (Princeton: Princeton University Press), p. 150.

70. This and subsequent two paragraphs from Smee, A. (1851), *The Process of Thought Adapted to Words and Language, Together with a Description of the Relational and Differential Machines* (London: Longman, Brown, Green, and Longmans), pp. xv, 2, 39, 40, 42–3, 45, 49–50. Boden, M. (2006), *Mind as Machine: A History of Cognitive Science*, 2 vols. (Oxford: Clarendon), vol. 1, p. 121, suggests that Smee was influenced by George Boole's *The Mathematical Analysis of Logic*, which was published in 1847. Smee's writings contain no reference to Boole or his ideas and historians of computing argue there was no connection – Buck, G. and Hunka, S. (1999), *IEEE Annals of the History of Computing* 21:21–7.

71. Aspray, W. (1990), *Computing Before Computers* (Ames: Iowa State University Press), pp. 108–10.

4. Function: 19th century

1. Liebknecht, K. (1908), *Karl Marx: Biographical Memoirs* (Chicago: Kerr), p. 64. The title of the group that met in 1850 is usually given either as the German Labourers' Educational Association, or the Communist Club.

2. Parssinen, T. (1974), *Journal of Social History* 7:1–20, p. 1.

3. Shuttleworth, S. (1989), in J. Christie and S. Shuttleworth (eds.), *Nature*

Transfigured: Science and Literature, 1700–1900 (Manchester: Manchester University Press), pp. 121–51; Boshears, R. and Whitaker, H. (2013), *Progress in Brain Research* 205:87–112.

4. McLaren, A. (1981), *Comparative Studies in Society and History* 23:3–22.
5. Clark and Jacyna (1987), pp. 222–3.
6. Gall, F. and Spurzheim, G. (1810), *Anatomie et physiologie du système nerveux en général, et du cerveau en particulier*, vol. 1 (Paris: Schoell), p. xvii.
7. Young, R. (1990), *Mind, Brain and Adaptation in the Nineteenth Century: Cerebral Localization and its Biological Context from Gall to Ferrier* (Oxford: Oxford University Press), p. 56.
8. Gall, F. and Spurzheim, G. (1812), *Anatomie et physiologie du système nerveux en général, et du cerveau en particulier*, vol. 2 (Paris: Schoell), p. 225; Gall, F. (1818), *Anatomie et physiologie du système nerveux en général, et du cerveau en particulier*, vol. 3 (Paris: Librairie Grècque-latine-allemande), pp. 307–22.
9. The word Gall used for 'pride' was 'hauteur', which also means 'height'.
10. Boring, E. (1950), *A History of Experimental Psychology* (Englewood Cliffs: Prentice-Hall), p. 53; Boshears and Whitaker (2013).
11. Spurzheim, J. (1815), *The Physiognomical System of Drs. Gall and Spurzheim* (London: Baldwin, Cradock, and Joy).
12. Gall (1818), p. xxix.
13. Ibid. and McLaren (1981).
14. Cooter, R. (1984), *The Cultural Meaning of Popular Science: Phrenology and the Organisation of Consent in Nineteenth-Century Britain* (Cambridge: Cambridge University Press).
15. Combe, G. (1836), *Testimonials on Behalf of George Combe, as a Candidate for the Chair of Logic in the University of Edinburgh* (Edinburgh: Anderson), p. 5; Parsinnen (1974), p. 1.
16. McLaren (1981).
17. Hegel, G. (2003), *The Phenomenology of Mind* (Mineola: Dover), pp. 175–98.
18. Napoleon (1824), *Profils des contemporains* (Paris: Pollet), p. 54.
19. All quotes from Roget, P. (1824b), *Supplement to the Fourth, Fifth, and Sixth Editions of the Encyclopaedia Britannica*, vol. 3 (Edinburgh: Constable) – entry on Cranioscopy.
20. Clark, J. and Hughes, T. (1980), *The Life and Letters of the Reverend Adam Sedgwick*, vol. 2 (Cambridge: Cambridge University Press), p. 83.
21. Parssinen (1974), p. 12.
22. Young (1990), p. 61.
23. Material from Flourens, P. (1842), *Recherches expérimentales sur les*

propriétés et les fonctions du système nerveux, dans les animaux vertébrés (Paris: Ballière), pp. 135, 131, 132, 244.

24. Swazey, J. (1970), *Journal of the History of Biology* 3:213–34.

25. Flourens, P. (1824), *Recherches expérimentales sur les propriétés et les fonctions du système nerveux, dans les animaux vertébrés* (Paris: Crevot), p. 122.

26. Luzzatti, C. and Whitaker, H. (2001), *Archives of Neurology* 58:1157–62.

27. Andral, G. (1840), *Clinique médicale, ou choix d'observations recueillies à l'Hôpital de la Charité*, vol. 5: *Maladies de l'encéphale* (Paris: Fortin, Masson), pp. 155, 523; Stookey, B. (1963), *Journal of the American Medical Association* 184:1024–9.

28. Finger (2000), p. 139.

29. Broca, P. (1861a), *Bulletins de la Société d'anthropologie de Paris* 2:139–204, 301–21, 441–6; LaPointe, L. (2013), *Paul Broca and the Origins of Language in the Brain* (San Diego: Plural Publishing).

30. Pearce, J. (2006), *European Neurology* 56:262–4; Schiller, F. (1979), *Paul Broca: Founder of French Anthropology, Explorer of the Brain* (Berkeley: University of California Press), p. 175.

31. Auburtin, E. (1863), *Considérations sur les localisations cérébrales et en particulier sur le siège de la faculté du langage articulé* (Paris: Masson et Fils), pp. 24–5.

32. Joynt, R. (1961), *Archives of Internal Medicine* 108:953–6; Schiller, F. (1963), *Medical History* 7:79–81.

33. Broca, P. (1861b), *Bulletins de la Société d'anthropologie de Paris* 2: 235–38, p. 238; Schiller (1979), p. 178.

34. Broca, P. (1861c), *Bulletins de la Société anatomique de Paris* 36:330–57.

35. Broca, P. (1861d), *Bulletins de la Société anatomique de Paris* 36:398–407.

36. Ibid., pp. 406–7.

37. Broca, P. (1863), *Bulletins de la Société d'anthropologie de Paris* 4:200–202, p. 202.

38. Dax, M. (1865), *Gazette hebdomodaire de médicine et de chirurgie* 17:259–60; Dax, M. G. (1865), *Gazette hebdomodaire de médicine et de chirurgie* 17:260–62; Finger, S. (1996), *Archives of Neurology* 53:806–13.

39. Dax, M. G. (1865), p. 262.

40. Broca, P. (1865), *Bulletins de la Société d'anthropologie de Paris* 6:377–93, p. 383.

41. Glickstein, M. (2014), *Neuroscience: A Historical Introduction* (Cambridge, MA: MIT Press), p. 278.

42. Rutten, G.-J. (2017), *The Broca–Wernicke Doctrine. A Historical and Clinical Perspective on Localization of Language Functions* (Cham, Switzerland: Springer).

43. Duval, A. (1864), *Bulletins de la Société d'anthropologie de Paris* 5:213–17, p. 215.

44. Bartholow, R. (1874a), *American Journal of the Medical Sciences* 134:305–13; Bartholow, R. (1874b), *British Medical Journal* 1(700):727.
45. Ferrier, D. (1876), *The Functions of the Brain* (London: Smith, Elder), p. 296; Harris, L. and Almerigi, J. (2009), *Brain and Cognition* 70:92–115.
46. Fritsch, G. and Hitzig, E. (1870), *Archiv für Anatomie, Physiologi und wissenschaftliche Medizin* 37:300–332 – for a translation, see Wilkins, R. (1963), *Journal of Neurosurgery* 20:904–16; Taylor, C. and Gross, C. (2003), *The Neuroscientist* 9:332–42; Hagner, M. (2012), *Journal of the History of the Neurosciences* 21:237–49.
47. Ferrier (1876), p. 80.
48. Wilkins (1963), p. 909.
49. Ibid., p. 916.
50. Ferrier (1876); Taylor and Gross (2003).
51. Quotes from Ferrier (1876), pp. 44–5, 124–5, 39, 40, 213, 130, 141–5.
52. Macmillan, M. (2000), *An Odd Kind of Fame: Stories of Phineas Gage* (London: MIT Press). Quotes from Ferrier (1876), pp. 231–2.
53. Ferrier, D. (1878a), *British Medical Journal* 1:443–7; Ferrier, D. (1878b), *The Localisation of Cerebral Function* (London: Smith, Elder).
54. Macmillan (2000), pp. 401–22, 414–15.
55. Ibid., pp. 314–33 contains a rogue's gallery of mistaken accounts.
56. Quotes from Ferrier (1876), pp. 288, 255–8.

5. Evolution: 19th century

1. Abercrombie, J. (1838), *Inquiries Concerning the Intellectual Powers and the Investigation of Truth* (London: Murray), p. 34.
2. https://www.biodiversitylibrary.org/title/50381#page/52/mode/1up.
3. Barrett, P., et al. (eds.) (2008), *Charles Darwin's Notebooks, 1836–1844: Geology, Transmutation of Species, Metaphysical Enquiries* (Cambridge: Cambridge University Press), p. 165.
4. Müller (1843), https://www.biodiversitylibrary.org/item/105993#page/53/mode/1up; Richards, R. (1987), *Darwin and the Emergence of Evolutionary Theories of Mind and Behavior* (Chicago: University of Chicago Press), p. 94; Swisher, C. (1967), *Bulletin of the History of Medicine* 41:24–43, p. 27.
5. Barrett et al. (2008), pp. 291, 614.
6. Partridge, D. (2015), *Biological Journal of the Linnean Society* 116:247–51.
7. Darwin, C. (2004), *The Descent of Man, and Selection in Relation to Sex* (London: Penguin), p. 17; Bizzo, N. (1992), *Journal of the History of Biology* 25:137–47.
8. Chadwick, O. (1975), *The Secularisation of the European Mind in the Nineteenth Century* (Cambridge: Cambridge University Press), p. 184.

9. Tyndall, J. (1875), *Popular Science Monthly*, February 1875, pp. 422–40, p. 438.

10. Harrington, A. (1987), *Medicine, Mind, and the Double Brain* (Princeton: Princeton University Press), p. 124.

11. Anonymous (1875), *Popular Science Monthly*, February 1875, pp. 501–4, p. 503. See Tyndall, J. (1874), *John Tyndall's Address Delivered Before the British Association Assembled at Belfast, with Additions* (London: Longmans, Green) and various articles in *Popular Science*, February 1875.

12. Finkelstein (2014), p. 165; Finkelstein, G. (2013), *Emil du Bois-Reymond: Neuroscience, Self, and Society in Nineteenth-Century Germany* (London: MIT Press).

13. Van Strien, M. (2015), *Annals of Science* 72:381–400, p. 387.

14. Richards (1987), pp. 176–9.

15. Ibid., p. 178; Wallace, A. (1871), *Contributions to the Theory of Natural Selection. A Series of Essays* (London: Macmillan).

16. Smith, C. (2010), *Journal of the History of the Neurosciences* 19:105–20, p. 118.

17. Lyell, C. (1863), *Geological Evidences of the Antiquity of Man* (London: John Murray); Cohen, C. (1998), in D. Blundell and A. Scott (eds.), *Lyell: The Past is the Key to the Present* (Bath: Geological Society), pp. 83–93.

18. Lyell (1863), p. 201; Richards, R. (2009), in J. Hodge and G. Radick (eds.), *The Cambridge Companion to Darwin* (Cambridge: Cambridge University Press), pp. 96–119, p. 106.

19. This and material in next two paragraphs from Darwin (2004), pp. 86, 231, 240.

20. Darwin (2004), p. 87.

21. This and subsequent quotes ibid., pp. 74, 88–9, 151; Smith (2010).

22. Huxley, T. (1874), *Nature* 6:362–6; Wallace, A. (1874), *Nature* 10:502–3; Wetterhan, I. (1874), *Nature* 6:438; Anger, S. (2009), *Victorian Review* 35:50–52.

23. Huxley (1874), p. 365.

24. This and subsequent quotes from Huxley (1898), pp. 237, 240, 244, 191; also Huxley (1874).

25. Richards (1987), pp. 352, 368.

26. This and next quote from Lloyd Morgan, C. (1900), *Animal Behaviour* (London: Edward Arnold), pp. 95, 93.

27. McGrath, L. (2014), *Journal of the Western Society for French History* 42:1–12, p. 1.

28. Maudsley, H. (1872), *The Lancet* 100:185–9, pp. 186–7.

29. Maudsley, H. (1883), *Body and Will* (London: Kegan Paul, Trench), pp. 101–2.

30. Hughlings Jackson, J. (1887), *Journal of Mental Science* 33:25–48, pp. 37–8.

6. Inhibition: 19th century

1. Diamond, S., et al. (1963), *Inhibition and Choice: A Neurobehavioral Approach to Problems of Plasticity in Behavior* (New York: Harper & Row); Smith. R. (1992a), *Inhibition: History and Meaning in the Sciences of Mind and Brain* (Berkeley: University of California Press).
2. Smith (1992a), pp. 80–81.
3. Ibid., p. 77.
4. Quotes from Sechenov, I. (1965), *Reflexes of the Brain* (Cambridge, MA: MIT Press), pp. 19, 86.
5. Young (1990), p. 205.
6. Sechenov (1965), p. 89.
7. Maudsley, H. (1867), *The Physiology and Pathology of the Mind* (New York: Appleton), p. 83.
8. Ferrier (1876), p. 287.
9. James, W. (1890), *Principles of Psychology*, 2 vols. (New York: Holt), vol. 2, p. 68.
10. Smith (1992a), pp. 132–3.
11. Diamond et al. (1963), p. 41.
12. Smith (1992a), p. 134.
13. McDougall, W. (1905), *Physiological Psychology* (London: Dent), p. 103.
14. Diamond et al. (1963), pp. 40, 45.
15. Ferrier (1876), p. 18.
16. Anstie, F. (1865), *Stimulants and Narcotics, Their Mutual Relations* (Philadelphia: Lindsay and Blakiston), pp. 86–7.
17. Smith, R. (1992b), *Science in Context* 5:237–63.
18. Hughlings Jackson (1887), p. 37.
19. Smith (1992a), p. 154.
20. Lloyd Morgan, C. (1896), *An Introduction to Comparative Psychology* (New York: Walter Scott), p. 182.
21. Morton, W. (1880), *Scientific American Supplement* 256:4085–6, p. 4085. For an exploration of the work of Charcot and his colleagues, and the significance of neurosyphilis in providing a biological model of mental illness, see Ropper, A. and Burrell, B. (2020), *How the Brain Lost Its Mind: Sex, Hysteria and the Riddle of Mental Illness* (London: Atlantic).
22. Goetz, C., et al. (1995), *Charcot: Constructing Neurology* (Oxford: Oxford University Press).
23. Heidenhain, R. (1899), *Hypnotism or Animal Magnetism: Physiological Observations* (London: Kegan Paul, Trench, Trübner), p. 46.
24. Smith (1992a), p. 129.

25. Fletcher, J. (2013), *Freud and the Scene of Trauma* (New York: Fordham University Press), p. 28.

26. Freud, S. (1963), in P. Rieff (ed.), *General Psychological Theory: Papers on Metapsychology* (New York: Collier), pp. 116–50, p. 125.

27. Crews, F. (2017), *Freud: The Making of an Illusion* (London: Profile), p. 448.

28. As might be expected, Crews (2017), pp. 435–51 is critical of this work, while Makari, G. (2008), *Revolution in Mind: The Creation of Psychoanalysis* (London: Duckworth), pp. 70–74 is sympathetic.

29. Todes, D. (2014), *Ivan Pavlov: A Russian Life in Science* (New York: Oxford University Press).

30. Helmholtz, H. von (1962), *Helmholtz's Treatise on Physiological Optics*, vol. 3 (New York: Dover), pp. 3, 4.

31. Ibid., pp. 4, 27.

32. Ibid., p. 14.

33. Ibid., p. 6.

34. Heidelberger, M. (1993), in D. Cahan (ed.), *Hermann von Helmholtz and the Foundations of Nineteenth-Century Science* (San Francisco: University of California Press), pp. 461–97, p. 493.

35. Cahan, D. (2018), *Helmholtz: A Life in Science* (Chicago: University of Chicago Press), p. 532.

36. Arbib, M. (2000), *Perspectives in Biology and Medicine* 43:193–216.

37. Meulders (2010), p. 145.

38. Sherrington, C. (1906), *The Integrative Action of the Nervous System* (New Haven: Yale University Press); Swazey, J. (1969), *Reflexes and Motor Integration: Sherrington's Concept of Integrative Action* (Cambridge, MA: Harvard University Press).

39. Sherrington (1906), pp. 7, 16, 181.

40. Ibid., p. 238.

41. Ibid., p. 55.

42. Ibid., pp. 65, 113, 187.

43. For material in this paragraph see ibid., pp. 308–31, 352, 393.

44. Bastian, H. (1880), *The Brain as an Organ of Mind* (New York: Appleton).

45. Sherrington (1906), p. 35.

46. Ferrier (1876), pp. 290, 294.

47. Sherrington (1906), p. 83. The figure is on p. 108.

7. Neurons: 19th to 20th century

1. Shepherd, G. (2016), *Foundations of the Neuron Doctrine*, 25th Anniversary Edition (Oxford: Oxford University Press).

2. Pannese, E. (1999), *Journal of the History of the Neurosciences* 8:132–40; Shepherd, G. (1999), *Journal of the History of the Neurosciences* 8:209–14.

3. This and subsequent quotes from Golgi, C. (1883), *The Alienist and Neurologist* 4:236–269, 383–416, pp. 396, 394, 401.

4. Cajal, S. (1937), *Memoirs of the American Philosophical Society* 8:1–638, p. 305.
5. Ibid., p. 321.
6. Cajal, S. (1909), *Histologie du système nerveux de l'homme et des vertébrés*, vol. 1 (Paris: Maloine), p. 29.
7. Ranvier, L.-A. (1878), *Leçons sur l'Histologie du système nerveux* (Paris: Savy), p. 131; Boullerne, A. (2016), *Experimental Neurology* 283B:431–45.
8. Shepherd (2016), p. 163.
9. Cajal (1937), pp. 356–7.
10. Ibid., p. 358.
11. Jones, E. (1999), *Journal of the History of the Neurosciences* 8:170–78; Bock, O. (2013), *Endeavour* 37:228–34.
12. Shepherd (2016), p. 189.
13. Ibid., p. 229.
14. López-Muñoz, F., et al. (2006), *Brain Research Bulletin* 70:391–405.
15. Golgi, C. (1967), in Nobel Foundation (ed.) *Nobel Lectures. Physiology or Medicine, 1901–1921* (Amsterdam: Elsevier), pp. 215, 216.
16. Cajal, S. (1894a), *Proceedings of the Royal Society of London* 55:444–68 – all translations are my own; Jones (1999).
17. Cajal (1894a), p. 444.
18. 'Units' can be found in Cajal (1894a), pp. 457, 465.
19. Ibid., p. 450.
20. Ibid., p. 465; Berlucchi, G. (1999), *Journal of the History of the Neurosciences* 8:191–201.
21. Shepherd (2016), pp. 203–10.
22. James (1890), vol. 2, p. 581.
23. Cajal (1894a), p. 452.
24. Otis, L. (2001), *Networking: Communicating with Bodies and Machines in the Nineteenth Century* (Ann Arbor: University of Michigan Press); Otis, L. (2002), *Journal of the History of Ideas* 63:105–28. These two works were extremely useful to me.
25. Cajal (1894a), pp. 466, 467.
26. Demoor, J. (1896), *Archives de Biologie* 14:723–52; Jones, E. (1994), *Trends in Neurosciences* 17:190–92; Berlucchi, G. (2002), *Journal of the History of the Neurosciences* 11:305–9.
27. Cajal (1894a), pp. 467–8. Recent English translations of Cajal's work have suggested that Cajal used a phrase from Max Nordau and described sense organs as 'true computing machines' – Cajal, S. (1999), *Texture of the Nervous System of Man and the Vertebrates* (Berlin: Springer), p. 8 – or 'computational devices' – Cajal, S. (1995), *Histology of the Nervous System* (Oxford: Oxford University Press). Nordau (1885), *Paradoxe* (Leipzig: Elischer Nachfolger), in fact used the phrase 'Zusammenfassung zahlreiche Organe' – 'a combination of numerous

organs'. It seems probable that a mistranslation by Cajal or others took place and has since been compounded. There is no evidence that Cajal viewed any part of the nervous system as any kind of computational device. For a full discussion of this point see theideaofthebrain.com.

28. Cajal, S. (1894b), *Les Nouvelles idées sur la structure du système nerveux chez l'homme et chez les vertébrés* (Paris: Reinwald), p. x.

29. Bergson, H. (1911), *Matter and Memory* (London: Allen and Unwin), pp. 19–20.

30. Keith, A. (1919), *The Engines of the Human Body* (London: Williams and Norgate), p. 259; Kirkland, K. (2002), *Perspectives in Biology and Medicine* 45:212–23.

31. Keith (1919), pp. 261–2.

32. Otis (2001), p. 67.

33. Robinson, J. (2001), *Mechanisms of Synaptic Transmission: Bridging the Gaps (1890–1990)* (Oxford: Oxford University Press), p. 21.

34. Foster, M. and Sherrington, C. (1897), *A Text Book of Physiology*, part III: *The Central Nervous System* (London: Macmillan), pp. 928–9.

35. Ibid., p. 969.

36. Material from Sherrington (1906), pp. 2, 3, 18.

37. Quotes are ibid., pp. 141, 155, 39.

38. Ibid., p. 39.

39. Valenstein, E. (2005), *The War of the Soups and the Sparks: The Discovery of Neurotransmitters and the Dispute Over How Nerves Communicate* (New York: Columbia University Press). See also Dupont, J.-C. (1999), *Histoire de la neurotransmission* (Paris: Presses Universitaires de France); Robinson (2001); Marcum, J. (2006), *Annals of Science* 63:139–56. I have not referred to the work of two Walters – Gaskell and Cannon – because it would have diverted attention from the central message.

40. Valenstein (2005), p. 6.

41. Ackerknecht, E. (1974), *Medical History* 18:1–8.

42. Valenstein (2005), p. 19.

43. Ibid., p. 22.

44. Ibid., p. 43.

45. Dale, H. (1914), *Journal of Pharmacology and Experimental Therapeutics* 6:147–90.

46. Loewi, O. (1960), *Perspectives in Biology and Medicine* 4:3–25, p. 17.

47. Valenstein (2005), p. 58.

48. Robinson (2001), pp. 63–7.

49. Valenstein (2005), pp. 59–60.

50. Ibid., p. 125; Eccles, J. (1976), *Notes and Records of the Royal Society of London* 30:219–30, p. 221.

51. Eccles (1976), p. 225; Brooks, C. and Eccles, J. (1947), *Nature* 159:760–64.

52. Brock, L., et al. (1952), *Journal of Physiology* 117:431–60, pp. 452, 455.

8. Machines: 1900 to 1950

1. Riskin (2016), pp. 296–304.
2. See for example Cohen, J. (1966), *Human Robots in Myth and Science* (London: Allen & Unwin); Mayor, A. (2018), *Gods and Robots: Myths, Machines, and Ancient Dreams of Technology* (Princeton: Princeton University Press).
3. Hill, A. (1927), *Living Machinery* (London: Bell); Herrick, C. (1929), *The Thinking Machine* (Chicago: University of Chicago Press). Disappointingly, Herrick had little to say about thinking machines.
4. Loeb, J. (1912), *The Mechanistic Conception of Life: Biological Essays* (Chicago: University of Chicago Press); Watson, J. (1913), *Psychological Review* 20:158–77.
5. Rignano, E. (1926), *Man Not a Machine: A Study of the Finalistic Aspects of Life* (London: Kegan Paul, Trench, Trübner); Needham, J. (1927), *Man a Machine* (London: Kegan Paul, Trench, Trübner).
6. Meyer, M. (1911), *The Fundamental Laws of Human Behaviour* (Boston: Badger), p. 39.
7. Russell, S. (1913), *Journal of Animal Behavior* 3:15–35, p. 17, note 5.
8. Ibid., p. 35.
9. Miessner, B. (1916), *Radiodynamics: The Wireless Control of Torpedoes and Other Mechanisms* (New York: Van Nostrand), p. 195.
10. Loeb, J. (1918), *Forced Movements, Tropisms and Animal Conduct* (London: Lippincott), pp. 68–9.
11. Miessner (1916), p. 199; Cordeschi, R. (2002), *The Discovery of the Artificial: Behavior, Mind and Machines Before and Beyond Cybernetics* (London: Kluwer).
12. Uexküll, J. von (1926), *Theoretical Biology* (London: Kegan Paul, Trench, Trübner).
13. Magnus, R. (1930), *Lane Lectures on Experimental Pharmacology and Medicine* (Stanford: Stanford University Press), p. 333.
14. Uexküll (1926), p. 273.
15. Lotka, A. (1925), *Elements of Physical Biology* (Baltimore: Williams & Wilkins), p. 342.
16. Hull, C. and Baernstein, H. (1929), *Science* 70:14–15; Baernstein, H. and Hull, C. (1931), *Journal of General Psychology* 5:99–106; Krueger, R. and Hull, C. (1931), *Journal of General Psychology* 5:262–9.
17. Krueger and Hull (1931), p. 267.
18. Baernstein and Hull (1931), p. 99.
19. Ross, T. (1933), *Scientific American* 148:206–8.
20. Ross, T. (1935), *Psychological Review* 42:387–93, p. 387.
21. Ross, T. (1938), *Psychological Review* 45:185–9, p. 138.
22. *Time*, 16 September 1935.

23. Bernstein, J. (1868), *Pflüger, Archiv für Physiologie* 1:173–207; Seyfarth, E.-A. (2006), *Biological Cybernetics* 94:2–8.
24. Bernstein, J. (1902), *Pflüger, Archiv für Physiologie* 92:521–62.
25. McComas, A. (2011), *Galvani's Spark: The Story of the Nerve Impulse* (Oxford: Oxford University Press); Campenot, R. (2016), *Animal Electricity: How We Learned That the Body and Brain are Electric Machines* (London: Harvard University Press).
26. Gotch, F. and Burch, G. (1899), *Journal of Physiology* 24:410–26.
27. Gotch, F. (1902), *Journal of Physiology* 28:395–416, p. 414.
28. Frank, R. (1994), *Osiris* 9:208–35.
29. https://tinyurl.com/Adrian-Nobel.
30. Hodgkin, A. (1979), *Biographical Memoirs of Fellows of the Royal Society* 25:1–73; Frank (1994); Garson, J. (2015), *Science in Context* 28:31–52.
31. Adrian, E. (1914), *Journal of Physiology* 47:460–74.
32. McComas (2011), pp. 73–4.
33. Forbes, A. and Thacher C. (1920), *American Journal of Physiology* 52:409–71, p. 468. Daly in the UK and Höber in Germany had similar ideas – Adrian, E. (1928), *The Basis of Sensation* (London: Christophers), p. 42.
34. Frank (1994), p. 218.
35. Hodgkin (1979), p. 25.
36. Ibid., p. 21.
37. Adrian, E. (1926a), *Journal of Physiology* 61:49–72; Adrian, E. (1926b), *Journal of Physiology* 62:33–51; Adrian, E. and Zotterman, Y. (1926a), *Journal of Physiology* 61:151–71; Adrian, E. and Zotterman, Y. (1926b), *Journal of Physiology* 61:465–83.
38. Frank (1994), p. 209.
39. Adrian, E. and Matthews, B. (1934), *Brain* 57:355–85, p. 355.
40. Ibid., p. 384. We still do not fully understand the origin of EEG – Cohen, M. (2017), *Trends in Neurosciences* 40:208–18.
41. This and subsequent quotes from Adrian (1928), pp. 6, 118–19, 120, 112.
42. Adrian, E. (1932), *The Mechanism of Nervous Action: Electrical Studies of the Neurone* (Philadelphia: University of Pennsylvania Press), p. 12.
43. Thomson, S. and Smith, H. (1853), *A Dictionary of Domestic Medicine and Household Surgery* (Philadelphia: Lippincott, Grambo), p. 291.
44. Adrian (1928), pp. 91, 100, 98.
45. Garson (2015), p. 46.

9. Control: 1930 to 1950

1. Smalheiser, N. (2000), *Perspectives in Biology and Medicine* 43:217–26, pp. 217–18.
2. Easterling, K. (2001), *Cabinet* 5, https://tinyurl.com/Easterling-Pitts; Gefter, A. (2015), *Nautilus* 21, https://tinyurl.com/Gefter-Pitts.

3. Chen, Z. (1999), in R. Wilson and F. Keil (eds.), *MIT Encyclopedia of Cognitive Science* (Cambridge, MA: MIT Press), pp. 650–52, p. 650.

4. Rashevsky eventually exasperated both the University of Chicago authorities and his principal funders, the Rockefeller Foundation. Dick Lewontin, the dean who shut down the group, recalled: 'What Rashevsky and his school failed to take into account was the conviction of biologists that real organisms were complex systems whose actual behaviour would be lost in idealizations. The work of the school was regarded as irrelevant to biology and was effectively terminated in the late 1960s, leaving no lasting trace.' Lewontin, R. (2003), *New York Review of Books*, 1 May.

5. This chapter is indebted to the work of Tara Abraham and Margaret Boden. Abraham, T. (2002), *Journal of the History of the Behavioral Sciences* 38:3–25; Abraham, T. (2004), *Journal of the History of Biology* 37:333–85; Abraham, T. (2016), *Rebel Genius: Warren S. McCulloch's Transdisciplinary Life in Science* (London: MIT Press); Boden (2006).

6. Rashevsky, N. (1936), *Psychometrika* 1:1–26, p. 1.

7. Kubie, L. (1930), *Brain* 53:166–77.

8. Pitts, W. (1942a), *Bulletin of Mathematical Biophysics* 4:121–9; Pitts, W. (1942b). *Bulletin for Mathematical Biophysics* 4:169–75.

9. Both dates are given by Abraham (2002), who wisely does not attempt to choose between them.

10. Abraham (2016); Magnus (1930).

11. Lettvin, J., et al. (1959), *Proceedings of the Institute of Radio Engineers* 47:1940–51, p. 1950. Eric Kandel described his work on *Aplysia* as a validation of Kant – Kandel, E. (2006), *In Search of Memory: The Emergence of a New Science of Mind* (New York: Norton), p. 202.

12. Hull, C. (1937), *Psychological Review* 44:1–32.

13. Arbib (2000), p. 199; Heims, S. (1991), *Constructing a Social Science for Postwar America: The Cybernetics Group, 1946–1953* (London: MIT Press), p. 38.

14. Heims (1991), pp. 40–41; Conway, F. and Siegelman, J. (2005), *Dark Hero of the Information Age: In Search of Norbert Wiener the Father of Cybernetics* (New York: Basic).

15. McCulloch, W. and Pitts, W. (1943), *Bulletin of Mathematical Biophysics* 5:115–33.

16. Arbib (2000), p. 207; Kay, L. (2001), *Science in Context* 14:591–614, p. 592.

17. McCulloch, W. (1965), *Embodiments of Mind* (Cambridge, MA: MIT Press).

18. Ibid., p. 9.

19. Arbib (2000), p. 199.

20. Quotes from McCulloch and Pitts (1943), pp. 122, 123, 120.

21. Masani, P. (1990), *Norbert Wiener 1894–1964* (Basel: Birkhaüser Verlag); Kay (2001); Abraham (2004); Piccinini, G. (2004), *Synthèse* 141:175–215; Koch, C. (1999), *Biophysics of Computation: Information Processing in Single Neurons* (New York: Oxford University Press); for a recent example of an AND gate embodied in a single cell, see Dobosiewicz, M., et al. (2019), *eLife* 8:e50566.

22. Heims, S. (1980), *John von Neumann & Norbert Weiner: From Mathematics to the Technologies of Life and Death* (London: MIT Press), pp. 192–9.

23. Quotes from von Neumann, J. (1993), *IEEE Annals of the History of Computing* 15:27–43, pp. 33, 37, 38.

24. Conway and Siegelman (2005).

25. Abraham (2016), p. 89.

26. Rosenblueth, A., et al. (1943), *Philosophy of Science* 10:18–24, p. 20.

27. Craik, K. (1943), *The Nature of Explanation* (Cambridge: Cambridge University Press), p. 52; Zangwill, O. (1980), *British Journal of Psychology* 71:1–16.

28. Craik (1943), p. 53.

29. Ibid., p. 61.

30. Collins, A. (2012), *Interdisciplinary Science Reviews* 37:254–68.

31. Craik (1943), p. 115.

32. Adrian, E. (1947), *The Physical Background of Perception* (Oxford: Clarendon Press), pp. 93–4.

33. Turing, A. (1937), *Proceedings of the London Mathematical Society* 42:230–65.

34. McCulloch and Pitts (1943), p. 129.

35. Von Neumann, J. (1951), in L. Jeffress (ed.), *Cerebral Mechanisms in Behavior: The Hixon Symposium* (London: Hafner), pp. 1–41, p. 32.

36. Soni, J. and Goodman, R. (2017), *A Mind at Play: How Claude Shannon Invented the Information Age* (London: Simon and Schuster), p. 107.

37. Hodges, A. (2012), *Alan Turing: The Enigma* (London: Vintage), p. 251.

38. Heims (1991), p. 20.

39. Masani (1990), pp. 243–5.

40. Organised by the Hixon Fund, this is often called the Hixon Conference or Hixon Meeting.

41. Quotes from von Neumann (1951), pp. 10, 20, 24, 34.

42. Quotes from McCulloch, W. (1951), in L. Jeffress (ed.), *Cerebral Mechanisms in Behavior: The Hixon Symposium* (New York: Wiley), pp. 45–57, p. 55.

43. Conway and Siegelman (2005), pp. 199, 169.

44. Wiener, N. (1948), *Cybernetics: or, Control and Communication in the Animal and the Machine* (New York: Technology Press), p. 124.

45. Pias, C. (ed.) (2016), *Cybernetics: The Macy Conferences 1946–1953* (Zurich: Diaphenes), pp. 171–202.

46. Von Neumann, J. (1958), *The Computer and the Brain* (New Haven: Yale University Press), p. 82.

47. Olby, R. (1994), *The Path to the Double Helix: The Discovery of DNA* (New York: Dover), p. 354.

48. Pias (2016), p. 128.

49. Conway and Siegelman (2005), pp. 217–29.

50. Husbands, P. and Holland, O. (2008), in P. Husbands, et al. (eds.), *The Mechanical Mind in History* (London: MIT Press), pp. 91–148; Pickering, A. (2010), *The Cybernetic Brain: Sketches of Another Future* (London: University of Chicago Press).

51. Husbands and Holland (2008), pp. 116–17; Husbands, P. and Holland, O. (2012), *Interdisciplinary Science Reviews* 37:237–53.

52. Hodges (2012), p. 251. From Hodges's interview with Turing's student Robin Gandy.

53. Pias (2016), pp. 474–9.

54. Soni and Goodman (2017), p. 204. For the full film, see: https://www.youtube.com/watch?v=vPKkXibQXGA.

55. Pias (2016), p. 478.

56. Soni and Goodman (2017), p. 205.

57. Pias (2016), p. 346.

58. I know that doesn't make sense. It's a Doctor Who joke.

59. Paul Mandel, 'Deux ex Machina', *The Harvard Crimson*, 5 May 1950.

60. Riskin (2016), p. 321.

61. https://www.youtube.com/watch?v=wQE82derooc.

62. Pias (2016), pp. 593–619; Dupuy, J.-P. (2009), *On the Origins of Cognitive Science: The Mechanization of the Mind* (London: MIT Press), pp. 148–50.

63. Ashby's book explains less than his title promises – Ashby, R. (1952), *Design for a Brain* (London: Chapman & Hall). Explorations of the Homeostat range from the useful (e.g. Cariani, P., 2009, *International Journal of General Systems* 38:139–54) to the infuriatingly obscure (Dupuy, 2009).

64. Dupuy (2009); Boden (2006), vol. 1, pp. 222–32.

65. Ryle, G. (1949), *The Concept of Mind* (London: Hutchinson); Turing, A. (1950), *Mind* 59:433–60.

66. Turing (1950), p. 442.

67. Ibid., p. 455.

68. MacKay, D. (1951), *British Journal for the Philosophy of Science* 2:105–21, p. 120.

69. Laslett, P. (ed.) (1950), *The Physical Basis of Mind* (Oxford: Blackwell); Young, J. (1951), *Doubt and Certainty in Science: A Biologist's Reflection on the Brain* (Oxford: Clarendon).

70. Young (1951), pp. 50–51.

71. Sherrington, C. (1940), *Man on his Nature* (Cambridge: Cambridge University Press), p. 225.
72. Smith, R. (2001), *Science in Context* 14:511–39.
73. Sherrington (1940), p. 357.
74. McCulloch and Pitts (1943), p. 132.

Present

1. Fields, R. (2018), *Journal of Neuroscience* 38:9311–17; Carandini, M. (2019), *Neuron* 102:732–4.
2. Hughes, J. and Söderqvist, T. (1999), *Endeavour* 23:1–2.

10. Memory: 1950 to today

1. Eccles, J. and Feindel, F. (1978), *Biographical Memoirs of Fellows of the Royal Society* 24:473–513; Lewis, J. (1981), *Something Hidden: A Biography of Wilder Penfield* (Toronto: Doubleday).
2. Penfield, W. (1952), *Archives of Neurology and Psychiatry* 67:178–91, p. 178.
3. Lashley, K. (1950), *Symposia of the Society for Experimental Biology* 4:454–82; Bruce, D. (2001), *Journal of the History of the Neurosciences* 10:308–18.
4. Lashley (1950), pp. 477–8.
5. Penfield (1952), p. 185.
6. Ibid., p. 196.
7. Penfield, W. (1954), in J. Delafresnaye (ed.), *Brain Mechanisms and Consciousness* (Oxford: Blackwell Scientific), pp. 284–304, p. 306.
8. Higgins, J., et al. (1956), *Archives of Neurology and Psychiatry* 76:399–419; Jacobs, J., et al. (2012), *Journal of Cognitive Neuroscience* 24:553–63.
9. Penfield, W. (1975), *The Mystery of the Mind: A Critical Study of Consciousness and the Human Brain* (Princeton: Princeton University Press), explains his change of view.
10. Penfield, W. and Boldrey, E. (1937), *Brain* 60:389–443.
11. Pogliano, C. (2012), *Nuncius* 27:141–62.
12. Penfield, W. and Rasmussen, T. (1950), *The Cerebral Cortex of Man* (New York: Macmillan).
13. Penfield described a homunculus in the thalamus although he accepted it made 'no pretence to detailed accuracy' – Penfield, W. and Jasper, H. (1954), *Epilepsy and the Functional Anatomy of the Human Brain* (New York: Little, Brown), p. 159. A later researcher remarked of Penfield's thalamic homunculus that 'any scientific significance of these displays is difficult to discern' – Schott, G. (1993), *Journal of Neurology, Neurosurgery and Psychiatry* 56:329–33, p. 331.
14. Hebb, D. (1949), *The Organization of Behavior: A Neuropsychological Theory* (London: Chapman & Hall); Brown, R. and Milner, P. (2003), *Nature Reviews Neuroscience* 4:1013–19.

15. Hebb (1949), p. xiii.
16. Quotes and material from ibid., pp. 12, 62, 70, 76, 197, 166.
17. Corkin, S. (2013), *Permanent Present Tense: The Man with No Memory, and What He Taught the World* (London: Allen Lane); Dittrich, L. (2016), *Patient H. M. – A Story of Memory, Madness, and Family Secrets* (London: Chatto & Windus).
18. Scoville, W. and Milner, B. (1957), *Journal of Neurology, Neurosurgery and Psychiatry* 20:11–21, p. 11.
19. Milner, B., et al. (1968). *Neuropsychologia* 6:215–34, p. 217.
20. In 2018 Milner celebrated her 100th birthday; she carried on working well into her nineties.
21. Scoville and Milner (1957).
22. Shepherd, G. (2010), *Creating Modern Neuroscience: The Revolutionary 1950s* (Oxford: Oxford University Press), p. 173. I found this excellent book to be invaluable.
23. Dittrich (2016), p. 233.
24. Milner et al. (1968); Dittrich (2016).
25. Scoville and Milner (1957).
26. Annese, J., et al. (2014), *Nature Communications* 5:3122.
27. Dittrich (2016), p. 230.
28. Tolman, E. (1949), *Psychological Review* 55:189–208. Alarmingly, the results Tolman based his theory on could not be replicated at the time. See for example Gentry, G., et al. (1948), *Journal of Comparative and Physiological Psychology* 41:312–18.
29. O'Keefe, J. (2014), *The Nobel Prizes 2014*, pp. 275–307.
30. O'Keefe, J. and Dostrovsky, J. (1971), *Brain Research* 34:171–5, p. 174.
31. Yartsev, M. and Ulanovsky, N. (2013), *Science* 340:367–72.
32. Hafting, T., et al. (2005), *Nature* 436:801–6; Moser, E., et al. (2008), *Annual Review of Neuroscience* 31:69–89.
33. O'Keefe (2014); Moser, E. (2014), https://www.nobelprize.org/prizes/medicine/2014/edvard-moser/lecture; Moser, M.-B. (2014), https://www.nobelprize.org/prizes/medicine/2014/may-britt-moser/lecture.
34. Maguire, E., et al. (1998), *Science* 280:921–4; Maguire, E., et al. (2000), *Proceedings of the National Academy of Sciences USA* 97:4398–403. For a failure to find such an effect in everyday drivers, see Weisberg, S., et al. (2019), *Cortex* 115:280–93.
35. Butler, W., et al. (2019), *Science* 363:1447–52; Baraduc, P., et al. (2019), *Science* 363:635–9.
36. Omer, D., et al. (2018), *Science* 359:218–24; Danjo, T., et al. (2018), *Science* 359:213–18.
37. Wilson, M. and McNaughton, B. (1994), *Science* 265:676–9.
38. Ólafsdóttir, H., et al. (2015), *eLife* 4:e06063; Stachenfeld, K., et al. (2017), *Nature Neuroscience* 20:1643–53.

39. Schuck, M. and Niv, Y. (2019), *Science* 364:eaaw5181; Liu, Y., et al. (2019), *Cell* 178:640–52.

40. Eichenbaum, H. (2016), *Learning and Behavior* 44:209–22, p. 213.

41. Lisman, J., et al. (2017), *Nature Neuroscience* 20:1434–47.

42. Brodt, S., et al. (2018), *Science* 362:1045–8.

43. Teyler, T. and DiScenna, P. (1986), *Behavioral Neuroscience* 100:147–54.

44. Tanaka, K., et al. (2018), *Science* 361:392–7.

45. Igarashi, K., et al. (2014), *Nature* 510:143–7.

46. Eichenbaum, H., et al. (1983), *Brain* 106:459–72.

47. Dahmani, L., et al. (2018), *Nature Communications* 9:4162; Bao, X., et al. (2019), *Neuron* 102:1066–75.

48. Knierim, J. (2015), *Current Biology* 25:R1116–R1121.

49. Kandel (2006), p. 134.

50. Hodgkin, A. and Huxley, A. (1952), *Proceedings of the Royal Society of London B* 140:177–83.

51. Kandel (2006), p. 147.

52. Hesse, R., et al. (2019), https://www.biorxiv.org/content/10.1101/631556v1; Asok, A., et al. (2019), *Trends in Neuroscience* 42:14–22.

53. McConnell, J., et al. (1959), *Journal of Comparative and Physiological Psychology* 52:1–5; Travis, G. (1981), *Social Studies of Science* 11:11–32.

54. Morange, M. (2006), *Journal of Bioscience* 31:323–7.

55. Byrne, W., et al. (1966), *Science* 153:658–9.

56. Malin, D. and Guttman, H. (1972), *Science* 178:1219–20.

57. Ungar, G., et al. (1972), *Nature* 238:198–202.

58. Stewart, W. (1972), *Nature* 238:202–9.

59. Wilson, D. (1986), *Nature* 320:313–14.

60. Irwin, L. (2007), *Scotophobin: Darkness at the Dawn of the Search for Memory Molecules* (Plymouth: Hamilton); Setlow, B. (1997), *Journal of the History of the Neurosciences* 6:181–92.

61. Nye, M. (1980), *Historical Studies in the Physical Sciences* 11:125–56.

62. Shomrat, T. and Levin, M. (2013), *Journal of Experimental Biology* 216:3799–810.

63. Bliss, T. and Lømo, T. (1973), *Journal of Physiology* 232:331–56.

64. Lømo, T. (2017), *Acta Physiologica* 222:e12921.

65. Cooke, S. and Bliss, T. (2006), *Brain* 129:1659–73.

66. Bliss, T. and Collingridge, G. (1993), *Nature* 361:31–9.

67. Cooke and Bliss (2006).

68. Nabavi, S., et al. (2014), *Nature* 511:348–52; Titley, H., et al. (2017), *Neuron* 95:19–32.

69. Ryan, T., et al. (2015), *Science* 348:1007–13.

70. Tonegawa, S., et al. (2018), *Nature Reviews Neuroscience* 19:485–98.

71. Crick, F. (1982), *Trends in Neuroscience* 5:44–6.

72. Roberts, T., et al. (2010), *Nature* 463:948–52; Hayashi-Takagi, A., et al. (2015), *Nature* 525:333–8.
73. Adamsky, A., et al. (2018), *Cell* 174:59–71.
74. Other forms of learning are available – see Tonegawa et al. (2018).
75. Han, J., et al. (2009), *Science* 323:1492–6.
76. Ramirez, S., et al. (2013), *Science* 341:387–91.
77. Redondo, R., et al. (2014), *Nature* 513:426–30.
78. Ramirez, S., et al. (2015), *Nature* 522:335–9.
79. Vetere, G., et al. (2019), *Nature Neuroscience* 22:933–40.
80. Saunders, B., et al. (2018), *Nature Neuroscience* 21:1072–83.
81. Phelps, E. and Hofmann, G. (2019) *Nature* 572:43–50.
82. Liu, X., et al. (2014), *Philosophical Transactions of the Royal Society of London: B* 369:20130142.
83. Poo, M.-M., et al. (2016), *BMC Biology* 14:40.

11. Circuits: 1950 to today

1. Hubel, D. and Wiesel, T. (2005), *Brain and Visual Perception: The Story of a 25-Year Collaboration* (Oxford: Oxford University Press), p. 60; Hubel, D. and Wiesel, T. (1959), *Journal of Physiology* 148:574–91; Hubel, D. and Wiesel, T. (2012), *Neuron* 75:182–4.
2. Barlow, H. (1953), *Journal of Physiology* 119:69–88.
3. Lorente de Nó, R. (1938), *Journal of Neurophysiology* 1:207–44.
4. Mountcastle, V. (1957), *Journal of Neurophysiology* 20:408–34.
5. Lettvin et al. (1959); Maturana, H., et al. (1960), *Journal of General Physiology* 43:129–76.
6. Spinelli, D., et al. (1968), *Experimental Neurology* 22:75–84; Cayco-Gajic, N. and Sweeney, Y. (2018), *Journal of Neuroscience* 38:6442–4.
7. Blakemore, C. and Cooper, G. (1970), *Nature* 228:477–8.
8. Hebb (1949), p. 31.
9. Gross, C. (2002b), *The Neuroscientist* 8:512–18; for a perceptive exploration of the history and philosophical underpinnings of the grandmother cell, see Barwich, A.-S. (2019) *Frontiers in Neuroscience* 13:1121.
10. Konorski, J. (1967), *Integrative Action of the Brain: A Multidisciplinary Approach* (Chicago: University of Chicago Press); Gross (2002b).
11. Gross, C., et al. (1972), *Journal of Neurophysiology* 35:96–111.
12. Gross, C., et al. (1969), *Science* 166:1303–6; Gross, C. (1998), *Brain, Vision, Memory: Tales in the History of Neuroscience* (London: MIT Press).
13. Perrett, D., et al. (1982), *Experimental Brain Research* 47:329–42; Kendrick, K. and Baldwin, B. (1987), *Science* 236:448–50.
14. Kendrick and Baldwin (1987), p. 450.
15. Quian Quiroga, R., et al. (2005), *Nature* 435:1102–7.

16. Koch, C. (2012), *Consciousness: Confessions of a Romantic Reductionist* (London: MIT Press), p. 65.
17. Quian Quiroga, R., et al. (2008), *Trends in Cognitive Science* 12:87–91.
18. Waydo, S., et al. (2006), *Journal of Neuroscience* 26:10232–4.
19. Yuste, R. (2015), *Nature Reviews Neuroscience* 16:487–97, p. 488.
20. Goodale, M. and Milner, A. (1992), *Trends in Neuroscience* 15:20–25.
21. Milner, A. (2017), *Experimental Brain Research* 235:1297–308.
22. Vargas-Irwin, C., et al. (2015), *Journal of Neuroscience* 35:10888–97.
23. Saur, D., et al. (2008), *Proceedings of the National Academy of Sciences USA* 105:18035–40.
24. Barlow, H. (1972), *Perception* 1:371–94; Barlow, H. (2009), *Perception* 38:795–807.
25. Crick, F. (1958), *Symposia of the Society of Experimental Biology* 12:138–63.
26. Boden (2006), vol. 2, p. 1206.
27. James (1890), vol. 1, p. 179.
28. Barlow (1972), p. 390.
29. Ibid., p. 381.
30. Barlow (2009), p. 797.
31. White, J., et al. (1986), *Philosophical Transactions of the Royal Society of London: B* 314:1–340.
32. White J. (2013), in The *C. elegans* Research Community (eds.), *WormBook*, https://tinyurl.com/mindofworm.
33. Crick, F. and Jones, E. (1993), *Nature* 361:109–10.
34. Felleman, D. and Van Essen, D. (1991), *Cerebral Cortex* 1:1–47.
35. Sporns O., et al. (2005), *PLoS Computational Biology* 1:e42, p. 245; Hagmann, P. (2005), 'From Diffusion MRI to Brain Connectomics' (PhD Thesis, Lausanne: EPFL), doi:10.5075/epfl-thesis-3230; Seung, S. (2012), *Connectome: How the Brain's Wiring Makes Us Who We Are* (Boston: Houghton Mifflin Harcourt).
36. Morabito, C. (2017), *Nuncius* 32:472–500.
37. Swanson, L. and Lichtman, J. (2016), *Annual Review of Neuroscience* 39:197–216, p. 197.
38. Bardin, J. (2012), *Nature* 483:394–6.
39. Smith, S., et al. (2015), *Nature Neuroscience* 18:1565–7.
40. Ingalhalikar, M., et al. (2014), *Proceedings of the National Academy of Sciences USA* 111:823–8; Joel, D. and Tarrasch, R. (2014), *Proceedings of the National Academy of Sciences USA* 111:E637; Cahill, L. (2015), *Proceedings of the National Academy of Sciences USA* 111:577–8.
41. Morgan, J. and Lichtman, J. (2013), *Nature Methods* 10:494–500, p. 497.
42. Economo, M., et al. (2016), *eLife* 5:e10566.
43. Wolff, S. and Ölveczky, B. (2018), *Current Opinion in Neurobiology* 49:84–94; Winnubst. J., et al. (2019), *Cell*, 179:268–81

44. Erö, C., et al. (2018), *Frontiers in Neuroinformatics* 12:00084.

45. Bargmann, C. (2013), *Bioessays* 34:458–65, p. 464.

46. White (2013).

47. Swanson and Lichtman (2016), p. 198.

48. Bargmann, C. and Marder, E. (2013), *Nature Methods* 10:483–90.

49. Shimizu, K. and Stopfer, M. (2013), *Current Biology* 23:R1026–R1031.

50. Ohyama, T., et al. (2015), *Nature* 520:633–9.

51. Morgan, J. and Lichtman, J. (2019), https://www.biorxiv.org/content/10.1101/683276v1

52. Sasaki, T., et al. (2012), *Proceedings of the National Academy of Sciences USA* 109:20720–5.

53. Mu, Y., et al. (2019), *Cell* 178:27–43.

54. Savtchouk I. and Volterra, A. (2018), *Journal of Neuroscience* 38:14–25; Fiacco, T. and McCarthy, K. (2018), *Journal of Neuroscience* 38:3–13.

55. Fitzsimonds, R., et al. (1997), *Nature* 388:439–48.

56. Bullock, T., et al. (2005), *Science* 310:791–2.

57. Yuste (2015).

58. Harvey, C., et al. (2012), *Nature* 484:62–8.

59. Yuste (2015), p. 494.

60. Buzsáki, G. (2010), *Neuron* 68:362–85; Buzsáki, G. (2019), *The Brain from Inside Out* (New York: Oxford University Press).

61. Saxena, S. and Cunningham, J. (2019), *Current Opinion in Neurobiology* 55:103–11.

62. For a simple explanation of low-dimensional manifolds, see Richard Gao's blog post: https://tinyurl.com/manifold-explanation.

63. Gallego, J., et al. (2017), *Neuron* 94:978–84; Gonzalez, W., et al. (2019), *Science* 365:821–5; Oby, E., et al. (2019), *Proceedings of the National Academy of Sciences* 116:15210–5.

64. Nassim, C. (2018), *Lessons from the Lobster: Eve Marder's Work in Neuroscience* (Cambridge, MA: MIT Press).

65. Delcomyn, F. (1980), *Science* 210:492–8; Marder, E. and Bucher, D. (2001), *Current Biology* 11:R986–R996.

66. Selverston, A. (1980), *Behavioral and Brain Sciences* 3:535–40.

67. Nusbaum, N., et al. (2017), *Nature Reviews Neuroscience* 18:389–403.

68. Turrigiano, G., et al. (1994), *Science* 264:974–7.

69. Stern, S., et al. (2017), *Cell* 171:1649–62.

70. Nassim (2018), p. 163.

71. Prinz, A., et al. (2004), *Nature Neuroscience* 7:1345–52; Calabrese, R. (2018), *Trends in Neurosciences* 41:488–91.

72. Sakurai, A. and Katz, P. (2017), *Current Biology* 27:1721–34.

73. Bargmann and Marder (2013).

74. Hassenstein, B. and Reichardt, W. (1956), *Zeitschrift Für Naturforschung:*

B 11:513–24; Barlow, H. and Levick, W. (1965), *Journal of Physiology* 178:477–504; Chi, K. (2016), *Nature* 531:S16–S17.

75. Takemura, S.-Y, et al. (2017), *eLife* 6:e24394.
76. Bargmann and Marder (2013); Motta, A., et al. (2019), *Science* 366:eaay3134.
77. Haley, J., et al. (2018), *eLife* 7:e41877; Bhattacharya, A., et al. (2019), *Cell* 176:1174–89.
78. Kato, S., et al. (2015), *Cell* 163:656–69.
79. Ryan, K., et al. (2016), *eLife* 5:e16962.
80. Wang, X., et al. (2018), *Science* 361:eaat5691.
81. Moffitt, J., et al. (2018), *Science* 362:eaau5324; Tasic, B., et al. (2018), *Nature* 563:72–8.
82. Economo, M., et al. (2018), *Nature* 563:79–84.
83. Mountcastle, V. (1998), *Perceptual Neuroscience: The Cerebral Cortex* (Cambridge, MA: Harvard University Press), p. 366.
84. Ohyama et al. (2015); Miroschnikow, A., et al. (2018), *eLife* 7:e40247.
85. https://tinyurl.com/Fly-brain-quote.
86. Vladimirov, N., et al. (2018), *Nature Methods* 15:1117–25; Hanchate, N., et al. (2019), https://www.biorxiv.org/content/10.1101/454835v1; Kunst, M., et al. (2019), *Neuron* 103:21–38.
87. Laurent, G. (2016), *e-Neuroforum* 7:54–5.
88. Tosches, M., et al. (2018), *Science* 360:881–8.
89. Mars, R., et al. (2018), *eLife* 7:e35237.
90. Laurent (2016), p. 55.
91. Morgan and Lichtman (2013).
92. Ibid., p. 497.
93. Marr, D. (1982), *Vision* (London: W. H. Freeman), p. 15.

12. Computers: 1950 to today

1. Boden (2006); Abbott, L. (2008), *Neuron* 60:489–95; Gerstner, W., et al. (2012), *Science* 338:60–65.
2. Rochester, N., et al. (1956), *IRE Transactions on Information Theory* 2:80–93.
3. Selfridge, O. (1959), in *Symposium on the Mechanisation of Thought Processes* (London: HMSO), pp. 513–26, p. 516.
4. Grainger, J., et al. (2008), *Trends in Cognitive Sciences* 12:381–7.
5. Boden (2006), vol. 2, p. 899.
6. Rosenblatt, F. (1958), *Psychological Review* 65:386–408.
7. Rosenblatt, F. (1959), *Two Theorems of Statistical Separability in the Perceptron* (Buffalo: Cornell Aeronautical Laboratory), p. 424.
8. See photo in Rosenblatt, F. (1961), *Principles of Neurodynamics: Perceptrons and the Theory of Brain Mechanisms*. Report no. 1196-G-8, 15 March 1961 (Buffalo: Cornell Aeronautical Laboratory).

9. *New York Times*, 7 July 1958.
10. McCorduck, P. (1979), *Machines Who Think: A Personal Inquiry into the History and Prospects of Artificial Intelligence* (San Francisco: W. H. Freeman), p. 87.
11. Rosenblatt (1961), p. 28.
12. Cowan, J. (1967), *Nature* 213:237.
13. Minsky, M. and Papert, S. (1969), *Perceptrons: An Introduction to Computational Geometry* (Cambridge, MA: MIT Press); Boden (2006), vol. 2, p. 915.
14. Olazaran, M. (1996), *Social Studies of Science* 26:611–59 argues that the impact of Minsky and Papert's critique has been exaggerated.
15. Marr (1982), pp. 13–14.
16. Ibid., p. xvii.
17. Glennerster, A. (2007), *Current Biology* 17:R397–R399; Frisby, J. and Stone, J. (2012), *Perception* 41:1040–52; Stevens, K. (2012), *Perception* 41:1061–72.
18. Marr (1982), p. 361.
19. Frisby, J. and Stone, J. (2010), *Seeing: The Computational Approach to Biological Vision* (Cambridge, MA: MIT Press), p. 548. John Frisby tried vainly to explain Marr's ideas to me when I was a psychology student at the University of Sheffield. The failure was all mine.
20. Marr (1982), p. 27.
21. Marr, D. (1976), *Cold Spring Harbor Symposia on Quantitative Biology* 40:647–62, p. 653; Marr, D. and Hildreth, E. (1980), *Proceedings of the Royal Society: Biological Sciences* 207:187–217; Martinez-Conde, S., et al. (2018), *Trends in Neurosciences* 41:163–5.
22. Greene, M. and Hansen, B. (2018), *PLoS Computational Biology* 14:e1006327.
23. Stevens (2012), p. 1071.
24. Chang, L. and Tsao, D. (2017), *Cell* 169:1013–28.
25. Landi, S. and Freiwald, W. (2017), *Science* 357:591–5.
26. Abbott, A. (2018), *Nature* 564:176–9, p. 179.
27. Kadipasaoglu, C., et al. (2017), *PLoS One* 12:e0188834.
28. Ponce, C., et al. (2019), *Cell* 177:999–1009.
29. Bashivan, P., et al. (2019), *Science* 364:eaav9436.
30. Carrillo-Reid, L., et al. (2019), *Cell* 178:447–57; Marshel, J., et al. (2019), *Science* 365:eaaw5202.
31. Rumelhart, D., et al. (eds.) (1986), *Parallel Distributed Processing: Explorations in the Microstructure of Cognition*, vol. 1: *Foundations*; vol. 2: *Psychological and Biological Models* (Cambridge, MA: MIT Press); Anderson, J. and Rosenfeld, E. (eds.) (1998), *Talking Nets: An Oral History of Neural Networks* (Cambridge, MA: MIT Press).

32. Sejnowski, T. (2018), *The Deep Learning Revolution* (London: MIT Press), p. 118.
33. Crick, F. (1989), *Nature* 337:129–32, p. 130.
34. Crick, F. (1994), *The Astonishing Hypothesis: The Scientific Search for the Soul* (New York: Charles Scribner's Sons), p. 186.
35. Sejnowski, T. and Rosenberg, C. (1987), *Complex Systems* 1:145–68.
36. Rumelhart, D. and McClelland, J. (1986), in D. Rumelhart, et al. (eds.), *Parallel Distributed Processing: Explorations in the Microstructure of Cognition*, vol. 1: *Foundations* (Cambridge, MA: MIT Press), pp. 216–71.
37. Le, Q., et al. (2016), https://ai.google/research/pubs/pub38115.
38. Hochreiter, S. and Schmidhuber, J. (1997), *Neural Computation* 9:1735–80; LeCun, Y., et al. (2015), *Nature* 521:436–44.
39. Banino, A., et al. (2018), *Nature* 557:429–33.
40. Rajalingham, R., et al. (2018), *Journal of Neuroscience* 38:7255–69; Gangopadhyay, P. and Das, J. (2019), *Journal of Neuroscience* 39:946–8.
41. Marcus, G. (2015), in G. Marcus and J. Freeman (eds.), *The Future of the Brain: Essays by the World's Leading Neuroscientists* (Oxford: Princeton University Press), pp. 204–15, p. 206.
42. Hassabis, D., et al. (2017), *Neuron* 95:245–58.
43. Silver, D., et al. (2016), *Nature* 529:484–9.
44. O'Doherty, J., et al. (2003), *Neuron* 38:329–37.
45. Caron, S., et al. (2013), *Nature* 497:113–17.
46. Aso, Y., et al. (2014), *Elife* 3:e04577.
47. Thum, A. and Gerber, B. (2019), *Current Opinion in Neurobiology* 54:146–54.
48. Ullman, S. (2019), *Science* 363:692–3. For a call for systems neuroscientists to pay more attention to the results from deep learning programs, see Richards, B. (2019), *Nature Neuroscience* 22:1761–70.
49. Sejnowksi and Rosenberg (1987), p. 157.
50. Hutson, M. (2018), https://tinyurl.com/AI-alchemy. For some idea of the irritated response from the audience, see Sejnowski, T. (2015), *Daedalus* 144:123–32, p. 122.
51. https://tinyurl.com/Hinton-quote.
52. Crick (1989). In 1963 he had published an article on breakthroughs in molecular genetics entitled 'The Recent Excitement in the Coding Problem'.
53. Ibid., p. 130.
54. Ibid., p. 132.
55. Husbands, P., et al. (1998), *Connection Science* 10:185–210.
56. Lillicrap, T., et al. (2016), *Nature Communications* 7:13276.
57. LeCun et al. (2015).
58. Wilson, M. and Bower, J. (1992), *Journal of Neurophysiology* 67:981–95.

59. Bower, J. (1994), in J. Bower and D. Beeman (eds.), *The Book of GENESIS: Exploring Realistic Neural Models with the GEneral NEural SImulation System* (New York: Springer-Verlag/TELOS), pp. 195–202, p. 196.

60. Markram, H., et al. (2011), *Procedia Computing Science* 7:39–42, p. 40.

61. Kandel, E., et al. (2013), *Nature Neuroscience* 14:659–66, p. 659; Hill, S. (2015), in G. Marcus and J. Freeman (eds.), *The Future of the Brain: Essays by the World's Leading Neuroscientists* (Oxford: Princeton University Press), pp. 111–24.

62. Dudai, Y. and Evers, K. (2014), *Neuron* 84:254–61; Serban, M. (2017), *Progress in Brain Research* 233:129–48.

63. Tiesinga, P., et al. (2015) *Current Opinion in Neurobiology* 32:107–14.

64. Frégnac, Y. and Laurent, G. (2014), *Nature* 513:27–9. For a critical account of the Human Brain Project by Leonid Schneider, see https://tinyurl.com/Schneider-HBP; for the view of computational neuroscientist Mark Humphries, see https://tinyurl.com/Humphries-HBP.

65. Markram, H., et al. (2015), *Cell* 163:456–92; Ramaswamy, S., et al. (2015), *Frontiers in Neural Circuits* 9:44; Reimann, M., et al. (2015), *Frontiers in Computational Neuroscience* 9:28.

66. Markram et al. (2015), p. 483.

67. Fan, X. and Markram, H. (2019) *Frontiers in Neuroinformatics* 13:32.

68. https://tinyurl.com/EdYong-HBP.

69. Eliasmith, C., et al. (2012), *Science* 338:1202–5.

70. Quotes from Chi (2016).

71. Seth, A. (2015), in T. Metzinger and J. Windt (eds.), *Open MIND* (Frankfurt: MIND Group), pp. 1–24; Clark, A. (2016), *Surfing Uncertainty: Prediction, Action and the Embodied Mind* (Oxford: Oxford University Press).

72. Gregory, R. (1980), *Philosophical Transactions of the Royal Society of London: B* 290:192–7.

73. Frith, C. (2007), *Making Up the Mind: How the Brain Creates Our Mental World* (London: Wiley-Blackwell).

74. Friston, K. (2009), *Trends in Cognitive Sciences* 13:293–301, p. 293.

75. Friston, K. (2003), *Neural Networks* 116:1325–52.

76. Friston (2009), p. 294.

77. Seth, A. and Tsakiris, M. (2018), *Trends in Cognitive Sciences* 22:969–81.

78. Gregory, R. (1983), *Perception* 12:233–8.

79. Clark (2016), Seth and Tsakiris (2018).

80. Pascual-Leone, A. and Walsh, V. (2001), *Science* 292:510–12. In a related phenomenon, known as 'blindsight', clinically blind subjects can nevertheless precisely guess the location of visual stimuli – Weiskrantz, L., et al. (1974), *Brain* 97:709–28.

81. Knill, D. and Pouget, A. (2004), *Trends in Neurosciences* 27:712–19, p. 712. Within four years, Pouget and his colleagues were able to provide such data, which has since been repeated by many studies. Beck, J., et al. (2008), *Neuron* 60:1142–52.

82. Sohn, H., et al. (2019) *Neuron* 104:458–470.

83. Collett, T. and Land, M. (1978), *Journal of Comparative Physiology* 125:191–204.

84. Fabian, S., et al. (2018), *Journal of the Royal Society Interface* 15:20180466; Mischiati, M., et al. (2015), *Nature* 517:333–8; Dickinson, M. (2014), *Current Biology* 25:R232–R234.

85. Cannon, W. (1927), *American Journal of Psychology* 39:106–24; Cannon, W. (1931), *Psychological Review* 38:281–95.

86. Dalgleish, T. (2004), *Nature Reviews Neuroscience* 5:582–9; Adolphs, R. and Anderson, D. (2018), *The Neuroscience of Emotion: A New Synthesis* (Princeton: Princeton University Press).

87. Hess, W. (1958), *The Functional Organization of the Diencephalon* (New York: Grune & Stratton).

88. Marzullo, T. (2017), *Journal of Undergraduate Neuroscience Education* 15:R29–R35, p. R33.

89. *New York Times*, 17 May 1965.

90. Delgado, J. (1965), *International Review of Neurobiology* 6:349–449; Horgan, J. (2005), *Scientific American*, October 2005; Keiper, A. (2006), *New Atlantis* Winter 2006:4–41.

91. Frank, L. (2018), *The Pleasure Shock: The Rise of Deep Brain Stimulation and Its Forgotten Inventor* (New York: Dutton); Baumeister, A. (2000), *Journal of the History of the Neurosciences* 9:262–78.

92. Moan, C. and Heath, R. (1972), *Journal of Behavior Therapy and Experimental Psychiatry* 3:23–30.

93. Bishop, M., et al. (1963), *Science* 140:394–6.

94. Olds, J. and Milner, P. (1954), *Journal of Comparative and Physiological Psychology* 47:419–27.

95. Olds, J. (1958), *Science* 127:315–24.

96. https://www.defense.gov/Explore/News/Article/Article/1164793/darpa-funds-brain-stimulation-research-to-speed-learning/.

97. Reardon, S. (2017), *Nature* 551:549–50.

98. Chen, S. (2019), *Science* 365:456–7.

99. Donoghue, J. (2015), in G. Marcus and J. Freeman (eds.), *The Future of the Brain: Essays by the World's Leading Neuroscientists* (Oxford: Princeton University Press), pp. 219–33.

100. Hochberg, L., et al. (2012), *Nature* 485:372–5.

101. https://tinyurl.com/Cathy-coffee.

102. Ajiboye, A., et al. (2017), *The Lancet* 389:1821–30.

103. George, J., et al. (2019) *Science Robotics* 4:eaax2352.
104. Penaloza, C. and Nishio, S. (2018), *Science Robotics* 3:eaat1228.
105. Dobelle, W. (2000), *ASAIO Journal* 46:3–9.
106. Akbari, H., et al. (2019), *Scientific Reports* 9:874; Anumanchipalli, G., et al. (2019), *Nature* 568:493–8.
107. Gilbert, F., et al. (2019), *Science and Engineering Ethics* 25:83–96, pp. 87–8.
108. Drew, L. (2019), *Nature* 571:S19–S21. There can also be problems with cochlear implants, in particular in developing countries, where the cost of repairing broken implants can be prohibitive and may lead to the devices being removed. Non-technological responses to disability may be more appropriate – see the case study by Friedner, M., et al. (2019), *New England Journal of Medicine* 381:2381–4.

13. Chemistry: 1950 to today

1. Hofmann, A. (1979), *Journal of Psychedelic Drugs* 11:53–60. LSD is an abbreviation from the German – Lyserg-Saure-Diathylamid.
2. Ban, T. (2006), *Dialogues in Clinical Neuroscience* 8:335–44.
3. Material in this paragraph from Healy, D. (2004), *The Creation of Psychopharmacology* (Cambridge, MA: Harvard University Press), pp. 91, 99.
4. Rose, S. (2005), *The 21st Century Brain: Explaining, Mending and Manipulating the Mind* (London: Cape), pp. 221–42.
5. Osmond, H. and Smythies, J. (1952), *Journal of Mental Science* 98:309–15.
6. Barber, P. (2018), *Psychedelic Revolutionaries: Three Medical Pioneers, the Fall of Hallucinogenic Research and the Rise of Big Pharma* (London: Zed).
7. Hoffer, A., et al. (1954), *Journal of Mental Science* 100:29–45, p. 39; Smythies, J. (2002), *Neurotoxicity Research* 4:147–50.
8. Twarog, B. and Page, I. (1953), *American Journal of Physiology* 175:157–61; Shore, P., et al. (1955), *Science* 122:284–5; Costa, E., et al. (1989), *Annual Review of Pharmacology and Toxicology* 29:1–21.
9. Brodie, B., et al. (1955), *Science* 122:968; Brodie, B. and Shore, P. (1957), *Annals of the New York Academy of Sciences* 66:631–42.
10. Loomer, H., et al. (1957), *Psychiatric Research Reports* 8:129–41.
11. Ban (2006).
12. Cade, J. (1949), *Medical Journal of Australia* 1949–2:349–51, p. 350.
13. Schou, M., et al. (1954), *Journal of Neurology, Neurosurgery and Psychiatry* 17:250–60.
14. Harrington, A. (2019), *Mind Fixers: Psychiatry's Troubled Search for the Biology of Mental Illness* (London: Norton); Brown, A. (2019), *Lithium: A Doctor, a Drug, and a Breakthrough* (New York: Liveright).
15. Dupont (1999), p. 207.
16. Baumeister, A. (2011), *Journal of the History of the Neurosciences* 20:106–22.

17. Frank (2018), p. 251.
18. Kety, S. (1959), *Science* 129:1528–32, 1590–96.
19. Ibid., p. 1593.
20. The word was first used in an unsigned book review in *The Lancet*, 2 September 1961, p. 530, where it was used as an adjective.
21. Carlsson, A. (2001), *Science* 294:1021, p. 1021.
22. Burgen, A. (1964), *Nature* 204:412.
23. In 1998 Robert Furchgott, Louis Ignarro and Ferid Murad won the Nobel Prize for demonstrating the role of nitric oxide.
24. Snyder, S. (2018), in D. Linden (ed.), *Think Tank: Forty Neuroscientists Explore the Biological Roots of Human Experience* (London: Yale University Press), pp. 88–93.
25. Dupont (1999), p. 227.
26. Zhu, S., et al. (2018), *Nature* 559:67–72.
27. Leng, G. (2018), *The Heart of the Brain: The Hypothalamus and Its Hormones* (London: MIT Press).
28. Ibid.
29. Pert, C. and Snyder, S. (1973), *Science* 179:1011–14.
30. Hughes, J., et al. (1975), *Nature* 258:577–80.
31. Simantov, R. and Snyder, S. (1976), *Proceedings of the National Academy of Sciences USA* 73:2515–19.
32. Pollin, W. (1979), *Science* 204:8; Snyder, S. (2017), *Annual Review of Pharmacology and Toxicology* 57:1–11.
33. Jones, E. and Mendell, L. (1999), *Science* 284:739.
34. Vander Weele, C., et al. (2018), *Nature* 563:397–401.
35. Salinas-Hernández, X., et al. (2018), *eLife* 7:e38818; Mohebi, A., et al. (2019), *Nature* 570:65–70.
36. Handler, A., et al. (2019), *Cell* 178:60–75.
37. Leshner, A. (1997), *Science* 278:45–7.
38. Nutt, D., et al. (2015), *Nature Reviews Neurosciences* 16:305–12.
39. Lüscher, C. and Malenka, R. (2011), *Neuron* 69:650–63; Sulzer, D. (2011), *Neuron* 69:628–49.
40. Volkow, N., et al. (2016), *New England Journal of Medicine* 374:363–71.
41. *Observer*, 4 March 2018.
42. Koepp, M., et al. (1998), *Nature* 393:266–8.
43. Kirk, S. and Kutchins, H. (1992), *The Selling of DSM: The Rhetoric of Science in Psychiatry* (New York: Aldine de Gruyter); Decker, H. (2013), *The Making of DSM-III. A Diagnostic Manual's Conquest of American Psychiatry* (New York: Oxford University Press); Stein, D., et al. (2010), *Psychological Medicine* 40:1759–65.
44. Andrews, P., et al. (2015), *Neuroscience and Biobehavioral Reviews* 51:164–88.
45. Howard, D., et al. (2019), *Nature Neuroscience* 22:343–52, p. 350.

46. Gøtsche, P. (2014), *The Lancet Psychiatry* 1:104–6; Nutt, D., et al. (2014), *The Lancet Psychiatry* 1:102–4.

47. This was the aim of McGrath, C., et al. (2013), *JAMA Psychiatry* 70:821–9.

48. Schildkraut, J. (1965), *American Journal of Psychiatry* 122:509–22; Coppen, A. (1967), *British Journal of Psychiatry* 113:1237–64, p. 1258.

49. Mendels, J. and Frazer, A. (1974), *Archives of General Psychiatry* 30:447–51.

50. Van Praag, H. and de Haan, S. (1979), *Psychiatry Research* 1:219–24.

51. The earliest reference I have found to this phrase is Lurie, S. (1991), *American Journal of Psychotherapy* 45:348–58. France, C., et al. (2007), *Professional Psychology: Research and Practice* 38:411–20 promise to get to the origin of the idea, but do not get as far back as 1991. The phrase soon penetrated into popular culture. In 1992, in Series 4 Episode 6 of the sitcom *Seinfeld*, Jerry Seinfeld says of another character, 'he's not crazy, he has a chemical imbalance'.

52. Leo, J. and Lacasse, J. (2008), *Society* 45:35–45. For an exploration of how psychiatry embraced this idea, and ripostes by leading psychiatrists, see Lacasse, J. and Leo, J. (2015), *The Behavior Therapist* 38:206–13, Pies, R. (2015), *The Behavior Therapist* 38:260–2, and Carlat, D. (2015), *The Behavior Therapist* 38:262–3. Lacasse and Leo reply to the replies on pp. 263–6 of the same issue; Pies continues to robustly argue that psychiatrists never really adopted the theory: https://tinyurl.com/imbalance-myth.

53. Fibiger, H. (2012), *Schizophrenia Bulletin* 38:649–50.

54. Rose, N. (2019), *Our Psychiatric Future: The Politics of Mental Health* (Cambridge: Polity).

55. De Kovel, C. and Francks, C. (2019), *Scientific Reports* 9:5986.

56. Border, R., et al. (2019), *American Journal of Psychiatry* 176:376–87.

57. Mitchell, K. (2015), in G. Marcus and J. Freeman (eds.), *The Future of the Brain: Essays by the World's Leading Neuroscientists* (Oxford: Princeton University Press), pp. 234–42; Mitchell, K. (2018), *Innate: How the Wiring of Our Brains Shapes Who We Are* (Oxford: Princeton University Press).

58. The PsychENCODE Consortium (2018), *Science* 362:1262–3.

59. Shorter, E. and Healy, D. (2007), *Shock Therapy: A History of Electroconvulsive Treatment in Mental Illness* (New Brunswick, NJ: Rutgers University Press); Hirshbein, L. (2012), *Journal of the History of the Neurosciences* 21:147–69.

60. Plath, S. (2005), *The Bell Jar* (London: Faber & Faber), p. 138.

61. Leiknes, K., et al. (2012), *Brain and Behavior* 2:283–345.

62. Pollan, M. (2018), *How to Change Your Mind: The New Science of Psychedelics* (London: Allen Lane).

63. Preller, K., et al. (2019), *Proceedings of the National Academy of Sciences USA* 116:2743–8.

64. Deco, G., et al. (2018), *Current Biology* 28:3065–74.e6.

65. Berman, R., et al. (2000), *Biological Psychiatry* 47:351–4.

66. https://twitter.com/NIMHDirector/status/1103120788272697346.

67. https://www.wired.com/2017/05/star-neuroscientist-tom-insel-leaves-google-spawned-verily-startup/.

14. Localisation: 1950 to today

1. Uttal, W. (2001), *The New Phrenology: The Limits of Localizing Cognitive Processes in the Brain* (Cambridge, MA: MIT Press); Raichle, M. (2008), *Trends in Neurosciences* 32:118–26; Poldrack, R. (2018), *The New Mind Readers: What Neuroimaging Can and Cannot Reveal about Our Thoughts* (Princeton: Princeton University Press).

2. Beckmann, E. (2006), *British Journal of Radiology* 79:5–8, pp. 6–7.

3. Ter-Pogossian, M. (1992), *Seminars in Nuclear Medicine* 22:140–49.

4. Petersen, S., et al. (1988), *Nature* 331:585–9; Posner, M., et al. (1988), *Science* 240:1627–31.

5. Kanwisher quotes from Kanwisher, N. (2017), *Journal of Neuroscience* 37:1056–61, p. 1056.

6. Logothetis, N., et al. (2001), *Nature* 412:150–57.

7. Racine, E., et al. (2005), *Nature Reviews Neuroscience* 6:159–64.

8. Sajous-Turner, A., et al. (2019), *Brain Imaging and Behavior*, https://doi.org/10.1007/s11682-019-00155-y.

9. Rusconi, E. and Mitchener-Nissen, T. (2013), *Frontiers in Human Neuroscience* 7:594; Satel, S. and Lilienfeld, S. (2013), *Brainwashed: The Seductive Appeal of Mindless Neuroscience* (New York: Basic Books); Sahakian, B. and Gottwald, J. (2017), *Sex, Lies, and Brain Scans: How fMRI Reveals What Really Goes On in Our Minds* (Oxford: Oxford University Press).

10. Eklund, A., et al. (2016), *Proceedings of the National Academy of Sciences USA* 113:7900–905.

11. Brown, E. and Behrmann, M. (2017), *Proceedings of the National Academy of Sciences USA* 114:E3368–E3369; Cox, R., et al. (2017), *Proceedings of the National Academy of Sciences USA* 114:E3370–E3371; Eklund, A., et al. (2017), *Proceedings of the National Academy of Sciences USA* 114:E3374–E3375; Kessler, D., et al. (2017), *Proceedings of the National Academy of Sciences USA* 114:E3372–E3373.

12. Logothetis, N. (2008), *Nature* 453:869–78, pp. 876–7.

13. Poldrack, R. (2017), *Nature* 541:156.

14. Vaidya, A. and Fellows, L. (2015), *Nature Communications* 6:10120.

15. Vul, E., et al. (2009), *Perspectives in Psychological Science* 4:274–90.

16. Margulies, D. (2012), in S. Choudhury and J. Slaby (eds.), *Critical Neuroscience: A Handbook of the Social and Cultural Contexts of Neuroscience* (Oxford: Blackwell), pp. 273–85.

17. Bennett, C., et al. (2009), *NeuroImage* 47:S39–S40.

18. Poldrack, R., et al. (2017), *Nature Reviews Neuroscience* 18:115–26; Poldrack (2018).

19. Haxby, J., et al. (2001), *Science* 293:2425–30.

20. Kanwisher, N. (2010), *Proceedings of the National Academy of Sciences USA* 107:11163–70, p. 11165.

21. Kanwisher (2017), p. 1060.

22. https://tinyurl.com/macarthur-tweet; https://tinyurl.com/cardona-tweet; https://tinyurl.com/mitchell-tweet.

23. Poldrack, R. and Farah, M. (2015), *Nature* 526:371–9.

24. Logothetis (2008).

25. Dubois, J., et al. (2015), *Journal of Neuroscience* 35:2791–802.

26. Guest, O. and Love, B. (2017), *eLife* 6:e21397.

27. Bargmann, C. (2015), *Journal of the American Medical Association* 314:221–2.

28. Kashyap, S., et al. (2018), *Scientific Reports* 8:17063.

29. For critiques of research on sex differences, see Fine, C. (2010), *Delusions of Gender: The Real Science Behind Sex Differences* (London: Norton); Rippon, G. (2019), *The Gendered Brain: The New Neuroscience That Shatters the Myth of the Female Brain* (London: Bodley Head).

30. Simon Baron-Cohen argues that sex differences tip over into the autism spectrum with an 'extreme male brain' characterising autistic people of both sexes – Baron-Cohen, S. (2003), *The Essential Difference: Men, Women and the Extreme Male Brain* (London: Allen Lane).

31. Ritchie, S., et al. (2018), *Cerebral Cortex* 28:2959–75.

32. Knickmeyer, R., et al. (2014), *Cerebral Cortex* 24:2721–31.

33. Vidal, F. and Ortega, F. (2017), *Being Brains: Making the Cerebral Subject* (New York: Fordham University Press).

34. Huth, A., et al. (2016), *Nature* 532:453–8; Brennan, J. (2018), *Trends in Neurosciences* 41:770–2.

35. Damasio, H., et al. (1996), *Nature* 380:499–505.

36. Uttal (2001); Rose (2005); Nobre, A. and van Ede, F. (2020), *Journal of Neuroscience* 40:89–100.

37. Raichle, M., et al. (2001), *Proceedings of the National Academy of Sciences USA* 98:676–82.

38. Raichle, M. (2015), *Annual Review of Neuroscience* 38:433–47; Sormaz, M., et al. (2018), *Proceedings of the National Academy of Sciences USA* 115:9318–23; Kaplan, R., et al. (2016), *Current Biology* 26:686–91.

39. Fox, K., et al. (2018), *Trends in Cognitive Sciences* 20:307–24.

40. Frégnac, Y. (2017), *Science* 358:470–77, p. 472.
41. Lange, F. (1877), *History of Materialism and Criticism of Its Present Importance*, vol. 3 (London: Trübner), p. 137.
42. Quotes from Gregory, R. (1959), in *Symposium on the Mechanisation of Thought Processes* (London: HMSO), pp. 669–82, pp. 680, 664.
43. Gregory, R. (1961), in W. Thorpe and O. Zangwill (eds.), *Current Problems in Animal Behaviour* (Cambridge: Cambridge University Press), pp. 307–30; Gregory, R. (1981), *Mind in Science: A History of Explanations in Psychology and Physics* (London: Weidenfeld and Nicolson).
44. Gregory (1981), p. 84.
45. Friston, K. (2011), *Brain Connectivity* 1:13–36.
46. Paré, D. and Quirk, G. (2017), *npj Science of Learning* 2:6; Adolphs and Anderson (2018).
47. Pignatelli, M. and Beyeler, A. (2019), *Current Opinion in Behavioral Sciences* 26:97–106; Corder, G., et al. (2019), *Science* 363:276–81; Morrow, K., et al. (2019), *Journal of Neuroscience* 39:3663–75; Chen, P., et al. (2019), *Cell* 176:1206–21.e18.
48. Padmanabhan, K., et al. (2019), *Frontiers in Neuroanatomy* 12:115.
49. Baumann, O., et al. (2015), *Cerebellum* 14:197–220; Carta, I., et al. (2019), *Science* 363:eaav058.
50. Genon, S., et al. (2018), *Trends in Cognitive Sciences* 22:350–63.
51. Harrington, A. (1990), in Harrington, A. (ed.), *So Human a Brain: Knowledge and Values in the Neurosciences* (New York: Springer), pp. 247–325, p. 268; Butler, A. (2009), in L. Squire (ed.), *Encyclopedia of Neuroscience* (New York: Academic Press).
52. Pogliano, C. (2017), *Nuncius* 32:330–75, p. 352.
53. Koestler, A. (1967), *The Ghost in the Machine* (London: Hutchinson), p. 296.
54. Sagan, C. (1977), *The Dragons of Eden: Speculations on the Evolution of Human Intelligence* (London: Hodder and Stoughton). See also Holden, B. (1979), *Science* 204:1066–8.
55. Sagan (1977), p. 142. There are pages and pages of this stuff, which was absurd at the time.
56. MacLean, P. (1990), *The Triune Brain in Evolution: Role in Paleocerebral Functions* (New York: Plenum).
57. Reiner, A. (1990), *Science* 250:303–5.
58. Guillery, R. (1987), *Nature* 330:29.
59. Di Pellegrino, G., et al. (1992), *Experimental Brain Research* 91:176–80; Rizzolatti, G. and Craighero, L. (2004), *Annual Review of Neuroscience* 27:169–19; Hickok, G. (2014), *The Myth of Mirror Neurons: The Real Neuroscience of Communication and Cognition* (London: Norton).

60. Gallese, V. (2006), *Brain Research* 1079:15–24.
61. Ramachandran, V. (2011), *The Tell-Tale Brain: Unlocking the Mystery of Human Nature* (London: Norton), p. 125.
62. Mukamel, R., et al. (2010), *Current Biology* 20:750–56.
63. Grabenhorst, F., et al. (2019), *Cell* 177:986–8.
64. Feuillet, L., et al. (2007), *The Lancet* 370:262; Weiss, T., et al. (2020), *Neuron* 105:35–45.
65. Yu, F., et al. (2015), *Brain* 138:e353.
66. García, A., et al. (2017), *Frontiers in Aging Neuroscience* 8:335.
67. Otchy, T., et al. (2015), *Nature* 528:358–63.
68. Li, X., et al. (2018), *Journal of Neuroscience* 38:8549–62.
69. Allen, W., et al. (2019), *Science* 364:eaav3932.
70. Stringer, C., et al. (2019), *Science* 364:255; Steinmetz, N., et al. (2019), *Nature* 576:266–73.
71. Prior, H., et al. (2008), *PLoS Biology* 6:e202.
72. Maler, L. (2018), *Current Biology* 28:R213–R215.

15. Consciousness: 1950 to today

1. Miller, G. (2005), *Science* 309:79. Philosophers can take heart from the fact that science still does not have answers to the two key questions that philosophy has been worrying away at for millennia.
2. Sutherland, S. (1989), *International Dictionary of Psychology* (New York: Crossroad), p. 95.
3. Between 1969 and 2016, the Society for Neuroscience held only two small symposia on the topic – Storm, J., et al. (2017), *Journal of Neuroscience* 37:10882–93. Many neuroscientists – myself included – do not study brains of any kind, never mind consciousness.
4. Seth, A. (2017), in K. Almqvist and A. Haag (eds.), *The Return of Consciousness: A New Science on Old Questions* (Stockholm: Axel and Margaret Ax:son Johnson Foundation), pp. 13–37; Strawson, G. (2017), in K. Almqvist and A. Haag (eds.), *The Return of Consciousness: A New Science on Old Questions* (Stockholm: Axel and Margaret Ax:son Johnson Foundation), pp. 79–92.
5. Delafresnaye, J. (ed.) (1954), *Brain Mechanisms and Consciousness* (Oxford: Blackwell Scientific).
6. Marshall, L. (2004), *Biographical Memoirs* 84:251–69. They made their discovery by accident when they used too much power during stimulation – Moruzzi, G. and Magoun, H. (1949), *Electroencephalography and Clinical Neurophysiology* 1:455–73.
7. Magoun, H. (1954), in J. Delafresnaye (ed.), *Brain Mechanisms and Consciousness* (Oxford: Blackwell Scientific), pp. 1–20, p. 1.
8. Penfield (1954), pp. 286, 289.

9. Cobb, S. (1952), *Archives of Neurology and Psychiatry* 42:172–7, p. 176.
10. Fessard. A. (1954), in J. Delafresnaye (ed.), *Brain Mechanisms and Consciousness* (Oxford: Blackwell Scientific), pp. 200–235, p. 206; Tyč-Dumont, S., et al. (2012), *Journal of the History of the Neurosciences* 21:170–88.
11. Jung, R. (1954), in J. Delafresnaye (ed.), *Brain Mechanisms and Consciousness* (Oxford: Blackwell Scientific), pp. 310–44.
12. Penfield (1954), p. 304.
13. Delafresnaye (1954), p. 499.
14. Eccles, J. (1953), *The Neurophysiological Basis of Mind: The Principles of Neurophysiology* (Oxford: Oxford University Press), p. vi.
15. Eccles, J. (1951), *Nature* 168:53–7, p. 56.
16. Delafresnaye (1954), p. 501.
17. Smith, C. (2001), *Brain and Cognition* 46:364–72; Smith, C. (2014), in C. Smith and H. Whitaker (eds.), *Brain, Mind and Consciousness in the History of Neuroscience* (Dordrecht: Springer), pp. 255–72; Borck, C. (2017), *Nuncius* 32:286–329.
18. Penfield (1975), p. 114.
19. Ibid., pp. 80, 114.
20. Place, U. (1956), *British Journal of Psychology* 47:44–50.
21. Smart, J. (1959), *Philosophical Review* 68:141–56.
22. Miller, G. (1962), *Psychology: The Science of Mental Life* (New York: HarperCollins), p. 40.
23. Meyer, D. and Meyer, P. (1963), *Annual Review of Psychology* 14:155–74.
24. Lashley (1950).
25. Sperry, R. (1961), *Science* 133:1749–57.
26. Gazzaniga, M. (2015), *Tales from Both Sides of the Brain: A Life in Neuroscience* (New York: HarperCollins); Schechter, E. (2018), *Self-Consciousness and 'Split' Brains: The Mind's I* (Oxford: Oxford University Press).
27. Bogen, J. (2006), *The History of Neuroscience in Autobiography* 5:46–122, p. 90.
28. Gazzaniga (2015), pp. 35–7; https://vimeo.com/96626442.
29. Shen, H. (2014), *Proceedings of the National Academy of Sciences USA* 111:18097.
30. Sperry, R. (1966), in J. Eccles (ed.), *Brain and Conscious Experience* (New York: Springer), pp. 298–313, p. 304.
31. Gazzaniga, M. (2018), *The Consciousness Instinct: Unraveling the Mystery of How the Brain Makes the Mind* (New York: Farrar, Straus and Giroux), pp. 204–5.
32. Gazzaniga, M., et al. (1962), *Proceedings of the National Academy of Sciences USA* 48:1765–9; Gazzaniga, M., et al. (1963), *Neuropsychologia*

1:209–15; Gazzaniga, M., et al. (1965), *Brain* 88:221–36; Gazzaniga, M. and Sperry, R. (1967), *Brain* 90:131–48.

33. Gazzaniga, M., et al. (1987), *Neurology* 37:682–2.

34. Gazzaniga (2015), p. 90, https://vimeo.com/96627695.

35. Ibid., pp. 151, 153.

36. Pinto, Y., et al. (2017a), *Brain* 140:1231–7; Pinto, Y., et al. (2017b), *Brain* 140:e68; Volz, L. and Gazzaniga, M. (2017), *Brain* 140:2051–60; Volz, L., et al. (2018), *Brain* 141:e15; Corballis, M., et al. (2018), *Brain* 141:e46.

37. Miller, M., et al. (2010), *Neuropsychologia* 48:2215–20.

38. Steckler, C., et al. (2017), *Royal Society Open Science* 4:170172.

39. Gazzaniga (2018), p. 204.

40. Corballis, M. (2014), *PLoS Biology* 12:e1001767; Toga, A. and Thompson, P. (2003), *Nature Reviews Neuroscience* 4:37–48; Kliemann, D., et al. (2019), *Cell Reports* 29:2398–407.

41. Gazzaniga, M. (2014), *Proceedings of the National Academy of Sciences USA* 111:18093–4.

42. Gazzaniga (2018), p. 230.

43. Koch (2012), p. 20.

44. Crick, F. (1979), *Scientific American* 241:219–32.

45. Treisman, A. and Gelade, G. (1980), *Cognitive Psychology* 12:97–136; Crick, F. (1984), *Proceedings of the National Academy of Sciences USA* 81:4586–90.

46. Crick, F. and Koch, C. (1990), *Seminars in the Neurosciences* 2:263–175.

47. Ibid., p. 264.

48. Dennett, D. (1991), *Consciousness Explained* (London: Penguin), p. 255.

49. Crick (1994), p. 3.

50. Ibid., p. 259.

51. Crick, F. and Koch, C. (2003), *Nature Neuroscience* 6:119–26, p. 123.

52. Crick, F. and Koch, C. (2005), *Philosophical Transactions of the Royal Society: B* 360:1271–9, p. 1277.

53. Koch, C., et al. (2016), *Nature Reviews Neuroscience* 17:307–21; Jackson, J., et al. (2018), *Neuron* 99:1029–39.

54. Koch et al. (2016); Storm et al. (2017); van Vugt, B., et al. (2018), *Science* 360:537–42.

55. Boly, M., et al. (2017), *Journal of Neuroscience* 37:9603–13; Odegaard, B., et al. (2017), *Journal of Neuroscience* 37:9593–602.

56. Owen, A., et al. (2006), *Science* 313:1402; Stender, J., et al. (2014), *The Lancet* 384:514–22; Owen, A. (2019), *Neuron* 102:526–8.

57. Naci, L., et al. (2014), *Proceedings of the National Academy of Sciences USA* 111:14277–82; Casarotto, S., et al. (2016), *Annals of Neurology* 80:718–29; Massamini, M. and Tononi, G. (2018), *Sizing Up Consciousness: Towards an Objective Measure of the Capacity for Experience* (Oxford:

Oxford University Press); Demertzi, A., et al. (2019), *Science Advances* 5:eaat7603.

58. Crick, F. and Koch, C. (1998), *Cerebral Cortex* 8:97–107, p. 105.

59. Quian Quiroga, R., et al. (2008), *Proceedings of the National Academy of Sciences USA* 105:3599–604.

60. Gelbard-Sagiv, H., et al. (2018), *Nature Communications* 9:2057.

61. Crick, F. and Koch, C. (1995a), *Nature* 375:121–3; Crick, F. and Koch, C. (1995b), *Nature* 377:294–5; Pollen, D. (1995), *Nature* 377:293–4; Block, N. (1996), *Trends in Neurosciences* 19:456–9.

62. Fahrenfort, J., et al. (2017), *Proceedings of the National Academy of Sciences USA* 114:3744–9; Dehaene, S. (2014), *Consciousness and the Brain: Deciphering how the Brain Codes Our Thoughts* (New York: Penguin); Block, N. (2015), in G. Marcus and J. Freeman (eds.), *The Future of the Brain: Essays by the World's Leading Neuroscientists* (Oxford: Princeton University Press), pp. 161–76.

63. Olby, R. (2009), *Francis Crick: Hunter of Life's Secrets* (Cold Spring Harbor: Cold Spring Harbor Laboratory Press), p. 418.

64. Libet, B., et al. (1979), *Brain* 102:193–224.

65. Dominik, T., et al. (2018), *Consciousness and Cognition* 65:1–26.

66. Maoz, U., et al. (2019), *eLife* 8:e39787.

67. Frith, C. and Haggard, P. (2018), *Trends in Neurosciences* 41:405–7.

68. Libet, B. (1994), *Journal of Consciousness Studies* 1:119–26; Libet, B. (2006), *Progress in Neurobiology* 78:322–6, p. 324.

69. Koch et al. (2016).

70. Rangarajan, V., et al. (2014), *Journal of Neuroscience* 34:12828–36, p. 12831.

71. Jonas, J., et al. (2018), *Cortex* 99:296–310.

72. Parvizi, J., et al. (2013), *Neuron* 80:1359–67.

73. Churchland, P. (2013), *Neuron* 80:1337–8, p. 1337.

74. Ibid., p. 1338. For a summary of how consistent subjective experiences can be induced by stimulating different brain regions, see Fox, K., et al. (2020), *Nature Human Behaviour* 4:1039–52.

75. Chalmers, D. (1995), *Journal of Consciousness Studies* 2:200–219.

76. Nagel, T. (1974), *Philosophical Review* 83:435–50; Strawson (2017).

77. Nagel, T. (2017), in K. Almqvist and A. Haag (eds.), *The Return of Consciousness: A New Science on Old Questions* (Stockholm: Axel and Margaret Ax:son Johnson Foundation), pp. 41–6, p. 45.

78. Strawson (2017).

79. Dehaene, S. and Changeux, J.-P. (2011), *Neuron* 70:200–227; Dehaene (2014).

80. Dehaene (2014), p. 233.

81. For example, Edelman, G. and Tononi, G. (2000), *Consciousness: How*

Matter Becomes Imagination (London: Penguin), and Tononi, G., et al. (2016), *Nature Reviews Neuroscience* 17:450–61.

82. Tononi, G. (2008), *Biological Bulletin* 215:216–42 includes some frankly baffling diagrams.

83. Baluška, F. and Reber, A. (2019), *BioEssays* 2019:1800229.

84. Tononi, G. and Koch. C. (2015), *Philosophical Transactions of the Royal Society: B* 370:20140167; Koch (2012).

85. Pennartz, C. (2018), *Trends in Cognitive Sciences* 22:137–53; Morsella, E., et al. (2016), *Behavioral and Brain Sciences* 39:e168 (see also the critical discussion of their position that follows their article). See also Michael Grazziano's Attention Schema Theory, summarised in Webb, T. and Grazziano, M. (2015), *Frontiers in Psychology* 6:500.

86. For example Penrose, R. (1995), *Shadows of The Mind: A Search for the Missing Science of Consciousness* (London: Vintage).

87. Gazzaniga (2018).

88. Litt, A., et al. (2006), *Cognitive Science* 30:593–603.

89. See, for example, Dehaene (2014); Clark (2016); Shea, N. and Frith, C. (2019), *Trends in Cognitive Sciences* 23:560–71.

90. Tononi and Koch (2015), p. 10, claim that their integrated information theory 'predicts' the split-mind phenomenon in split-brain patients. Prediction is easy after the event. For details of the proposed tests of the two theories, see Reardon, S. (2019), *Science* 366:293.

91. Dehaene, S., et al. (2017), *Science* 358:486–92.

92. Sarasso, S., et al. (2015), *Current Biology* 25:3099–105.

93. Nieder, A., et al. (2020), *Science* 369:1626–9; Hesse, J. and Tsao, D. (2020), *eLife* 9:e58360.

94. Snaprud, P. (2018), *New Scientist*, 23 June 2018.

Future

1. Churchland, A. and Abbott, L. (2016), *Nature Neuroscience* 19:348–9. For a collective attempt to forecast the next fifty years of neuroscience, see Altimus, C., et al. (2020), *Journal of Neuroscience* 40:101–6.

2. Sporns, O. (2015), in G. Marcus and J. Freeman (eds.), *The Future of the Brain: Essays by the World's Leading Neuroscientists* (Oxford: Princeton University Press), pp. 90–99, p. 95.

3. *Science*, 27 October 2017.

4. Quotes from Frégnac (2017), pp. 471, 472.

5. Churchland, P. and Sejnowski, T. (1992), *The Computational Brain* (Cambridge, MA: MIT Press), p. 413.

6. Churchland and Abbott (2016), p. 346.

7. Pagán, O. (2019), *Philosophical Transactions of the Royal Society B* 374:20180383.

8. Ballard, D. (2015), *Brain Computation as Hierarchical Abstraction* (Cambridge, MA: MIT Press); Borthakur, A. and Cleland, T. (2019), *Frontiers in Neuroscience* 13:656.

9. Churchland and Abbott (2016), p. 349.

10. Abraham (2016), pp. 146–7.

11. Brette, R. (2019), *Behavioral and Brain Sciences* 42:e15; see also the responses to his article in the same journal.

12. Barlow, H. (1961), in W. Rosenblith (ed.), *Sensory Communication* (Cambridge, MA: MIT Press), pp. 217–34. Brette puts the figure at over 15,000. He used Google Scholar; I used Web of Knowledge. My own papers use the coding metaphor to explore how the neurons in a maggot's nose respond to smells – e.g. Hoare, D., et al. (2008), *Journal of Neuroscience* 28:9710–22; Grillet, M., et al. (2016), *Proceedings of the Royal Society B* 283:20160665.

13. Freeman, W. and Skarda, C. (1990), in J. McGaugh, et al. (eds.) *Third Conference, Brain Organization and Memory: Cells, Systems and Circuits* (New York: Guilford Press), pp. 375–80.

14. Buzsáki, G. (2019), *The Brain from Inside Out* (New York: Oxford University Press).

15. Arbib, M. (1972), *The Metaphorical Brain* (London: Wiley); Arbib, M. (1989), *The Metaphorical Brain 2* (London: Wiley); Keller, E. (1995), *Refiguring Life: Metaphors of Twentieth Century Biology* (New York: Columbia University Press); Brown, T. (2003), *Making Truth: Metaphor in Science* (Chicago: University of Illinois Press); Reynolds, A. (2018), *The Third Lens: Metaphor and the Creation of Modern Cell Biology* (Chicago: University of Chicago Press); Nicholson, D. (2019), *Journal of Theoretical Biology* 477:108–26; Olson, M., et al. (2019), *Trends in Ecology and Evolution* 34:605–15.

16. Kriegeskorte, N. and Diedrichsen, J. (2019), *Annual Review of Neuroscience* 42:407–32.

17. Cazé, R., et al. (2013), *PLoS Computational Biology* 9:e1002867; https://tinyurl.com/Humphries-blog.

18. Gregory (1981), p. 187.

19. Turkheimer, F. (2019), *Neuroscience and Biobehavioral Reviews* 99:3–10.

20. Daugman, J. (1990), in E. Schwartz (ed.), *Computational Neuroscience* (London: MIT Press), pp. 9–18; Gigerenzer, G. and Goldstein, D. (1996), *Creativity Research Journal* 9:131–44; Kirkland (2002); Borck (2012); Abrahams, N. (2018), *Humanity Journal* 8, https://novaojs.newcastle.edu.au/hass/index.php/humanity/article/download/49/53; Borck, C. (2012), in S. Choudhury and J. Slaby (eds.), *Critical Neuroscience: A Handbook of the Social and Cultural Contexts of Neuroscience* (London: Blackwell), pp. 113–33.

21. Brooks, R. (2015), in J. Brockman (ed.), *This Idea Must Die: Scientific Theories That Are Blocking Progress* (New York: HarperPerennial), pp. 295–8; Johansson, S. (1993), in H. Haken, et al. (eds.), *The Machine as Metaphor and Tool* (Berlin: Springer-Verlag), pp. 9–44, p. 38.

22. Carandini, M. (2015), in G. Marcus and J. Freeman (eds.), *The Future of the Brain: Essays by the World's Leading Neuroscientists* (Oxford: Princeton University Press), pp. 177–85, p. 179; Marcus (2015), p. 210.

23. Crick (1989), p. 132. Crick cited Churchland and Sejnowski (1988) as the origin of this insight; the comparison can be found on page 48 of their book. Brown, J. (2014), *Frontiers in Neuroscience* 8:349.

24. Jonas, E. and Kording, K. (2017), *PLoS Computational Biology* 13:e1005268, pp. 1, 18.

25. For a justification of such approaches see Einevoll, G., et al. (2019) *Neuron* 102:735–44; for a fascinating critique, by a modeller, see Mark Humphries' blog post, 'Why Model the Brain?': https://tinyurl.com/Humphries-Why.

26. Bartol, T., et al. (2015), *eLife* 4:e10778.

27. Abbott, L. (2006), in J. van Hemmen and T. Sejnowski (eds.), *23 Problems in Systems Neuroscience* (Oxford: Oxford University Press), pp. 423–31.

28. Chiel, H. and Beer, R. (1997), *Trends in Neurosciences* 20:553–7; Gomez-Marin, A. and Ghazanfar, A. (2019), *Neuron* 104:25–36, p. 34.

29. Sporns (2015), p. 99. For an exploration of how this applies to imagining studies of the human brain, see Nobre and van Ede (2019).

30. Dunn, T., et al. (2016), *eLife* 5:e12741.

31. Ormel, P., et al. (2018), *Nature Communications* 9:4167; Quadrato, G., et al. (2017), *Nature* 545:48–53; Giandomenico, S., et al. (2019), *Nature Neuroscience* 22:669–79; Velasco, C., et al. (2019), *Nature* 570:523–7.

32. Di Lullo, E. and Kriegstein, A. (2017), *Nature Reviews Neuroscience* 18:573–84; Pollen, A., et al. (2019), *Cell* 176:743–56; Ball, P. (2019), *How to Grow a Human: Adventures in Who We Are and How We Are Made* (London: Collins).

33. Cohen, J. (2018), *Science* 360:1284.

34. Farahany, N., et al. (2018), *Nature* 556:429–32.

35. Clarke, G., et al. (2013), *Molecular Psychiatry* 18:666–73; Jameson, K. and Hsiao, E. (2018), *Trends in Neurosciences* 41:413–14.

36. Adolphs and Anderson (2018).

37. Jasanoff, A. (2018), *The Biological Mind: How Brain, Body, and Environment Collaborate to Make Us Who We Are* (New York: Basic).

38. Sterling, P. and Laughlin, S. (2015), *Principles of Neural Design* (London: MIT Press).

39. Nummenmaa, L., et al. (2014), *Proceedings of the National Academy of*

Sciences USA 111:646–51; Nummenmaa, L., et al. (2018), *Proceedings of the National Academy of Sciences USA* 115:9198–203.

40. Keesy, I., et al. (2019), *Nature Communications* 10:1162.
41. Vosshall, L. (2007), *Nature* 450:193–7.
42. https://tinyurl.com/Patel-quote.
43. Perry, C. and Chittka, L. (2019), *Current Opinion in Neurobiology* 54:171–7; Buchanan, S., et al. (2015), *Proceedings of the National Academy of Sciences USA* 112:6700–705; Khuong, T., et al (2019), *Science Advances* 5:eaaw4099.
44. Krause, T., et al. (2019), *Current Biology* 29:1833–41.
45. Feinberg, T. and Mallat, J. (2016), *The Ancient Origins of Consciousness: How the Brain Created Experience* (London: MIT Press); Scholz, M., et al. (2018), https://www.biorxiv.org/content/10.1101/445643v1 – see also the comments.
46. Gutfreund, Y. (2017), *Trends in Neurosciences* 40:196–9.
47. Smith, A. (1978), *Animal Behaviour* 26:232–40.
48. Groothius, J., et al. (2019), *Arthropod Structure & Development* 51:41–51.
49. Webb, B. (2019), *Journal of Experimental Biology* 222:jeb188094; Calhoun, A., et al. (2019), *Nature Neuroscience* 22:2040–9.
50. Saxena and Cunningham (2019); Marques, J., et al. (2020), *Nature* 577:239–43.
51. Chettih, S. and Harvey, C. (2019), *Nature* 567:334–40.
52. Robinson, D. (1992), *Behavioral and Brain Sciences* 14:644–55. See also: Fetz, E. (1992), *Behavioral and Brain Sciences* 15:679–90.

PICTURE CREDITS

143 Cajal, S. (1894a), *Proceedings of the Royal Society of London* 55:444–68.
147 Keith, A. (1919), *The Engines of the Human Body* (London: Williams and Norgate).
163 Uexküll, J. von (1926), *Theoretical Biology* (London: Kegan Paul, Trench, Trubner).
165 Lotka, A. (1925), *Elements of Physical Biology* (Baltimore: Williams & Wilkins).
170 Adrian, E. and Zotterman, Y. (1926), *Journal of Physiology* 61:465–83.
172 Adrian, E. (1931), *Journal of Physiology* 72:132–51.
173 Adrian, E. (1928), *The Basis of Sensation* (London: Christophers).
181 McCulloch, W. and Pitts, W. (1943), *Bulletin of Mathematical Biophysics* 5:115–33.
198 Young, J. Z. (1951), *Doubt and Certainty in Science: A Biologist's Reflection on the Brain* (Oxford: Clarendon).
209 Penfield, W. (1952), *Archives of Neurology and Psychiatry* 67:178–91.
212 Penfield, W. and Rasmussen, T. (1950), *The Cerebral Cortex of Man* (New York: Macmillan).
235 Hubel, D. and Wiesel, T. (1962), *Journal of Physiology* 160:106–54.
249 Twitter. Used with permission of S. Scott.
262 Selfridge, O. (1959), in *Symposium on the Mechanisation of Thought Processes* (London: HMSO).
265 Rosenblatt, F. (1958), *Psychological Review* 65:386–408.
271 Ponce, C. et al. (2019), *Cell* 177: 999–1009.
342 Rare Books and Special Collections, University of Sydney Library.
387 Smith, A. (1978), *Animal Behaviour* 26:232–40.

Colour section credits

1.   Imaginary portrait of Galen from the sixteenth century.
2.   Galen's experiment on a pig, taken from the title page of a sixteenth-century collection of his works. (Science History Images/Alamy Stock Photo)
3.   Nicolaus Steno (1638–86). (Institute of Medical Anatomy, Copenhagen.)
4.   Portrait of Julien Offray de La Mettrie (1709–51) on the frontispiece of one of his books. (Harvard Art Museums/Fogg Museum, Gift of William Gray from the collection of Francis Calley Gray. Photo: ©President and Fellows of Harvard College)
5.   Self-portrait of Ramón y Cajal in his laboratory. (Courtesy Instituto Cajal del Consejo Superior de Investigaciones Cientificas, Madrid.)

6. The London performance of *RUR* in 1923.
7. A description of Seleno the Electric Dog from 1918.
8. Walter Pitts.
9. David Hubel (left) and Torsten Wiesel in 1981. (Harvard University Museum)
10. Eve Marder in the lab. (Photo courtesy of Brandeis University.)
11. Doris Tsao. (Photo courtesy of Doris Tsao.)
12. Cover of *Science* magazine from 1991, showing the use of fMRI to reveal brain activity.
13. Patient S3 (Cathy Hutchinson) controlling a robotic arm in Donoghue's lab, 2012. (Photo © Maciek Jasik)
14. Diagram from 'Could a Neuroscientist Understand a Microprocessor?' by Eric Jonas and Konrad Paul Kording (2017). (Copyright: © 2017 Jonas, Kording. Open access.)

INDEX

Note: The index covers the main text fully, but the endnotes (on page 393 onward) only very selectively. Italic page numbers indicate relevant illustrations; the suffix n, a footnote, and n followed by a number an endnote on the indicated pages.